A Photographic Atlas For The Botany Laboratory

Second Edition

Kent M. Van De Graaff

Samuel R. Rushforth
Brigham Young University

John L. Crawley

Morton Publishing Company
925 W. Kenyon, Unit 12
Englewood, Colorado 80110

*To biologists, conservationists,
and concerned people throughout the world
who actively strive to save nature.*

Copyright © 1994, 1995 by Morton Publishing Company

ISBN: 0-89582-302-0

10 9 8 7 6 5 4 3 2 1

All rights reserved. Permission in writing must be obtained from the publisher before any part of this work may be reproduced or transmitted in any form, or by any means, electronic or mechanical, including photocopying and recording or by any information storage or retrieval system.

Printed in the United States of America

Contents

Preface v

1 Cells and Tissues 1

2 Kingdom Monera 11
    Archaebacteria (methanogens, thermoacidophiles)
    Eubacteria (prokaryotes)

3 Kingdom Protista: Primarily Unicellular Organisms 18
    Chrysophyta (diatoms, golden algae)
    Pyrrhophyta (dinoflagellates)
    Rhizopoda (amoebas)
    Apicomplexa (sporozoans)
    Euglenophyta (euglenoids)
    Ciliophora (ciliates, *Paramecium*)

4 Kingdom Protista: Primarily Multicellular Organisms 25
    Algae
        Division Chlorophyta (green algae)
        Division Phaeophyta (brown algae, giant kelp)
        Division Rhodophyta (red algae)
    Protists Resembling Fungi
        Division Myxomycota (plasmodial slime molds)
        Division Acrasiomycota (cellular slime molds)
        Division Oomycota (water molds, white rusts, downy mildews)

5 Kingdom Fungi 51
    Division Zygomycota (conjugation fungi)
    Division Ascomycota (yeasts, morels, truffles)
    Division Basidiomycota (mushrooms, toadstools, rusts, smuts)
    Lichens

6 Kingdom Plantae: Division Bryophyta (Bryophytes) 68
    Class Hepaticae (liverworts)
    Class Anthocerotae (hornworts)
    Class Musci (mosses)

7 Kingdom Plantae: Seedless Vascular Plants 81
   Division Psilotophyta (whisk ferns)
   Division Lycophyta (club mosses, quillworts, spike mosses)
   Division Sphenophyta (horsetails)
   Division Pterophyta (ferns)

8 Kingdom Plantae: Gymnosperms (Exposed Seed Plants) 101
   Division Cycadophyta (cycads)
   Division Ginkgophyta (*Ginkgo*)
   Division Coniferophyta (conifers)
   Representative Herbarium Specimens of Conifers
   Division Gnetophyta (gnetophytes)

9 Kingdom Plantae: Angiosperms (Enclosed Seed Plants — Flowering Plants) 121
   Division Anthophyta (angiosperms: monocots and dicots)

■ Glossary of Terms 163

■ Index 167

Preface

Botany, a specialty of biology, is the study of plants. More than 95% of the Earth's biomass is composed of plants. Because plants are so abundant, visible, and necessary to life, everyone has some interest in and knowledge of botany.

Plants and flowers, for example, are welcomed into our homes for their beauty. Florists remind us to "Say it with flowers." Flowers are easily cared for in a vase of water, where they remain attractive and aromatic for a few days. In contrast, plants are not so easily cared for, but they can live and grow for more than just a few days. House plants need to be properly potted, watered regularly, and occasionally fertilized. By providing this care, we learn about the nature and requirements of plants as we become amateur botanists. Grass, trees, and shrubs landscaped around our houses provide us with beauty, comfort, shade, and a sense of being connected to nature.

Our existence depends on plants. From plants we are supplied with building materials, the oxygen we breathe, and the food we eat. Through photosynthesis, plants use sunlight to convert water and carbon dioxide to sugars, releasing oxygen as a byproduct.

Plants also provide us with the materials required for producing medicines and paper. Insights and discoveries in botany are occurring today at a rapid pace. What students will learn in a basic botany course will be of immeasurable value in understanding and making decisions about the ecological problems currently facing our world. A basic course also is essential as a secure foundation for planning advanced life science study.

Botany is a visually oriented science. *A Photographic Atlas for the Botany Laboratory, Second Edition,* provides clear photographs and drawings of tissues and organisms, similar to specimens seen in a botany laboratory. It is designed to accompany any botany (or biology) text or laboratory manual used in the classroom. In certain courses this atlas could serve as the laboratory manual.

This atlas provides a balanced visual representation of the major groups of botanical organisms. Care has been taken to construct completely labeled, informative figures. Parts of organisms are depicted clearly and accurately. The terminology used matches college botany texts.

Several dissections of plants are provided for students who have the opportunity to do similar dissections in the class laboratory. In addition, many photomicrographs, photos of living specimens, and herbarium collections are included. These figures enhance the student's understanding of plant structure and plant classification. Plants of significant economic importance for food, shelter, and medicines are highlighted.

Acknowledgments

Many professionals helped us prepare this atlas and have shared our enthusiasm for its value for botany students. We especially appreciate Brigham Young University for allowing us to photograph specimens in their herbarium and botanical greenhouses. Kaye H. Thorne and Thomas G. Black were most helpful in selecting specimens to include. Wilford M. Hess and James V. Allen provided many scanning electron micrographs, and we appreciate their generosity. The suggestions of H. Blaine Furniss were greatly appreciated. Ryan L. Van De Graaff assisted in many of the plant dissections.

The professional input by reviewers and numerous users were invaluable. Neil A. Harriman of the University of Wisconsin–Oshkosh was most generous in his meticulous critique of the manuscript. Others who offered especially helpful input include Frank W. Ewers, Patrick F. Fields, Dale M. J. Mueller, John W. Taylor, Brian Speer, Lawrence Virkaitis, Cecile Bochmer, and Anne S. Viscomi.

Christopher H. Creek and Roger Clarke rendered line art throughout the book. Jan Hall copyedited the manuscript. Joanne R. Saliger of Ash Street Typecrafters, Inc., was superb to work with and we appreciate her talent and interest in this project. We are indebted to Douglas Morton and the personnel at Morton Publishing Company for this opportunity and their encouragement and support.

Cells and Tissues

All organisms are comprised of one or more cells. Cells are the basic structural and functional units of organisms. A cell is a minute, membrane-enclosed, protoplasmic mass consisting of chromosomes surrounded by cytoplasm. Specific organelles are contained in the cytoplasm that function independently but in coordination with one another. Prokaryotic cells and eukaryotic cells are the two basic types.

Prokaryotic cells lack a membrane-bound nucleus, instead containing a single strand of *nucleic acid*. These cells contain few organelles. A rigid or semi-rigid *cell wall* provides shape to the cell outside the *cell (plasma) membrane*. Bacteria are examples of prokaryotic, single-celled organisms.

Eukaryotic cells contain a true *nucleus* with multiple chromosomes, have several types of specialized *organelles* and have a differentially permeable cell (plasma) membrane. Organisms comprised of eukaryotic cells include protozoa, fungi, algae, plants, and invertebrate and vertebrate animals.

Plant cells differ in some ways from other eukaryotic cells in that their cell walls contain *cellulose* for stiffness. Plant cells also contain vacuoles for water storage and membrane-bound *chloroplasts* with photosynthetic pigments for photosynthesis.

The *nucleus* is the large, spheroid body within the eukaryotic cell that contains the genetic material of the cell. The nucleus is enclosed by a double membrane called the *nuclear membrane*, or *nuclear envelope*. The *nucleolus* is a dense, non-membranous body in the nucleus composed of protein and RNA molecules. The chromatin are fibers of protein and DNA molecules. Prior to cellular division, the chromatin shortens and coils into rod-shaped *chromosomes*. Chromosomes consist of DNA and structural proteins called *histones*.

The *cytoplasm* of the eukaryotic cell is the medium between the nuclear membrane and the cell membrane. *Organelles* are small membrane-bound structures within the cytoplasm (other than the nucleus). The cellular functions carried out by organelles are referred to as *metabolism*. The structure and functions of the nucleus and principal plant organelles are listed on page 2. In order for cells to remain alive, metabolize, and maintain *homeostasis*, cells must have access to nutrients and respiratory gases, be able to eliminate wastes, and be in a constant, protective environment.

Tissues are groups of similar cells that perform specific functions. A flowering plant, for example, is composed of three tissue systems:

1. the *ground tissue system*, providing support, regeneration, respiration, photosynthesis, and storage;
2. the *vascular tissue system*, providing conduction of water, nutrients, and sugars through the plant;
3. the *dermal tissue system*, providing protection.

Organs are two or more tissue systems that carry out specific functions together. Examples of organs include flowers, leaves, stems, and roots.

The *organism* is the plant itself, which consists of all the organs functioning together to keep it alive, allow it to grow, and permit it to propagate.

The term *cell cycle* refers to how a multicellular organism develops, grows, and maintains and repairs body tissues. In the cell cycle, each new cell receives a complete copy of all genetic information in the parent cell, and the cytoplasmic substances and organelles to carry out hereditary instructions.

The cell cycle of a plant consists of growth, synthesis, mitosis, and cytokinesis. *Growth* is the increase in cellular mass as the result of metabolism. *Synthesis* is the production of DNA and RNA to regulate cellular activity. *Mitosis* is the exact duplication and division of chromosomes. *Cytokinesis* is the division of the cytoplasm that follows mitosis.

Unlike animal cells, plant cells have a rigid cell wall that does not cleave during cytokinesis. Instead, a new cell wall is constructed between the daughter cells. Furthermore, many land plants do not have centrioles for the attachment of spindles. The *microtubules* in these plants form a barrel-shaped anastral *spindle* at each pole. Mitosis and cytokinesis in plants occurs in basically the same sequence as these processes in animal cells.

Asexual reproduction is propagation of new organisms without sex; that is, the production of new individuals by processes that do not involve *gametes* (sex cells). Asexual reproduction occurs in a variety of microorganisms, plants, and animals, wherein a single parent produces offspring with characteristics identical to itself. Asexual reproduction is not dependent on the presence of other individuals. No egg or sperm is required. In asexual reproduction, all the offspring are genetically identical (except for mutants). Types of asexual reproduction and example organisms include:

1. *fission* — a single cell divides to form two separate cells (bacteria, protozoans, and other one-celled organisms);
2. *sporulation* — many cells are formed that may remain separate or join together in a cystlike structure (algae, fungi, protozoans);
3. *budding* — buds develop on the parent and then detach themselves (hydras, yeast, certain plants);
4. *fragmentation* — organisms break into two or more parts, and each part is capable of becoming a complete organism (flatworms, echinoderms, algae).

Sexual reproduction is propagation of new organisms through the union of genetic material from two parents. Sexual reproduction usually involves the fusion of haploid gametes (such as sperm and egg cells) during fertilization to form a zygote.

The major biological difference between sexual and asexual reproduction is that sexual reproduction produces genetic variation in the offspring. The combining of genetic material from the gametes produces offspring that are different from either parent and contain new combinations of characteristics. This may increase the ability of the species to survive environmental changes or to reproduce in new habitats. The only genetic variation that can arise in asexual reproduction comes from mutations.

TABLE 1.1.
Structure and function of components of plant eukaryotic cells.

Component	Structure	Function
Cell (plasma) membrane	Composed of protein and phospholipid molecules	Provides form to cell; controls passage of materials into and out of cell
Cell wall	Cellulose fibrils	Provides structure and rigidity to plant cell
Cytoplasm	Fluid to jelly-like substance	Serves as suspending medium for organelles
Endoplasmic reticulum	Interconnecting membrane-lined channels	Provides supporting framework for cell; enables cell transport
Ribosomes	Granules of nucleic acid	Synthesize protein
Mitochondria	Double-layered sacs with cristae	Produce ATP (cellular respiration)
Golgi apparatus	Flattened membrane-lined chambers	Synthesizes carbohydrates and packages molecules for secretion
Lysosomes	Membrane-surrounded sacs of enzymes	Digest foreign molecules and worn cells
Centrosome	Mass of two rod-like centrioles	Organizes spindle fibers and assists mitosis
Vacuoles	Membranous sacs	Store and excrete substances within the cytoplasm
Fibrils and microtubules	Protein strands	Support cytoplasm and transport materials
Cilia and flagella	Cytoplasmic extensions from cells	Move particles along cell surface, or move cell
Nucleus	Nuclear membrane, nucleolus, and chromatin (DNA)	Direct cell activity; forms ribosomes;
Chloroplast	Inner (grana) membrane within outer membrane	Carries out photosynthesis

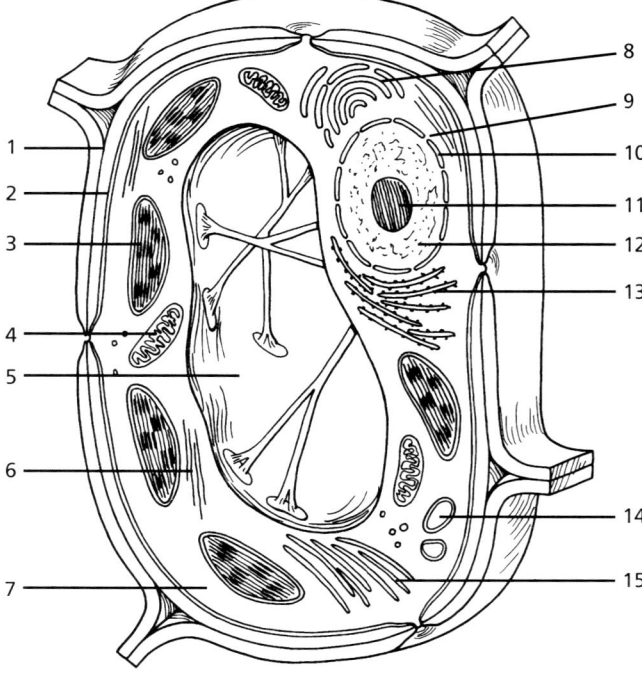

Figure 1.1 A prokaryotic cell.
1. Cell wall
2. Cell (plasma) membrane
3. Ribosomes
4. Circular molecule of DNA

Figure 1.2 A typical plant cell.
1. Cell wall
2. Cell (plasma) membrane
3. Chloroplast
4. Mitochondrion
5. Vacuole
6. Microfilament
7. Cytoplasm
8. Golgi apparatus (dictyosome)
9. Nuclear pore
10. Nuclear membrane (envelope)
11. Nucleolus
12. Chromatin
13. Rough endoplasmic reticulum
14. Vesicle
15. Smooth endoplasmic reticulum

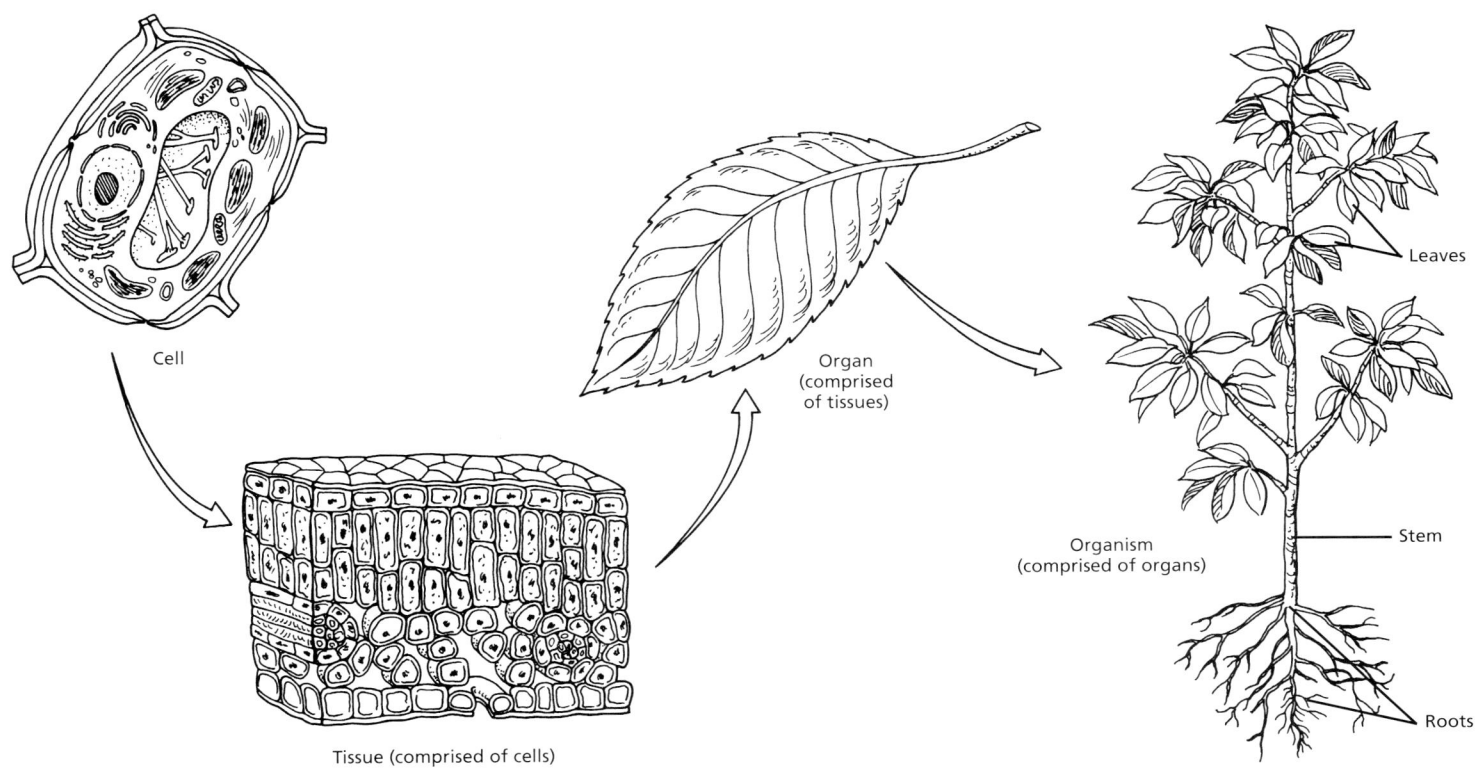

Figure 1.3 Structural levels of plant organization.

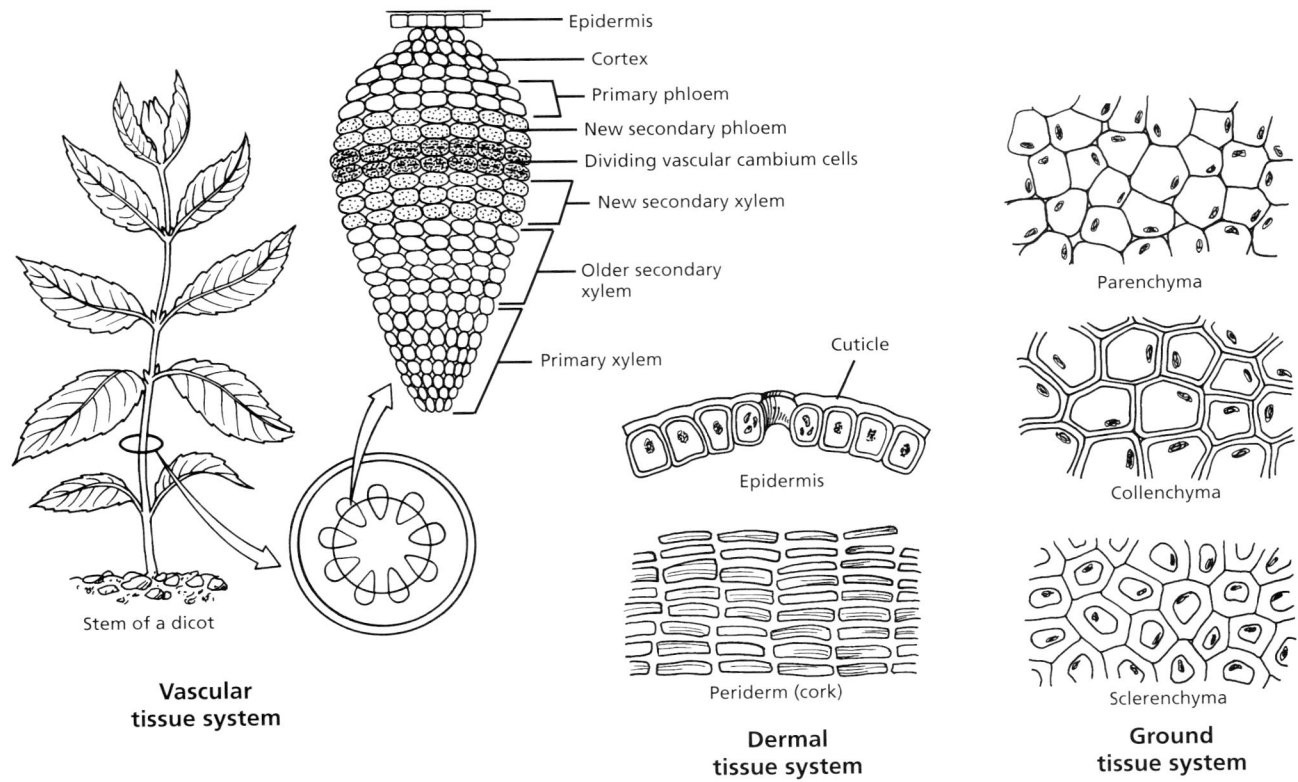

Figure 1.4 Examples of plant tissues.

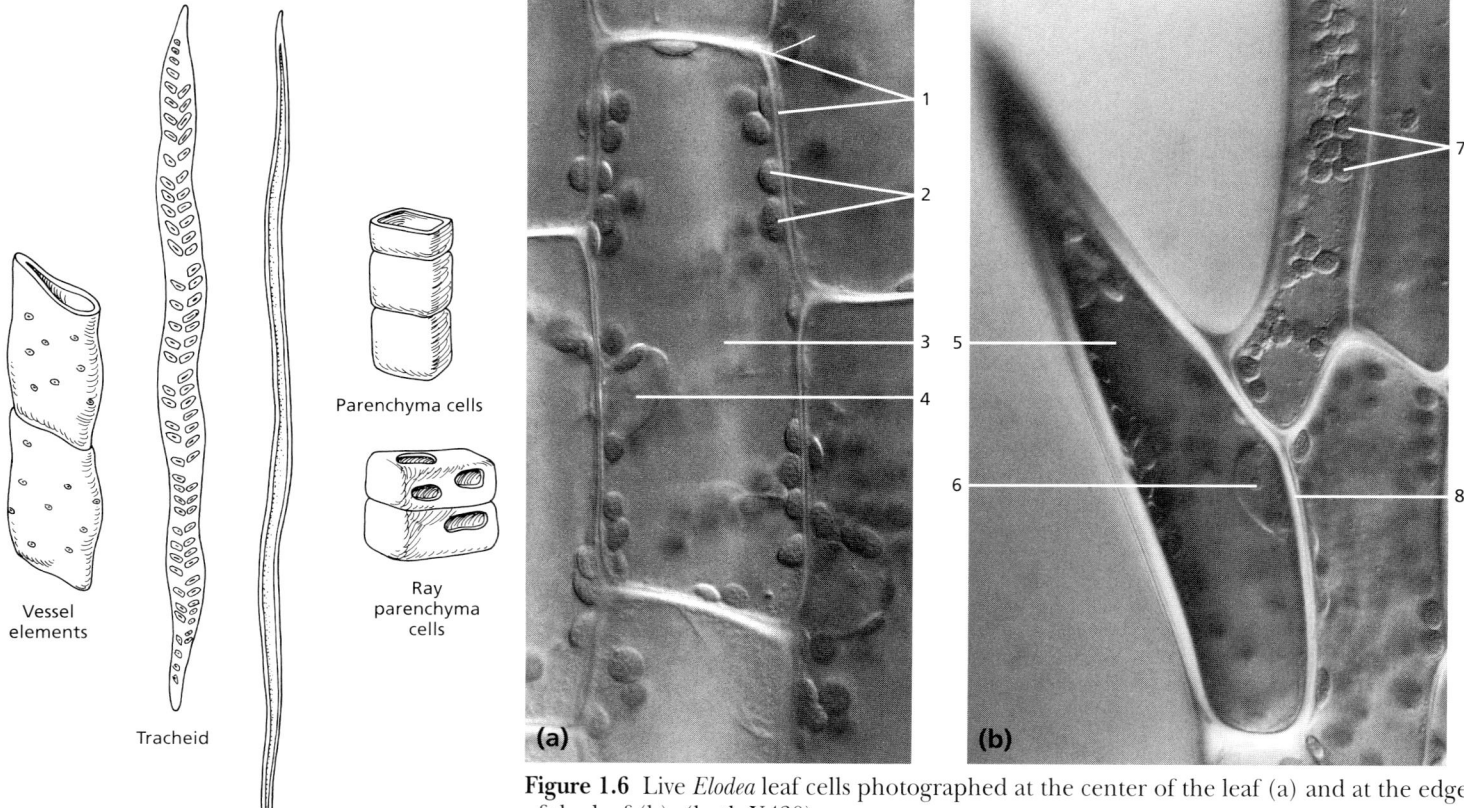

Figure 1.5 Examples of plant cells.

Figure 1.6 Live *Elodea* leaf cells photographed at the center of the leaf (a) and at the edge of the leaf (b). (both X430)

1. Cell wall
2. Chloroplasts
3. Vacuole
4. Nucleus
5. Spine-shaped cell on exposed edge of leaf
6. Nucleus
7. Chloroplasts
8. Cell wall

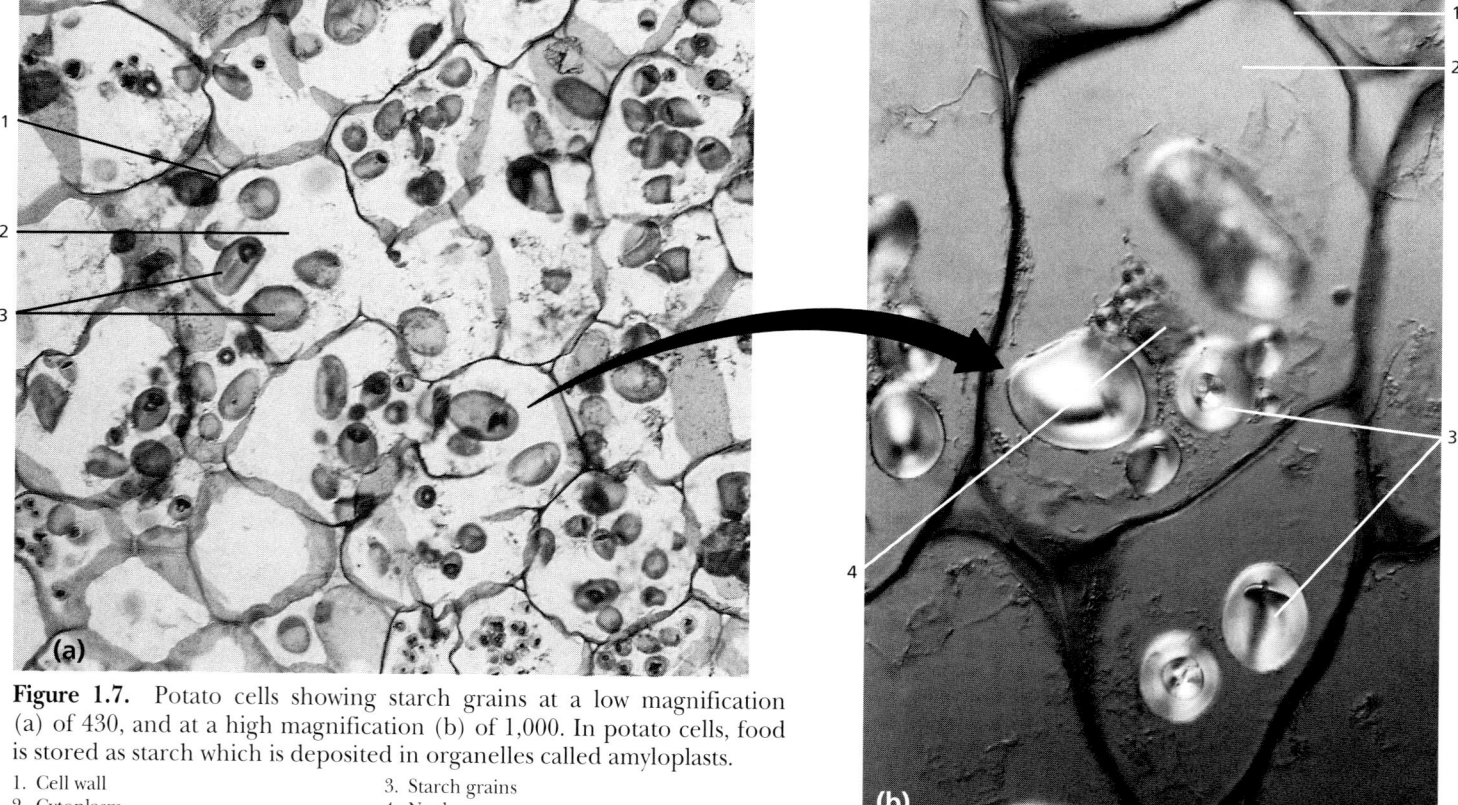

Figure 1.7. Potato cells showing starch grains at a low magnification (a) of 430, and at a high magnification (b) of 1,000. In potato cells, food is stored as starch which is deposited in organelles called amyloplasts.

1. Cell wall
2. Cytoplasm
3. Starch grains
4. Nucleus

Cells and Tissues

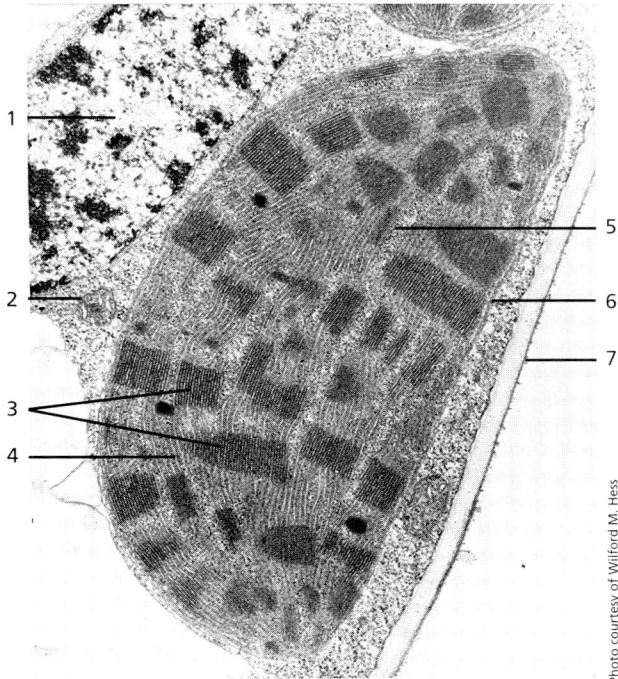

Figure 1.8 Electron micrograph of a portion of a sugar cane leaf cell.

1. Nucleus
2. Mitochondrion
3. Grana
4. Thylakoid membrane
5. Stroma
6. Chloroplast membrane
7. Cell wall

Figure 1.9 Fractured barley smut spore.

1. Cell wall
2. Cell membrane

Figure 1.10 Barley smut spore, fractured through the middle of the cell.

1. Nucleus
2. Vacuole
3. Cell wall
4. Cell membrane
5. Mitochondrion
6. Nuclear pores

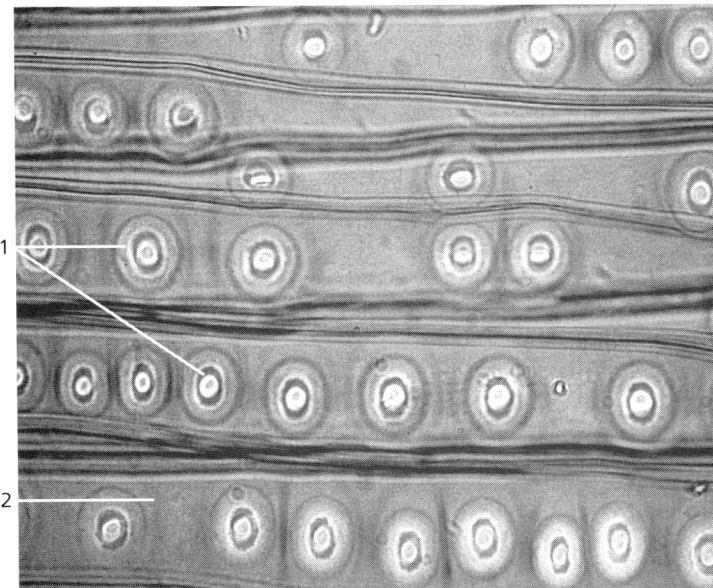

Figure 1.11 Longitudinal section through the xylem of a pine, *Pinus*, showing tracheid cells with prominent bordered pits.

1. Bordered pits
2. Tracheid cell

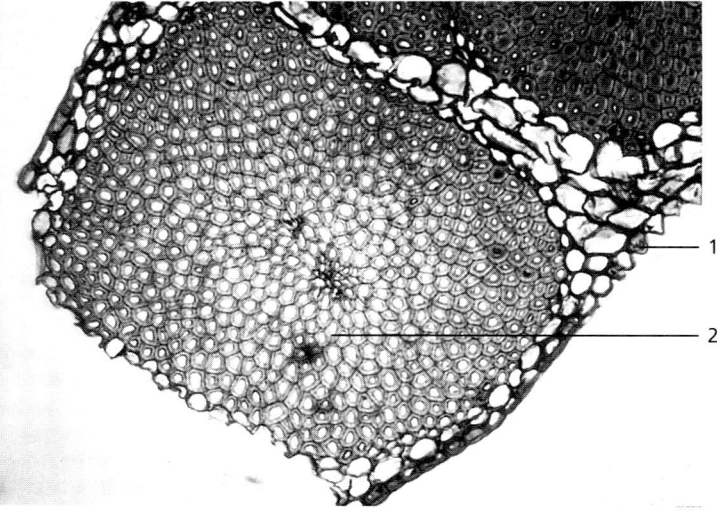

Figure 1.12 Transverse section through the leaf of a yucca, *Yucca brevifolia*. A bundle of leaf fibers (sclerenchyma) is evident at the edge of the leaf.

1. Epidermis
2. Bundle of sclerenchyma

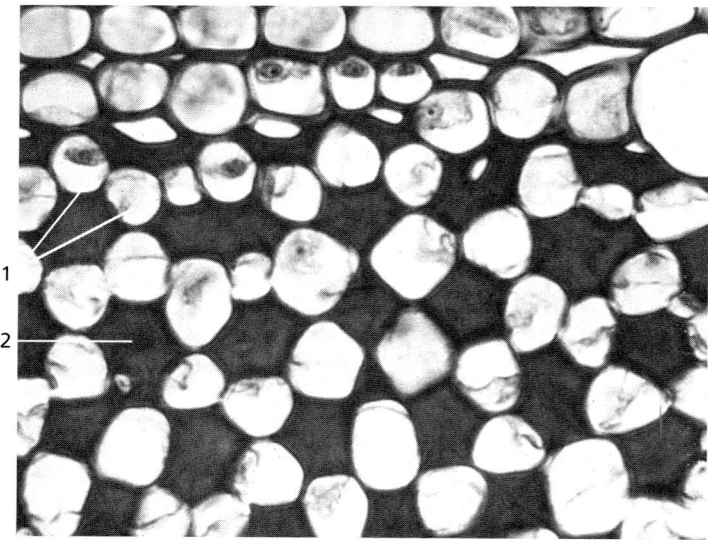

Figure 1.13 Collenchyma tissue from the stem of a begonia. Collenchyma tissue consists of thickened cells that form supportive strands beneath the epidermis in stems and petioles.

1. Cell lumen
2. Thickened cell walls

Figure 1.14 Longitudinal section through the xylem of a squash stem, *Cucurbita maxima*. The vessel elements shown here have several different patterns of wall thickenings.

1. Vessel elements
2. Parenchyma

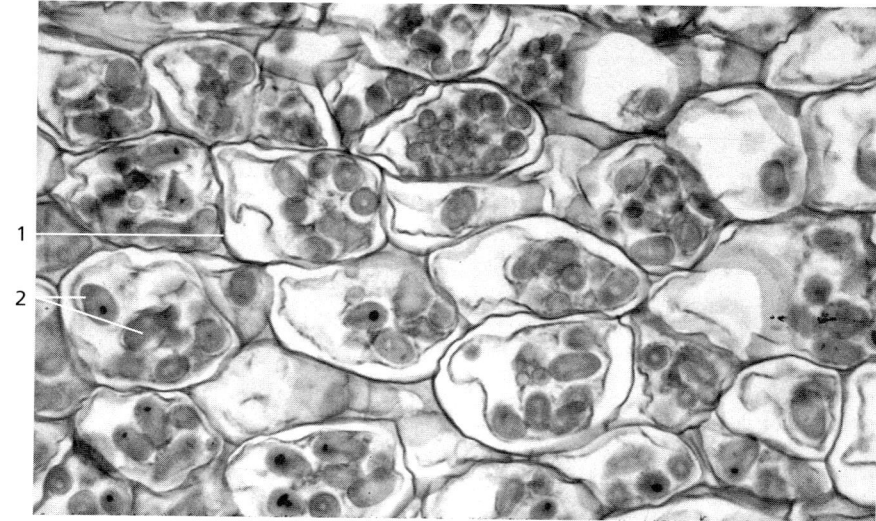

Figure 1.15 Section through the tuber of a potato, *Solanum tuberosum*, showing parenchyma cells containing numerous starch grains.

1. Cell wall
2. Starch grains

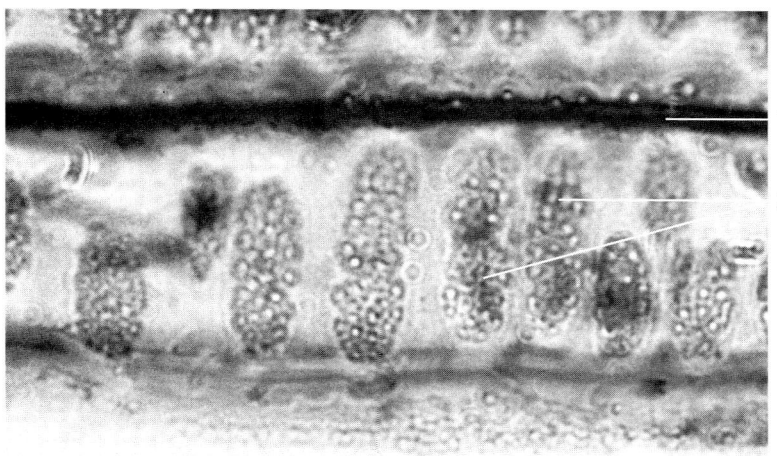

Figure 1.16 Close-up of sieve tube elements in the phloem of a grape, *Vitis vinifera*. Note the sieve plates on the sieve tube elements.

1. Cell wall
2. Sieve plates

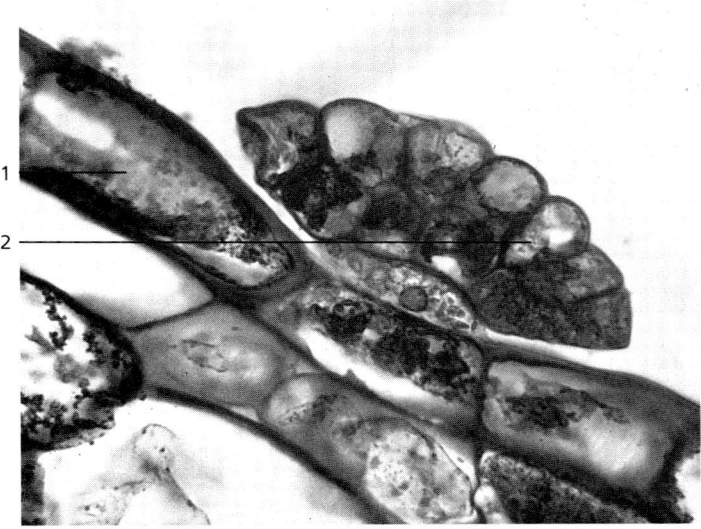

Figure 1.17 Section through a leaf of the venus flytrap, *Dionaea muscipula*, showing epidermal cells with an attached digestive gland. The gland is comprised of secretory parenchyma cells.

1. Epidermis
2. Gland

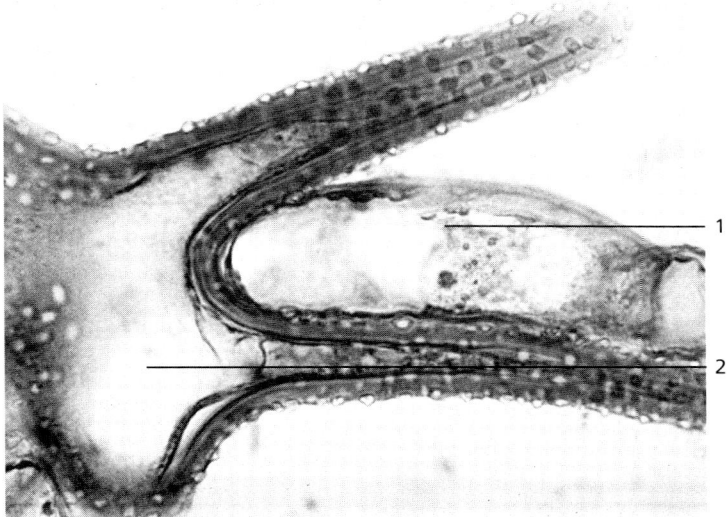

Figure 1.18 Astrosclereid in the petiole of a pondlily, *Nuphar.*

1. Parenchyma cell
2. Astrosclereid

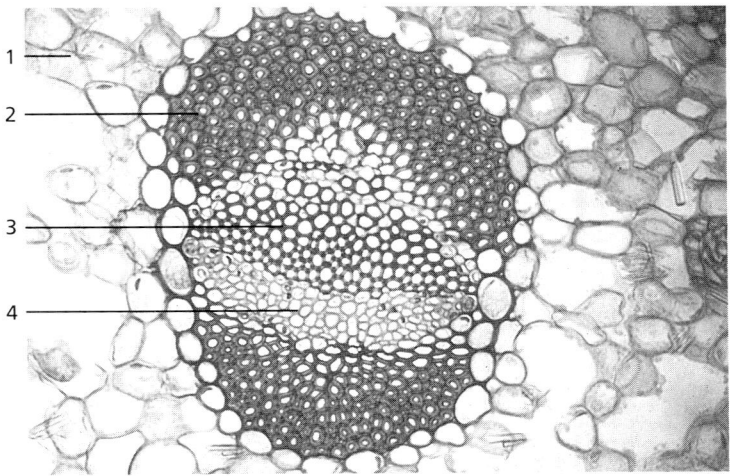

Figure 1.19 Transverse section through the leaf of a yucca, *Yucca brevifolia,* showing a vascular bundle (vein). Note the prominent sclerenchyma tissue forming caps on both sides of the bundle.

1. Leaf parenchyma
2. Leaf sclerenchyma (bundle sheath)
3. Xylem
4. Phloem

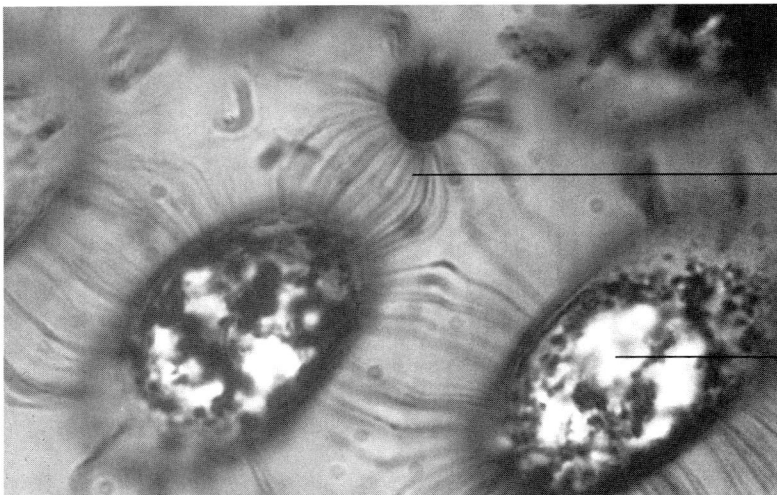

Figure 1.20 A section through the endosperm tissue of a persimmon, *Diospyros virginiana*. These thick-walled cells are actually parenchyma cells. Cytoplasmic connections, or plasmodesmata, are evident between cells.

1. Plasmodesmata
2. Cell lumen

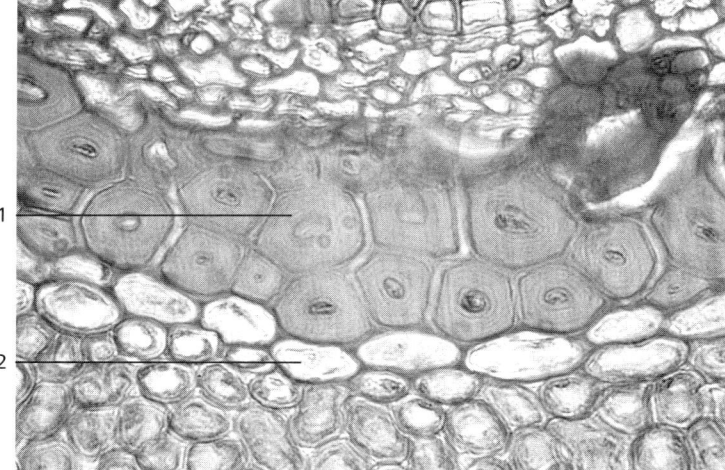

Figure 1.21 Transverse section through the stem of flax, *Linum*. Note the thick-walled fibers as compared to the thin-walled parenchyma cells.

1. Fibers
2. Parenchyma cell

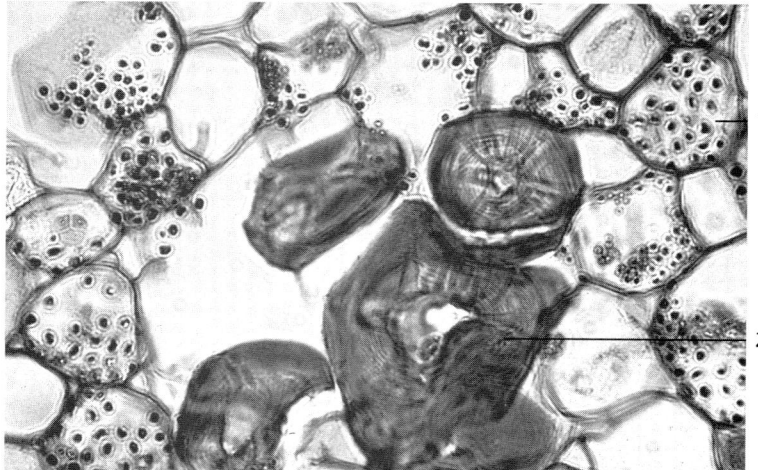

Figure 1.22 Section through the stem of a wax plant, *Hoya carnosa*. Thick-walled sclereids (stone cells) are evident.

1. Parenchyma cell
2. Sclereid (stone cell)

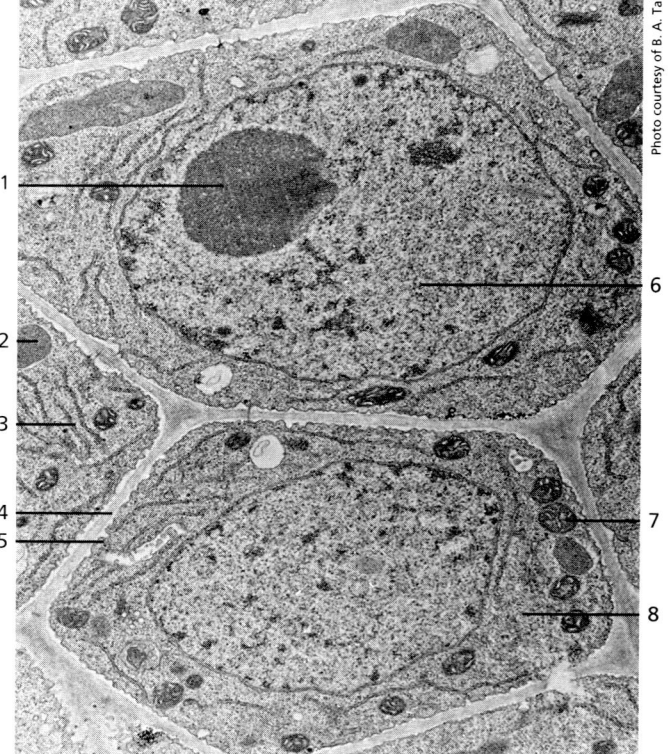

Figure 1.23 Electron micrograph of alfalfa root cells.

1. Nucleolus
2. Underdeveloped plastid
3. Endoplasmic reticulum
4. Cell wall
5. Cell membrane
6. Nucleus
7. Mitochondrion
8. Ribosomes

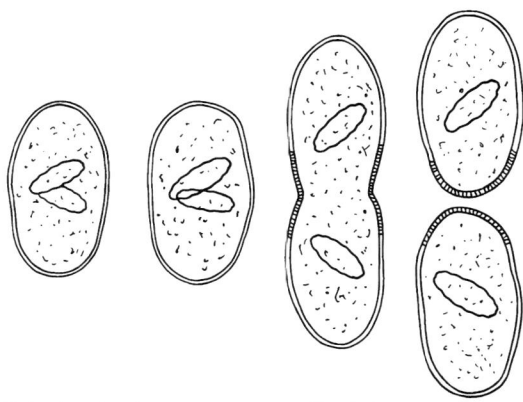

Fission

A single cell divides, forming two separate cells. Fission occurs in bacteria, protozoans, and other single-celled organisms.

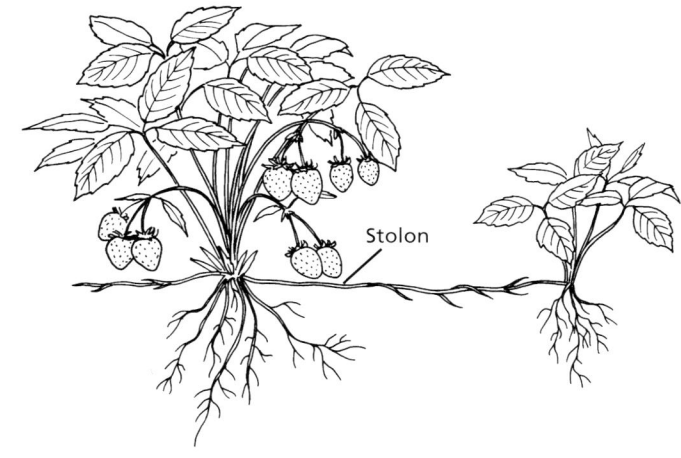

Vegetative propagation

A plant produces external stems, or runners. Budding occurs in a number of flowing plants, such as strawberries.

Figure 1.24 Types of asexual reproduction.

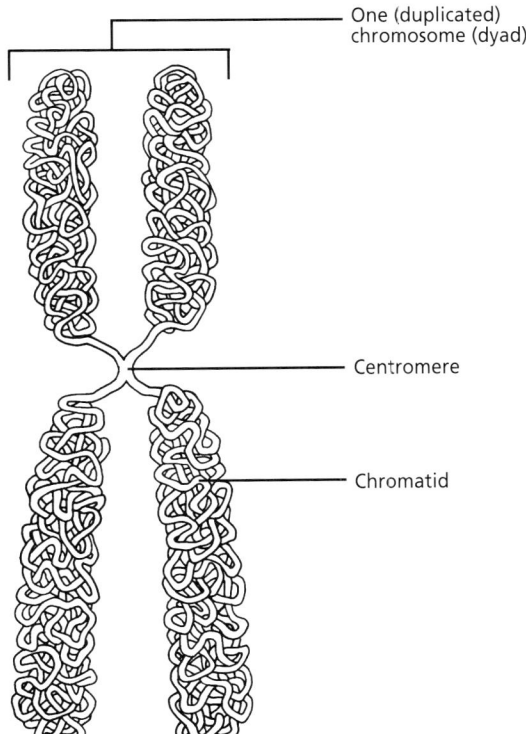

Figure 1.25 Each replicated chromosome consists of two identical sister chromatids attached at a constricted centromere.

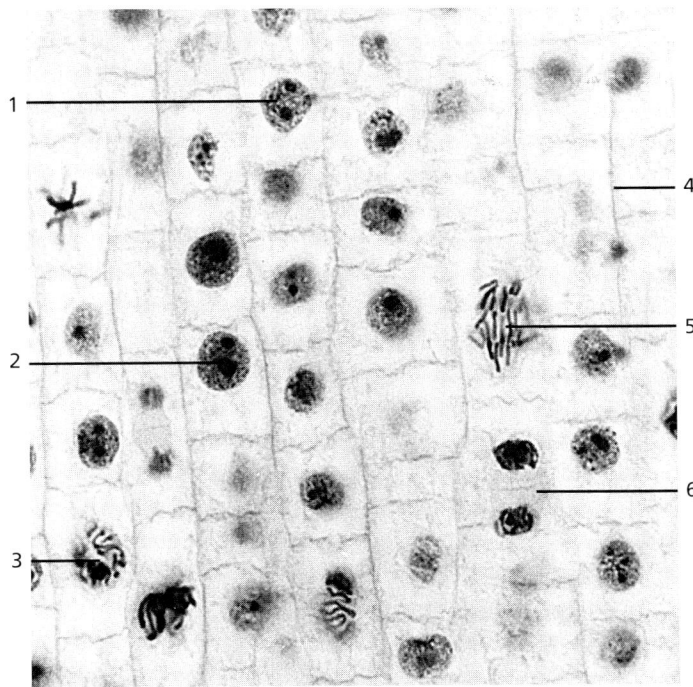

Figure 1.26 Cells in various stages of mitosis from an onion, *Allium*, root tip. (X100)

1. Interphase
2. Early prophase
3. Late prophase
4. Cell wall
5. Anaphase
6. Telophase

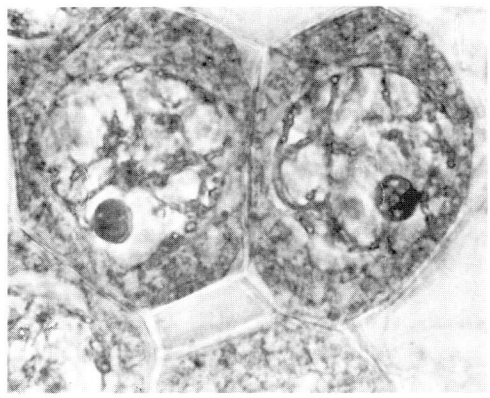

(a) **Prophase I** — Each chromosome consists of two chromatids joined by a centromere.

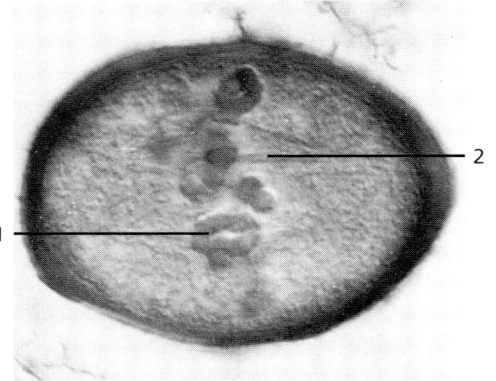

(b) **Metaphase I** — The chromosomes align at the equator with their homologous partner. During this stage, called synapsis, exchange (crossing over) occurs between the chromosomes.
1. Pairs of dyads   2. Spindle fibers

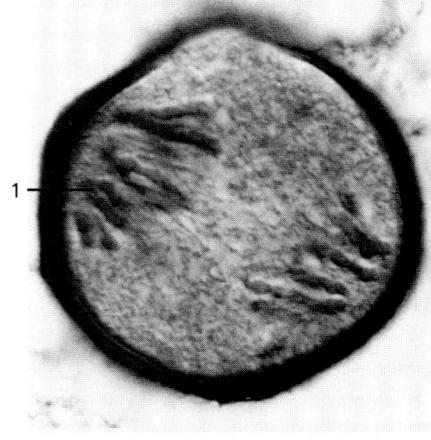

(c) **Anaphase I** — No division at the centromeres occurs as the chromatin separates, so both copies of a homologous chromosome go to one cell.
1. Dyad chromosome

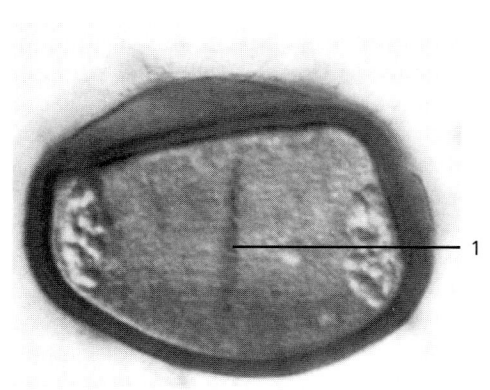

(d) **Telophase I** — The chromosomes lengthen and become less distinct. The cell plate (in some plants) forms between the forming cells.
1. Cell plate

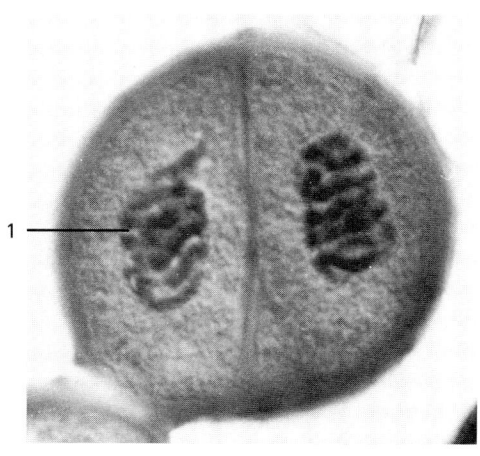
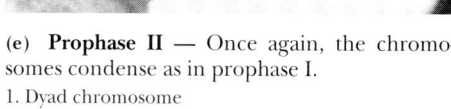

(e) **Prophase II** — Once again, the chromosomes condense as in prophase I.
1. Dyad chromosome

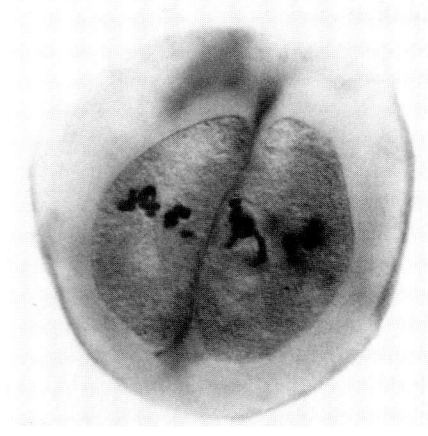

(f) **Metaphase II** — The dyad chromosomes align on the equator and the spindle fibers attach to the centromeres. This is similar to metaphase in mitosis.

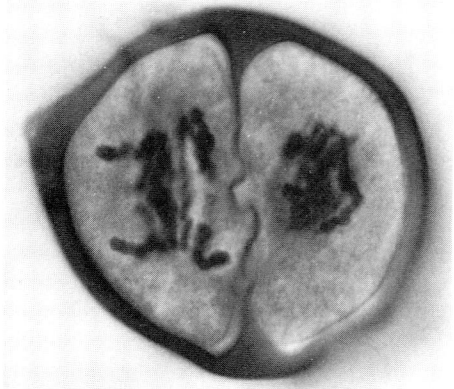

(g) **Anaphase II** — The chromatids separate and each is pulled to an opposite pole.

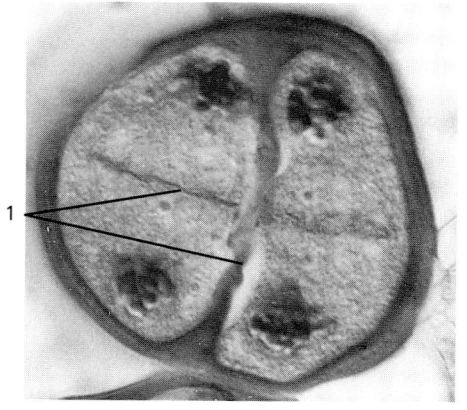

(h) **Telophase II** — Cell division is complete and cell walls of four haploid cells are formed.
1. Cell walls

Figure 1.27 Stages of meiosis in lily microspore mother cells. (X1000)

Kingdom Monera

The kingdom Monera (about 2,500 species of bacteria) contains all organisms comprised of prokaryotic cells. Prokaryotic cells were the first kinds of cells to evolve, probably about 3.5 billion years ago. The Archaebacteria and Eubacteria are the two divisions within the kingdom Monera.

Archaebacteria are adapted to a limited range of extreme conditions. The cell walls of Archaebacteria lack peptidoglycan (characteristic of Eubacteria). Archaebacteria have distinctive transfer RNAs and RNA polymerases. They include methanogens, typically found in swamps and marshes, and thermoacidophiles, found in acid hot springs and acidic soil.

Methanogens exist in oxygen-free environments and subsist on simple, inorganic compounds such as CO_2, acetate, and methanol. As their name implies, Methanobacteria produce methane gas as a byproduct of metabolism. These organisms are typically found in organic-rich mud and sludge, particularly that which contains fecal wastes.

The thermoacidophiles are resistant to hot temperatures and high acid concentrations. The plasma membrane of these organisms contains high amounts of saturated fats, and its enzymes and other proteins are able to withstand extreme conditions without denaturation. These microscopic organisms thrive in most hot springs and hot, acid soils.

In contrast to Archaebacteria, Eubacteria are considered the "true" bacteria. This division includes the Cyanobacteria (formerly known as blue-green algae) and a number of other diverse types (see Table 3.1). Cyanobacteria are photosynthetic bacteria that contain chlorophyll and release oxygen during photosynthesis. Some bacteria are obligate aerobes (require O_2 for metabolism) and others are facultative anaerobes

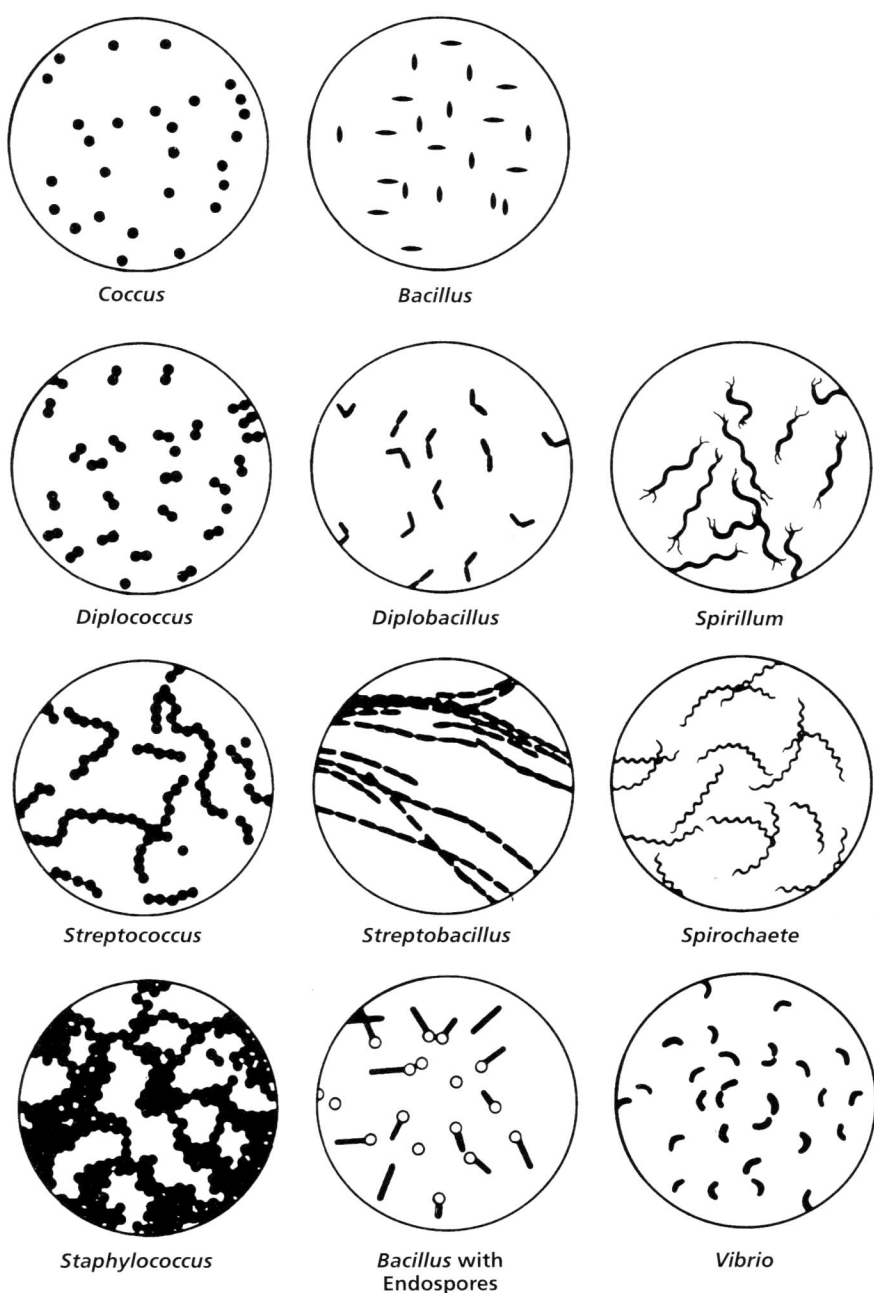

Figure 2.1 Various bacteria.

(indifferent to O_2 for metabolism). Most bacteria are heterotrophic saprophytes, which secrete enzymes to break down surrounding organic molecules into absorbable compounds.

Bacteria range between 1 and 10 μm in width or diameter. The morphological appearance may be spiral (spirillum), spherical (coccus), or rod-shaped (bacillus). Cocci and bacilli frequently form clusters or linear filaments, and may have cilia. Relatively few species of bacteria cause infection. Hundreds of species of non-pathogenic bacteria live on the human body and within the gastrointestinal (GI) tract. Those in the GI tract constitute a person's normal gut flora.

TABLE 2.1
Some Representatives of the Kingdom Monera

Categories	Representative Genera
Archaebacteria	
Methanogens	*Halobacterium, Methanobacterium*
Thermoacidophiles	*Thermoplasma, Sulfobolus*
Eubacteria	
Photosynthetic bacteria	
Cyanobacteria	*Anabaena, Oscillatoria, Spirulina, Nostoc*
Green bacteria	*Chlorobium*
Purple bacteria	*Rhodospirillum*
Gram-negative bacteria	*Proteus, Pseudomonas, Escherichia, Rhizobium, Desulfovibrio, Neisseria*
Gram-positive bacteria	*Streptococcus, Staphylococcus, Bacillus, Clostridium, Lactobacillus*
Spirochaetes	*Spirochaeta, Treponema*
Actinomycetes	*Actinomyces*
Rickettsias	*Rickettsiae, Chlamydia*
Mycoplasmas	*Mycoplasma*

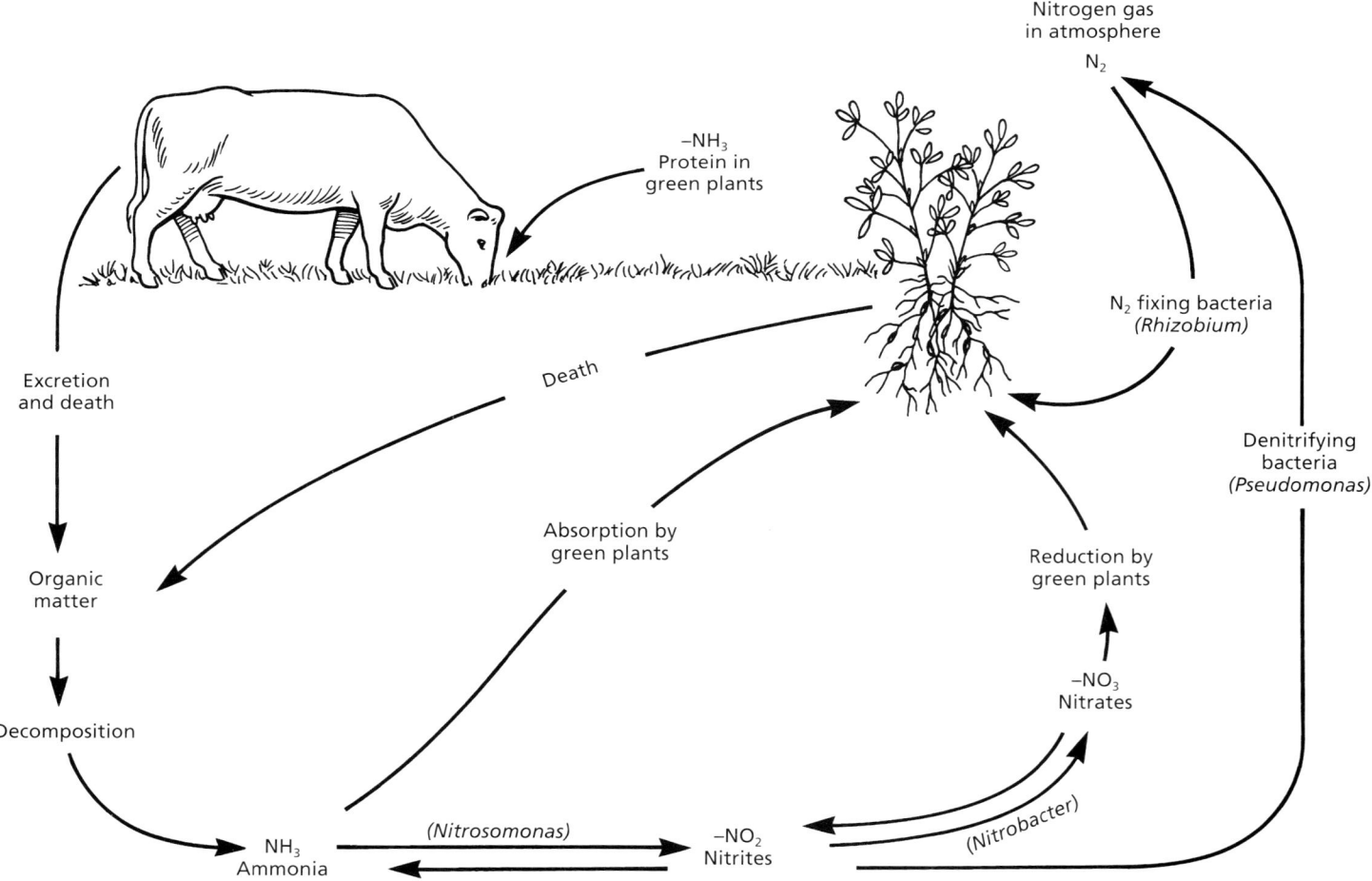

Figure 2.2 Nitrogen-fixing bacteria within the root nodules of legumes provide a usable source of nitrogen to plants.

Kingdom Monera

Figure 2.3 Cyanobacteria live in hot springs and hot streams, such as in the effluent from this geyser in Yellowstone National Park.
1. Mats of Cyanophyta

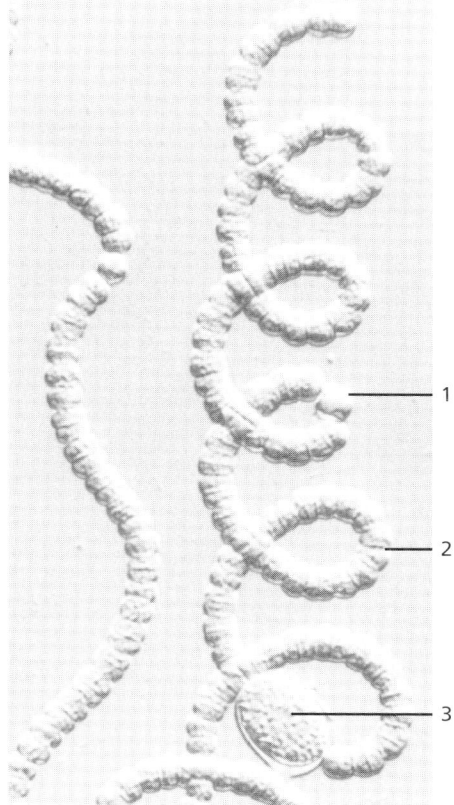

Figure 2.4 *Anabaena* filaments. This organism is a nitrogen-fixing cyanobacterium. Nitrogen fixation takes place within the heterocyst cells. (X430)
1. Heterocyst
2. Vegetative cell
3. Akinete (spore)

Figure 2.5 *Merismopedia*, a genus of cyanobacteria, is characterized by flattened colonies of cells. The cells are in a single layer and usually aligned into groups of two or four. (X430)

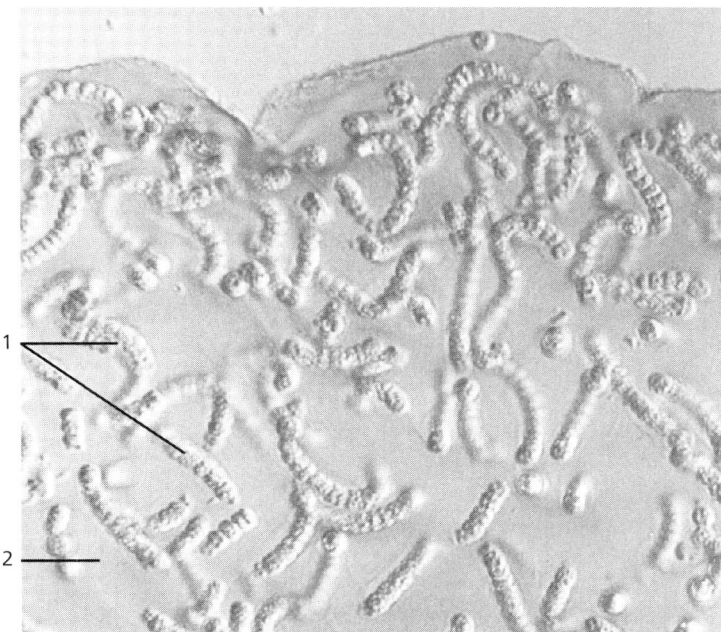

Figure 2.6 A colony of *Nostoc* filaments. Individual filaments secrete mucilage, which forms a gelatinous matrix around the filaments. (X430)
1. Filaments
2. Gelatinous matrix

Figure 2.7 *Oscillatoria* filaments. The only way this cyanobacterium can reproduce is by fragmentation of a filament. (X430)

Figure 2.8 A portion of a cylindrical filament of *Oscillatoria*. This cyanobacterium is common in most aquatic habitats. (X750)

1. Filament

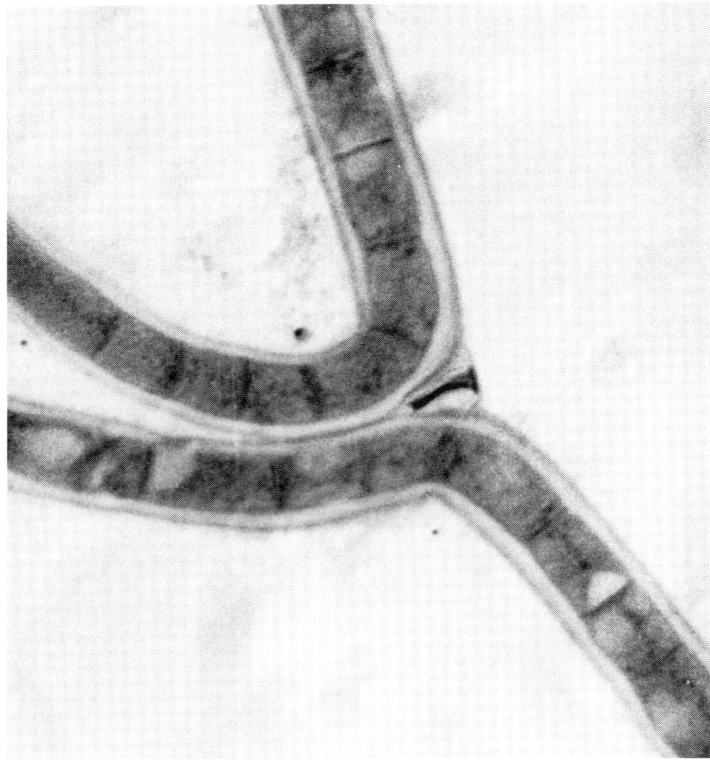

Figure 2.9 *Scytonema*, a cyanobacterium, is common on soil moistened from the spray of a waterfall or stream. Note the falsely branched filament typical of this genus. (about X500)

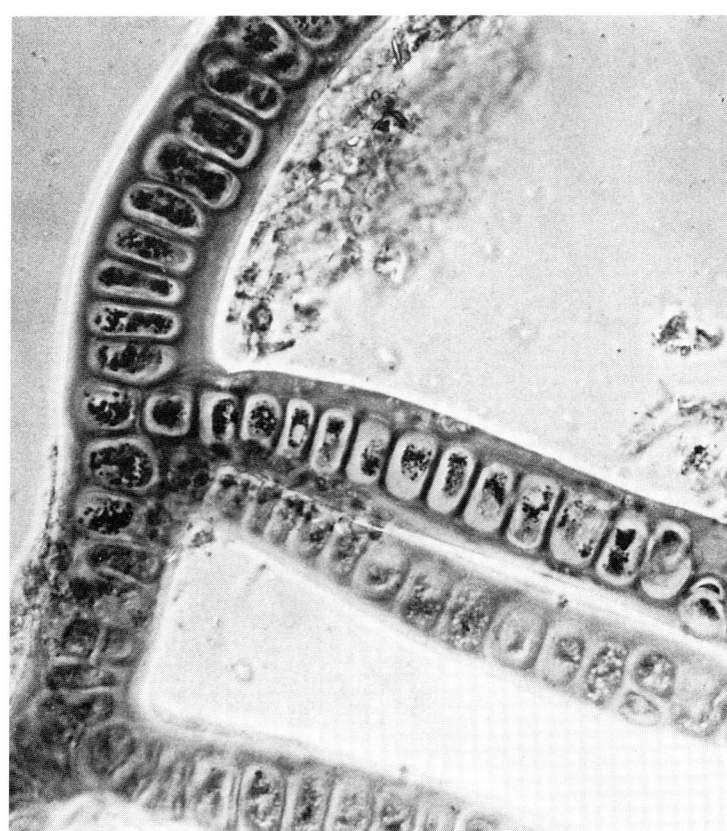

Figure 2.10 *Stigonema*, a cyanobacterium, has true branched filaments. (about X500)

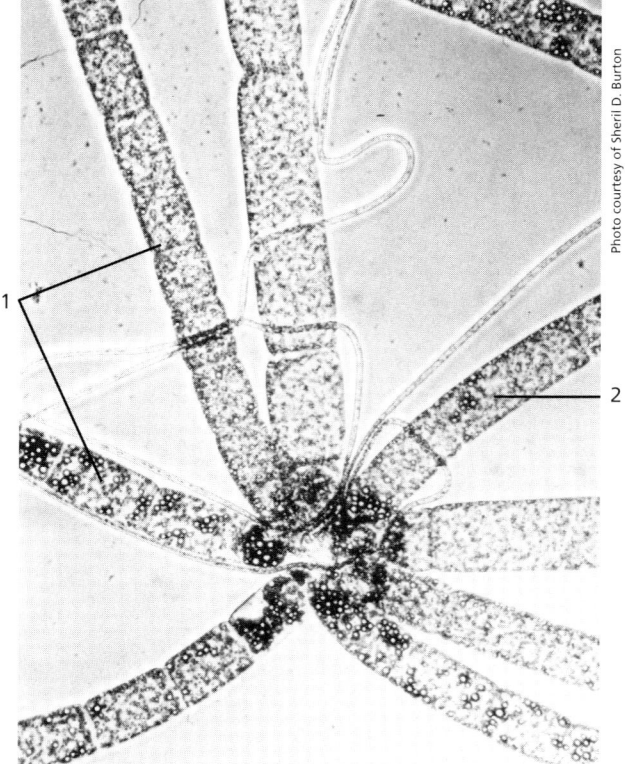

Figure 2.11 *Thiothrix*, a genus of bacteria that forms sulfur granules in its cytoplasm. These organisms obtain energy from the oxidation of H_2S. (X200)

1. Filaments 2. Sulfur granules

Kingdom Monera

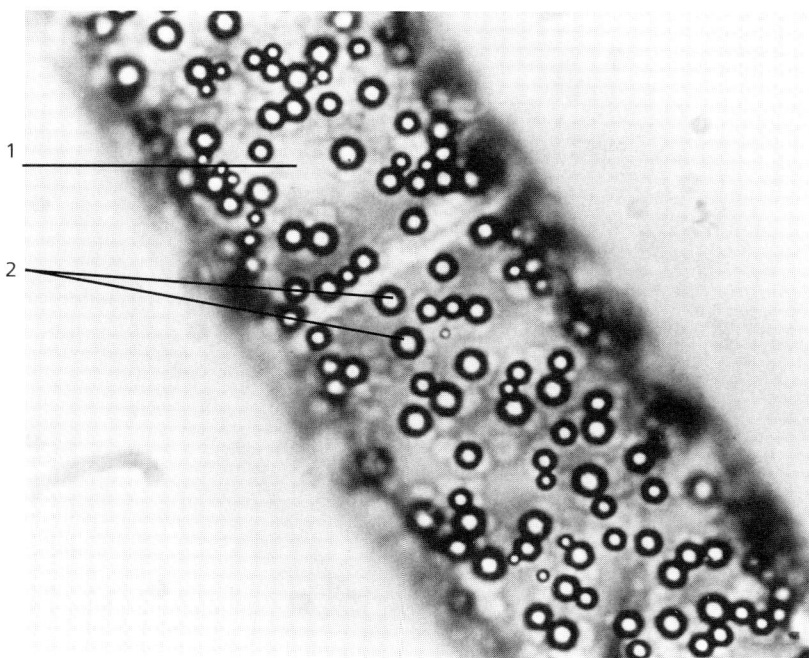

Figure 2.12 A magnified *Thiothrix* filament with sulfur granules in its cytoplasm. (X430)

1. Cytoplasm
2. Sulfur granules

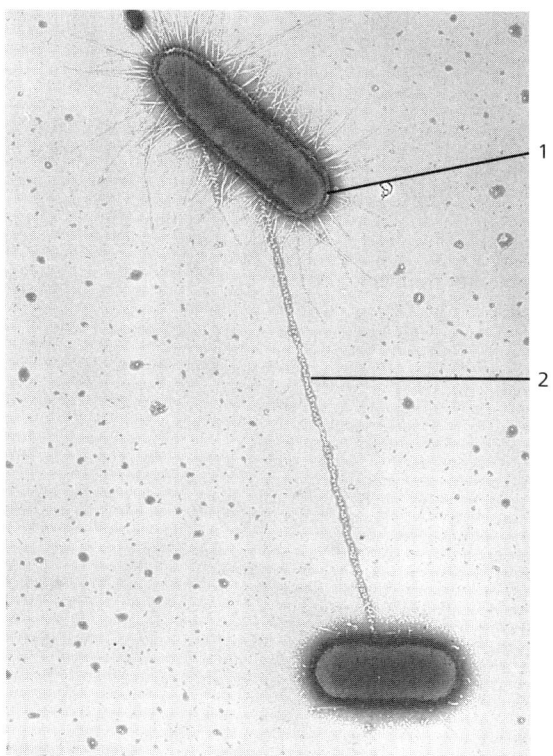

Figure 2.13 Conjugation of the bacterium *Escherichia coli*. In this process, genetic material is transferred through a conjugation tube from one cell to the other. (about X1700)

1. Bacterial cell
2. Conjugation tube

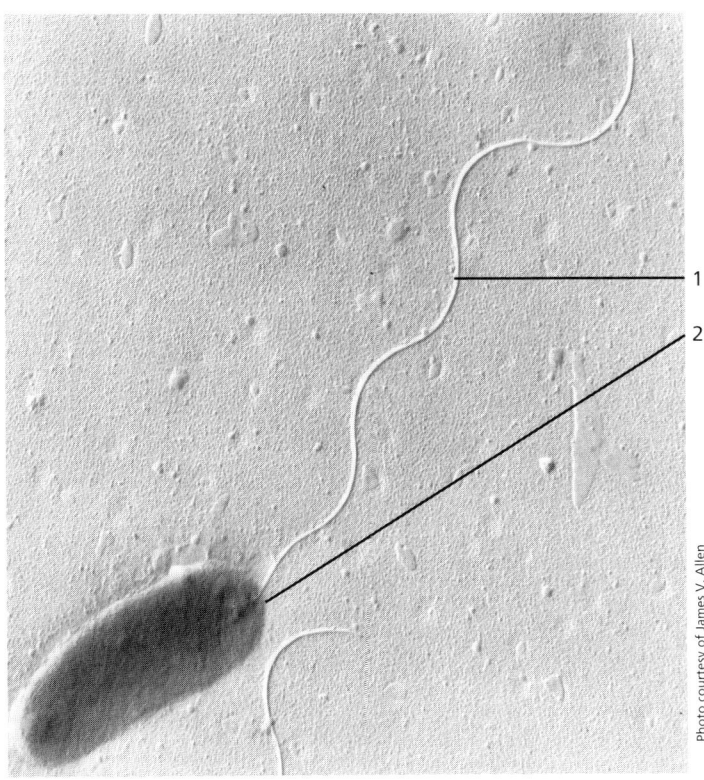

Figure 2.14 A flagellated bacterium, *Pseudomonas*. (about X2500)

1. Flagellum
2. Bacterial cell

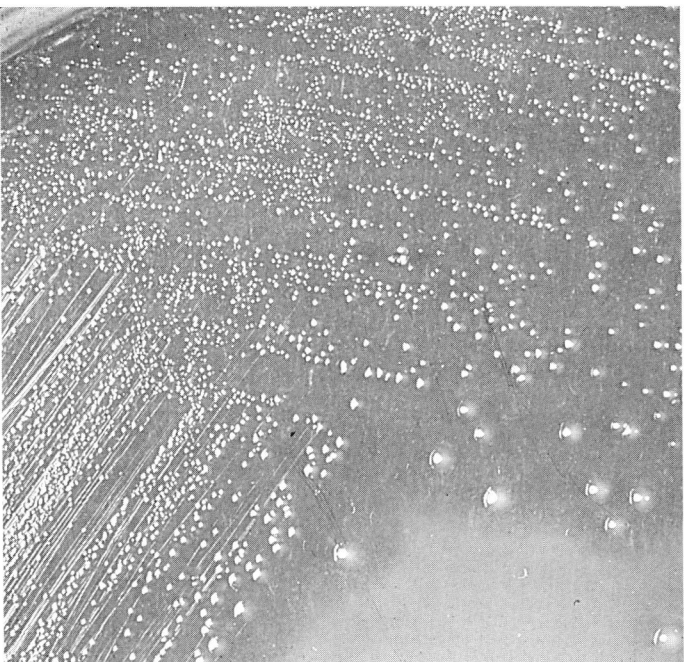

Figure 2.15 Colonies of *Streptococcus pyogenes* cultured on a nutrient agar plate. *S. pyogenes* causes strep throat and rheumatic fever. (X2)

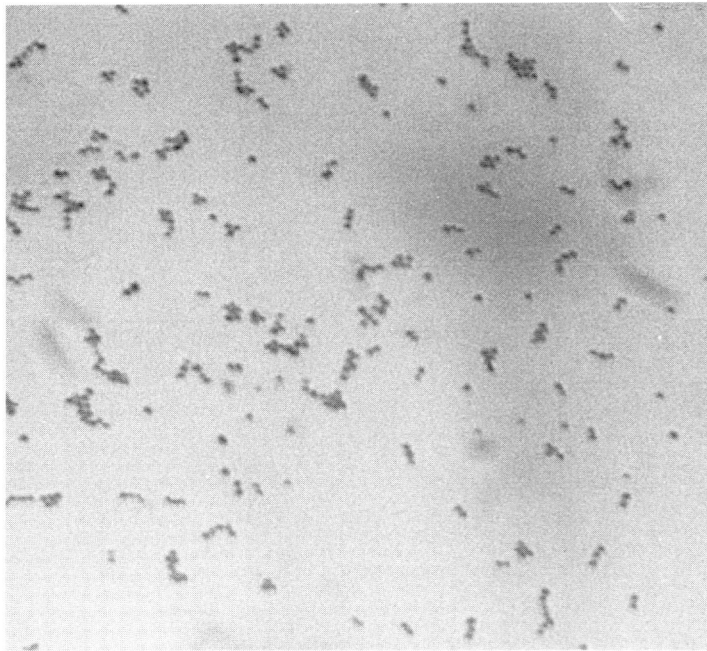

Figure 2.16 *Staphylococcus*, with cells arranged in irregular clusters. (X1000)

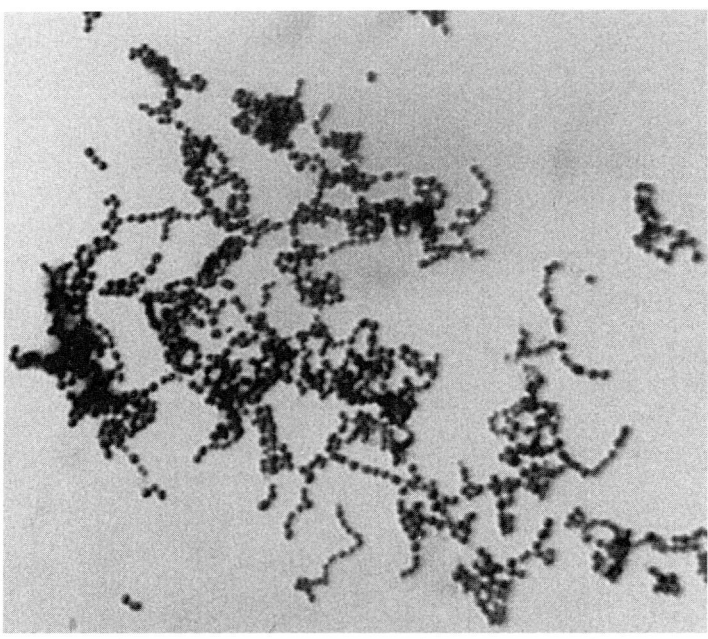

Figure 2.17 *Streptococcus pyogenes*, the organism that causes rheumatic fever. Note the chains of coccus-shaped bacteria. (X1000)

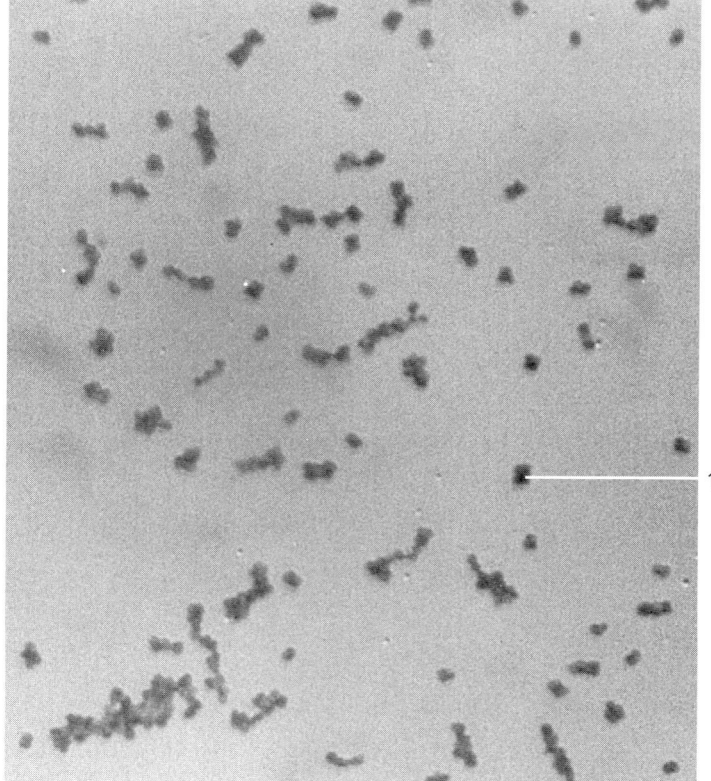

Figure 2.18 *Micrococcus luteus*. *Micrococcus* are gram-positive bacteria that are generally arranged as clusters or tetrads. (X1000)

1. Tetrad (4 cells)

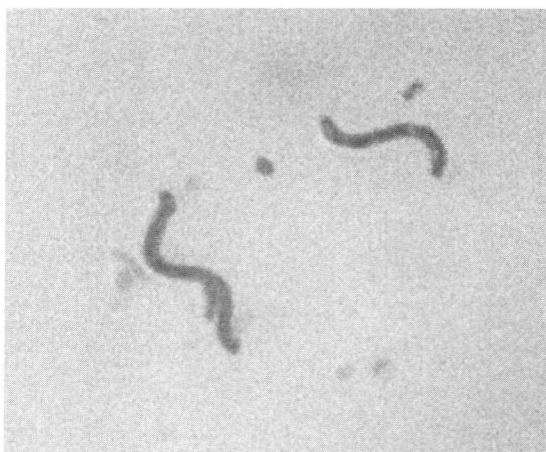

Figure 2.19 *Spirillum volutans*. Bacteria in the genus *Spirillum* are shaped like long rods twisted into rigid helices. They generally have multiple polar flagella. (X1000)

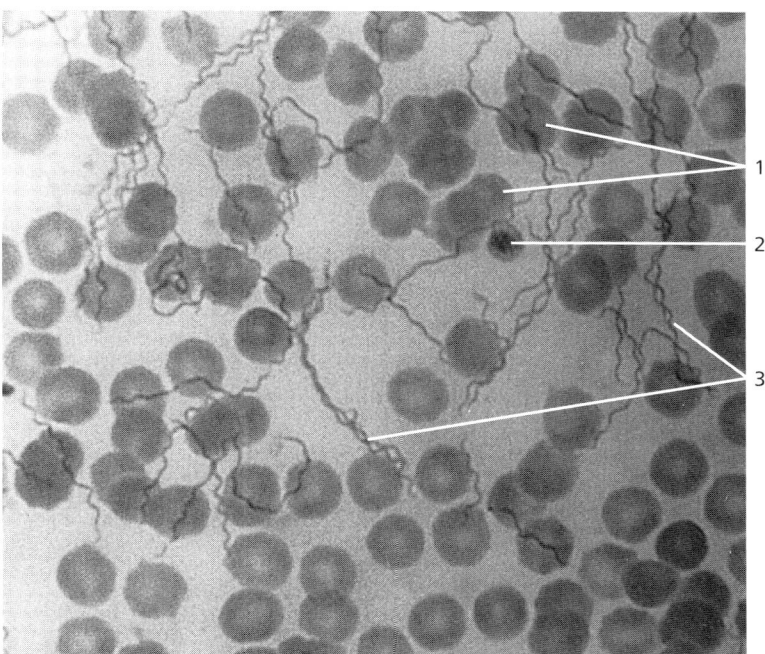
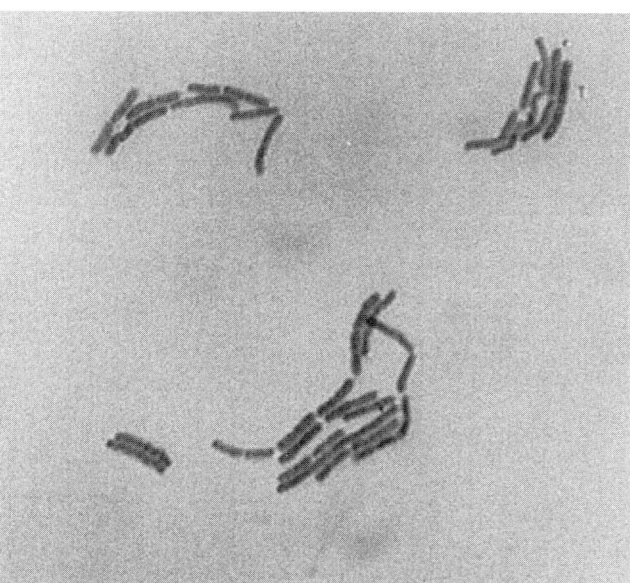

Figure 2.21 *Bacillus megaterium. Bacillus* is one of the few bacteria capable of producing endospores. This species of *Bacillus* generally remains in chains after it divides. (X1000)

Figure 2.20 A spirochete, *Borrelia recurrentis.* Spirochetes are flexible rods twisted into helical shapes. This species causes relapsing fever. (X1000)

1. Red blood cells
2. White blood cells
3. Spirochete

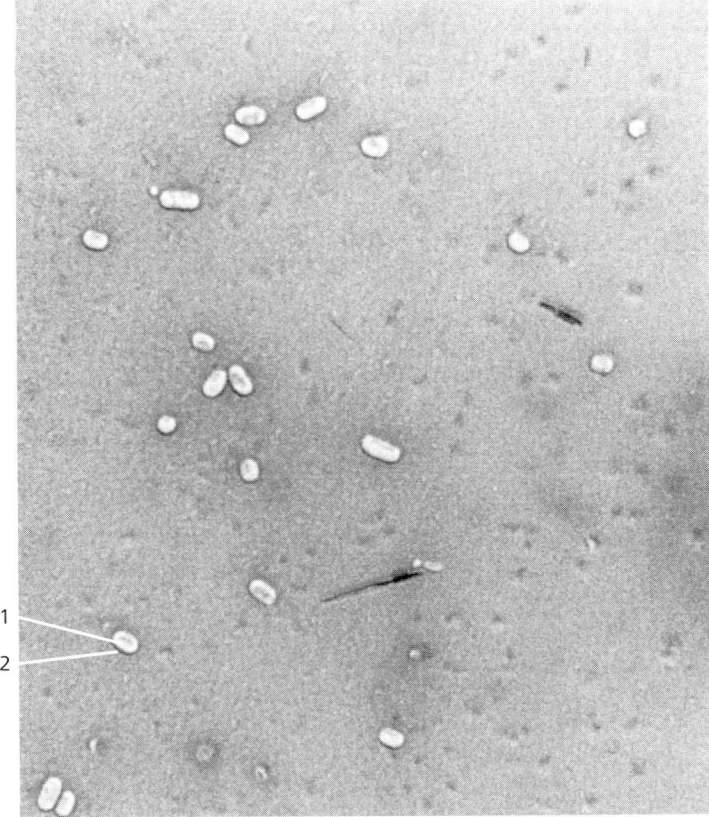

Figure 2.22 *Klebsiella pneumoniae* capsules. This bacterium is able to encapsulate, making itself more resistant to host defense mechanisms.

1. Cell   2. Capsule

Kingdom Protista: Primarily Unicellular Organisms

3

Most protists are unicellular, eukaryotic organisms, although some species are multicellular. Protists have a nucleus, mitochondria, chloroplasts, endoplasmic reticulae, and Golgi apparati. Protists are capable of meiosis and sexual reproduction; these processes evolved a billion or more years ago and occur in nearly all complex plants and animals.

Protists are abundant in aquatic habitats, and are important constituents of plankton. Plankton are communities of organisms that drift passively or swim slowly near the surface of ponds, lakes, and oceans. Plankton are a major source of food for other aquatic organisms. Photosynthetic protists are the primary food producers in aquatic ecosystems.

The unicellular algal protists include microscopic aquatic organisms within the divisions Chrysophyta and Pyrrhophyta. Chrysophyta are the yellow-green and golden-brown algae, and the diatoms. The cell wall of a diatom is composed largely of silica rather than cellulose. Some diatoms move in a slow, gliding way as cytoplasm glides through slits in the cell wall to propel the organism.

The Pyrrhophyta are single-celled, algae-like organisms, the most important of which are the dinoflagellates. In most species of dinoflagellates, the cell wall is formed of armor-like plates of cellulose. Dinoflagellates are motile, having two flagella, one encircling the organism in a transverse groove, and the other projecting to the posterior.

Protozoa are also protists. They are small ($2\mu m - 100 \mu m$), unicellular eukaryotic organisms that lack a cell wall. Movement of protozoa is due to flagella, cilia, or pseudopodia of various sorts. In feeding upon other organisms or organic particles, they use simple diffusion, pinocytosis (active transport), or phagocytosis. Although most Protozoa reproduce asexually, some species may also reproduce sexually during a portion of their life cycle. Most protozoa are harmless, although some are of immense clinical concern because they are parasitic and may cause human disease, including African sleeping sickness and malaria.

DIVISION CHRYSOPHYTA (diatoms, golden algae)

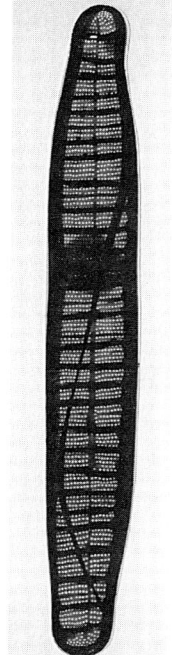
Diatoma

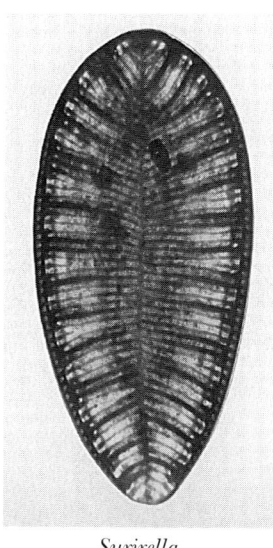
Cymbella

Surirella

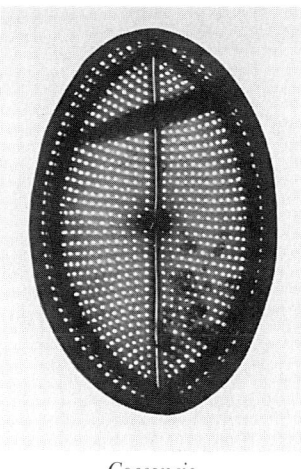
Cocconeis

Navicula

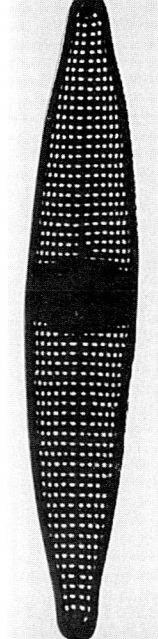
Synedra

Figure 3.1 Electron micrographs of several types of pennate diatoms.

CHAPTER 3 — Kingdom Protista: Primarily Unicellular Organisms

DIVISION CHRYSOPHYTA (diatoms, golden algae)

TABLE 3.1
Some Representatives of the Kingdom Protista: Primarily Unicellular Organisms

Taxa and Representative Kinds	Characteristics
Chrysophyta (diatoms and golden algae)	Diatom cell walls of silica, with two halves; plastids often golden
Pyrrhophyta (dinoflagellates)	Two flagella in grooves of wall; brownish plastids
Rhizopoda (amoebas)	Cytoskeleton of microtubules and microfilaments; amoeboid locomotion
Apicomplexa (sporozoa, *Plasmodium*)	Lack locomotor capabilities and contractile vacuoles; mostly parasitic
Euglenophyta (euglenoids)	Green flagellates lacking typical cell walls
Ciliophora (ciliates, *Paramecium*)	Use cilia to move and feed

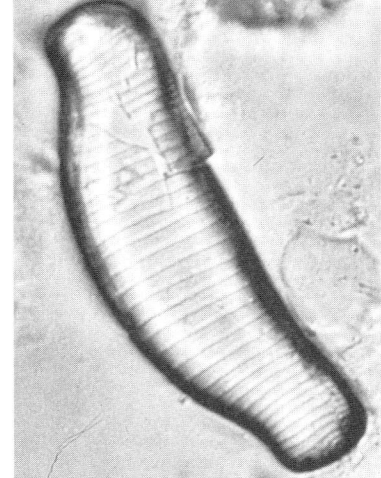
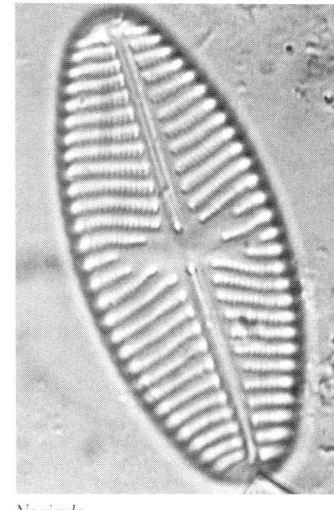
Eunotia   *Navicula*

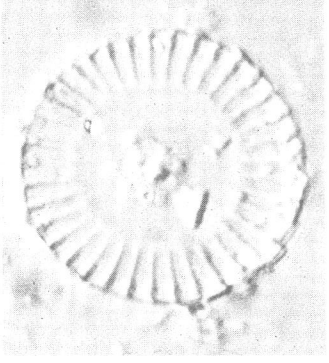
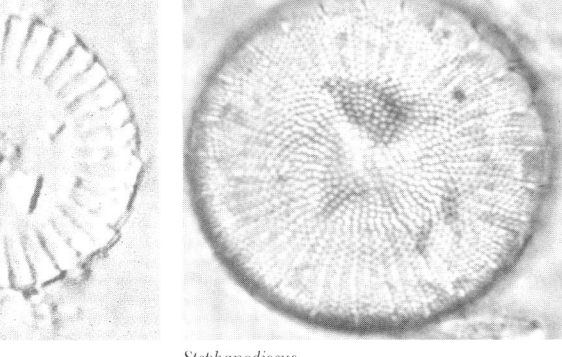
Cyclotella   *Stephanodiscus*

Figure 3.2 Examples of common freshwater diatoms. *Eunotia* species are often found in acidic waters common in forest ponds or bogs. *Navicula* is a large and widely distributed genus with species found on wet soil and in both marine and fresh-water habitats. *Cyclotella* and *Stephanodiscus* are centric (round in face view) diatoms, common in lake plankton.

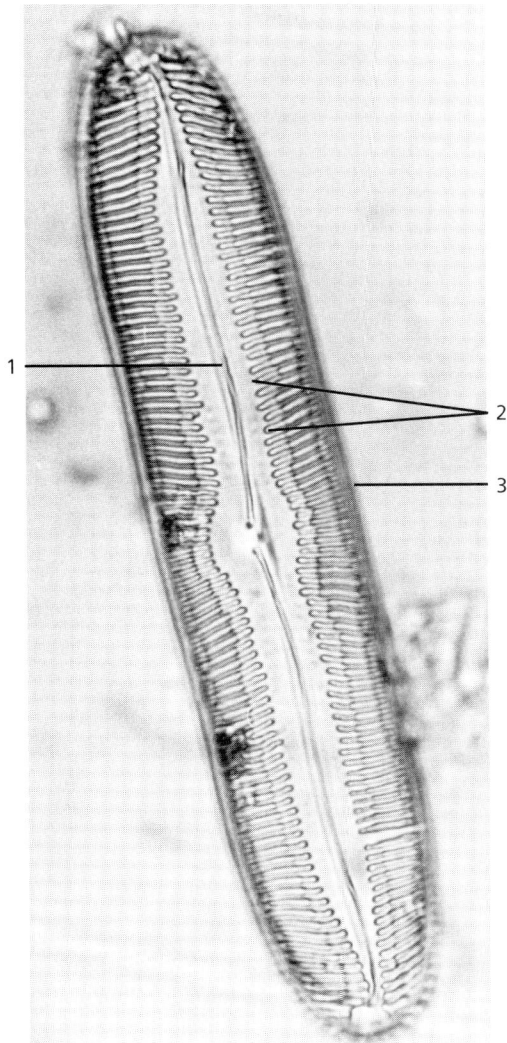

Figure 3.3 *Pinnularia*, a common diatom. (X430)
1. Raphe   2. Striae   3. Valve

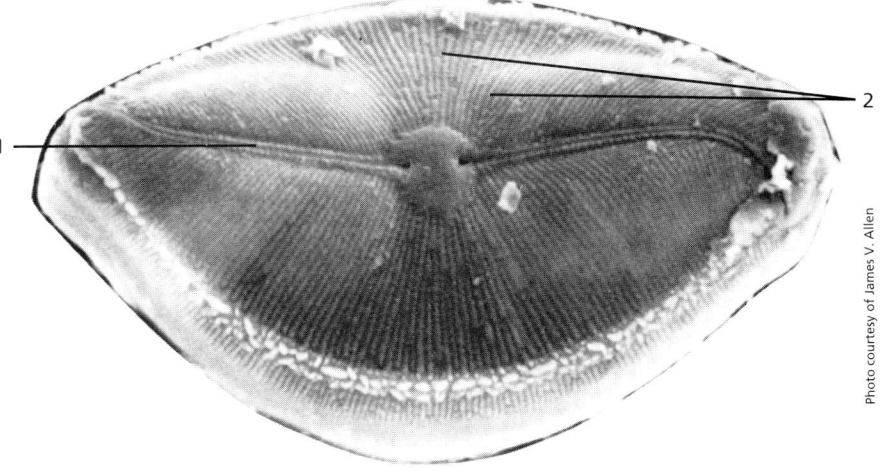

Figure 3.4 Scanning electron micrograph of the diatom *Achnanthes flexella*. (about X700)
1. Raphe   2. Striae

DIVISION CHRYSOPHYTA (diatoms, golden algae)

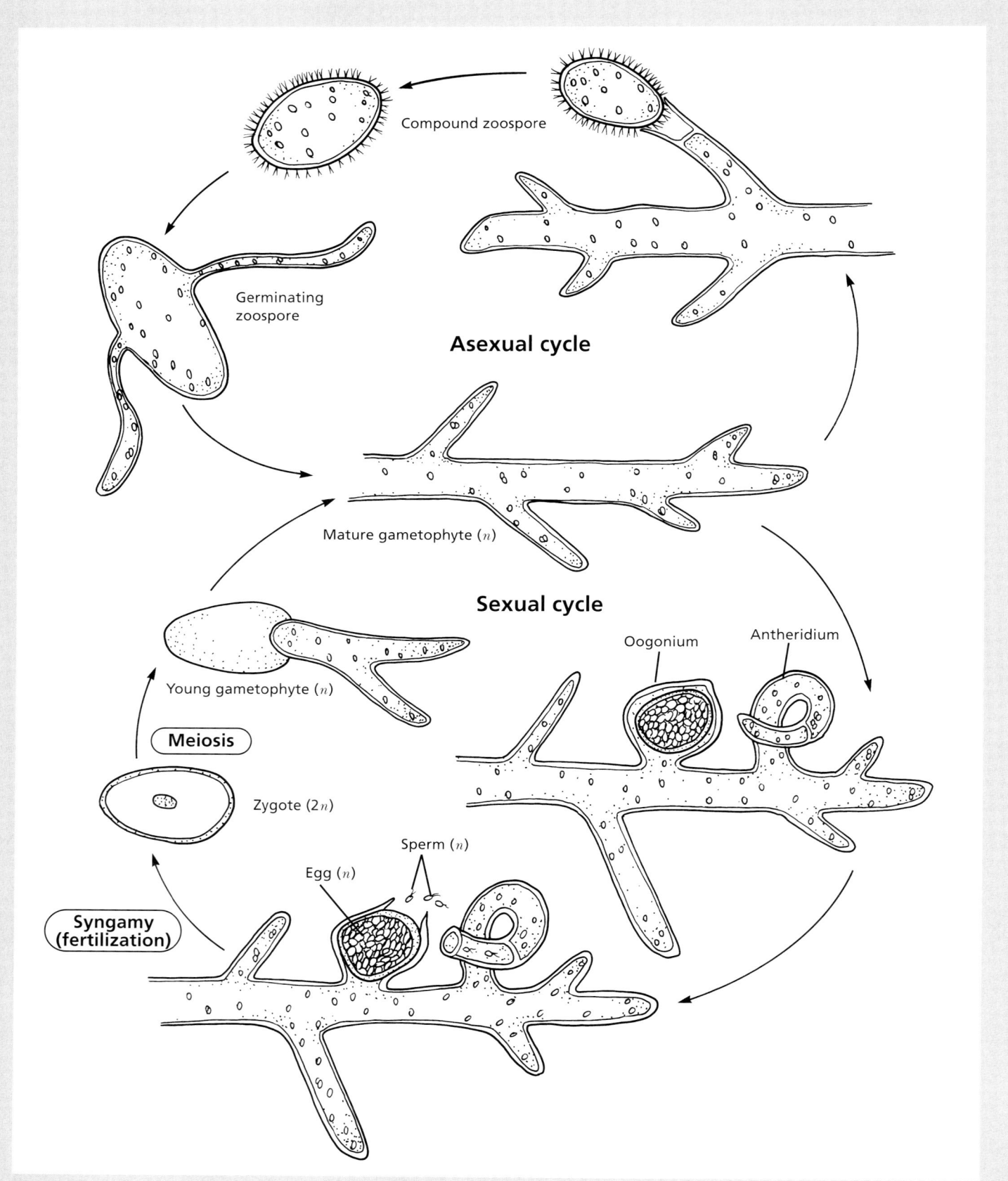

Figure 3.5 Life cycle of the "water felt," *Vaucheria*.

DIVISION CHRYSOPHYTA (diatoms, golden algae)

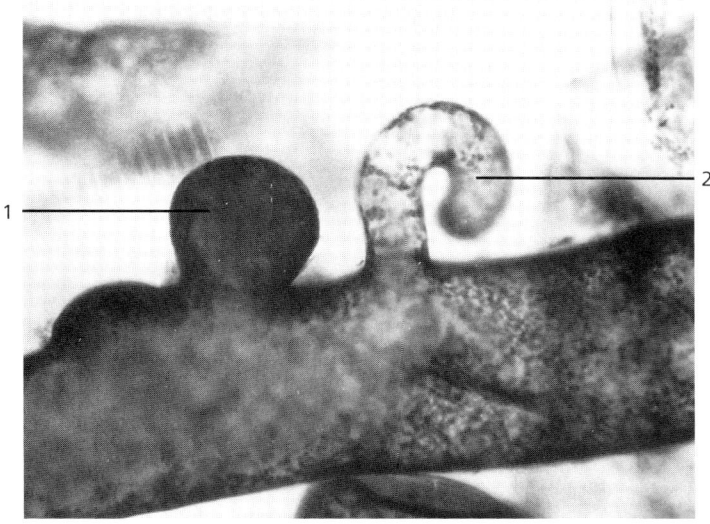

Figure 3.6 A filament with an immature gametangia of the "water felt," *Vaucheria*. *Vaucheria* is a chrysophyte that is widespread in freshwater and marine habitats. It is also found in the mud of brackish areas that periodically become submerged and then exposed to air. (X430)

1. Oogonium
2. Antheridium

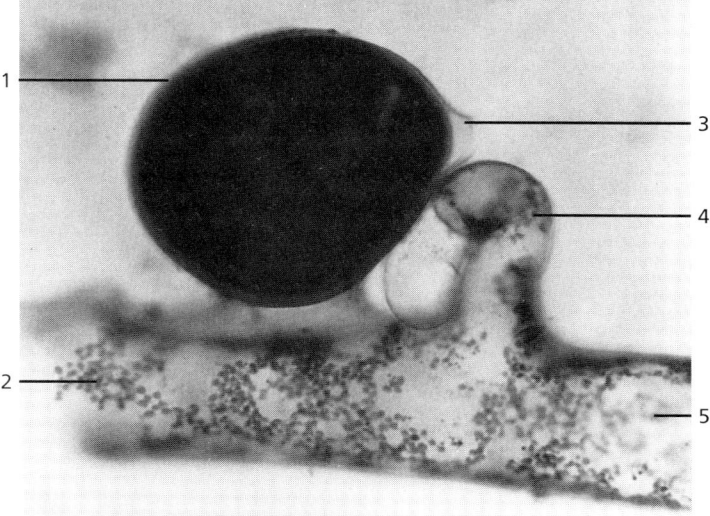

Figure 3.7 *Vaucheria*, with mature gametangia. (X430)

1. Oogonium
2. Chloroplasts
3. Fertilization pore
4. Antheridium
5. Coenocytic filament

DIVISION PYRRHOPHYTA (dinoflagellates)

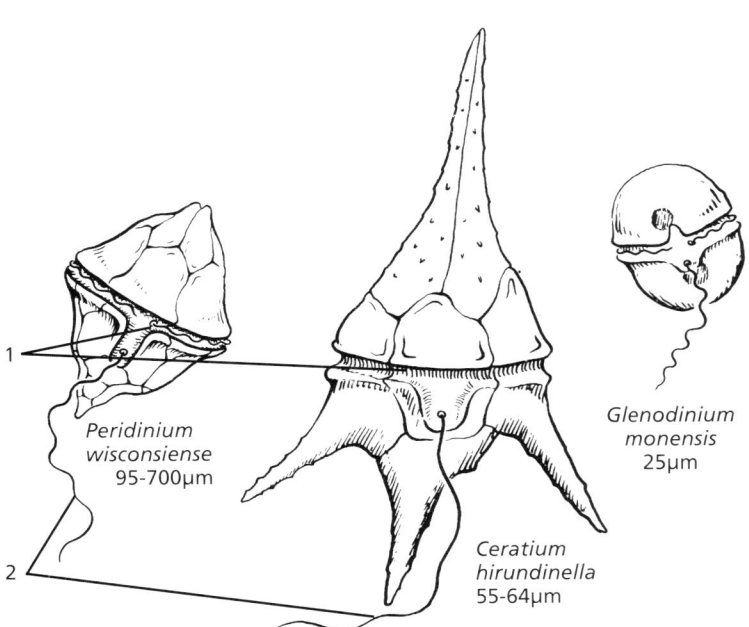

Figure 3.8 Representative dinoflagellates.
1. Girdles
2. Flagella

Figure 3.9 A diagram of the dinoflagellate, *Gymnodinium*. Lacking a cell wall, specimens within this genus are known as unarmored dinoflagellates. (about 50 μm)

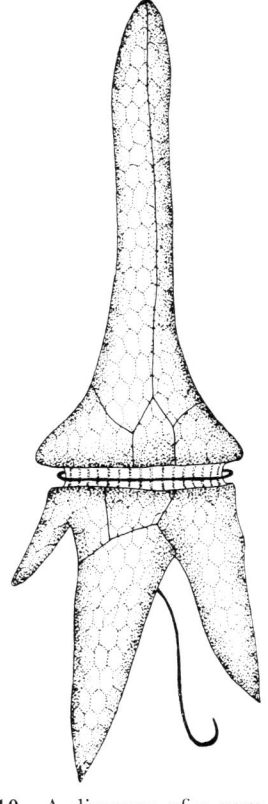

Figure 3.10 A diagram of a common freshwater dinoflagellate, *Ceratium hirundinella*. (about 60 μm)

DIVISION PYRRHOPHYTA (dinoflagellates)

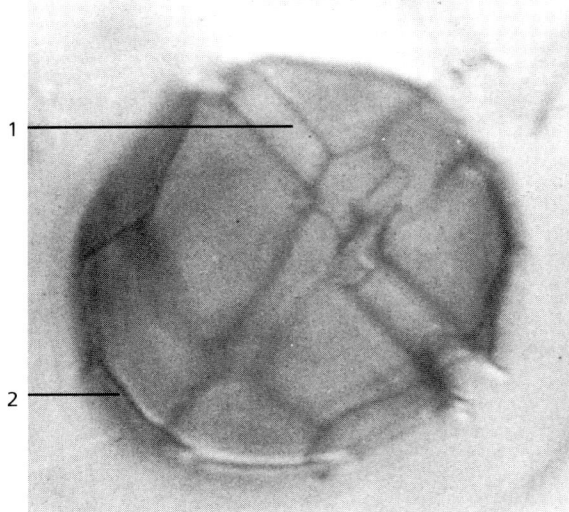

Figure 3.11 Photomicrograph of a dinoflagellate. The cell wall of many dinoflagellates is composed of overlapping plates of cellulose, evident in this photomicrograph. (X430)

1. Transverse groove
2. Wall of cellulose plates

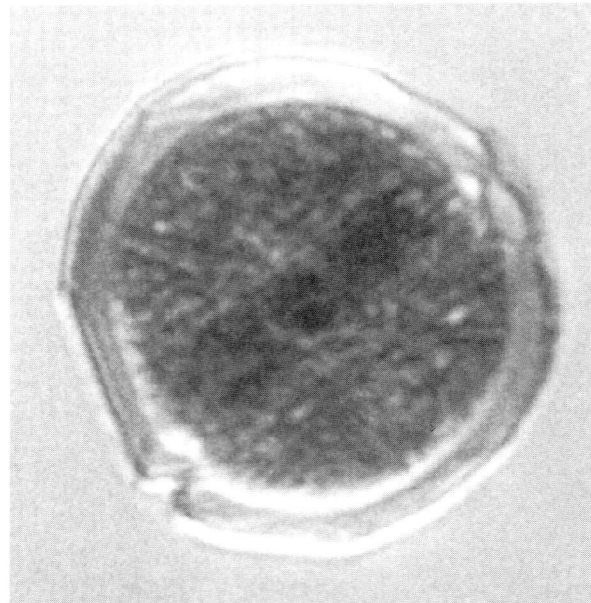

Figure 3.12 *Peridinium* species, a common fresh-water dinoflagellate.

DIVISION RHIZOPODA (amoebas)

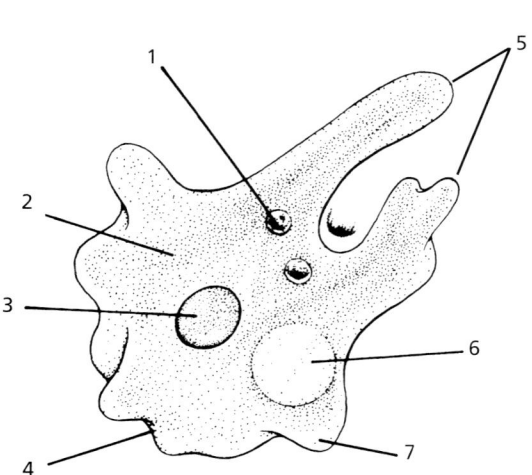

Figure 3.13 *Amoeba proteus* is a fresh-water protozoan that moves by forming cytoplasmic extensions called pseudopodia.

1. Food vacuole
2. Endoplasm
3. Nucleus
4. Cell membrane
5. Pseudopodia
6. Contractile vacuole
7. Ectoplasm

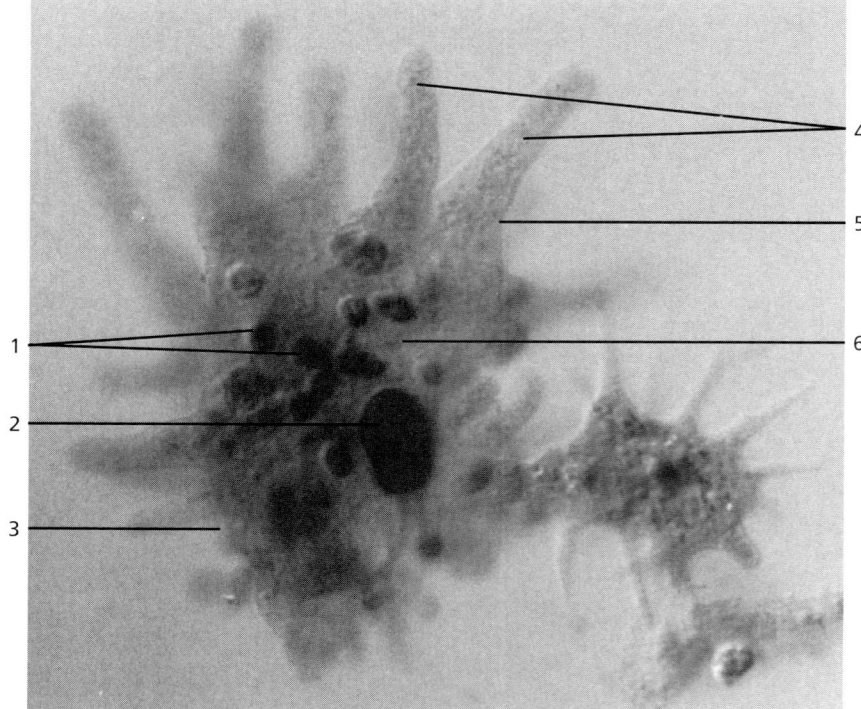

Figure 3.14 *Amoeba proteus.* (X160)

1. Food vacuoles
2. Nucleus
3. Cell membrane
4. Pseudopodia
5. Ectoplasm
6. Endoplasm

DIVISION APICOMPLEXA (sporozoans)

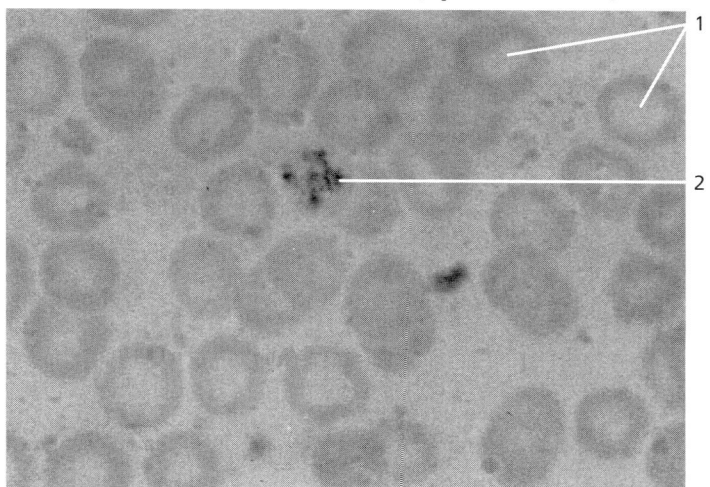

Figure 3.15 *Plasmodium vivax*. This protozoan causes malaria, transmitted by the female mosquito. (X1000)
1. Red blood cells
2. Merozites in red blood cell

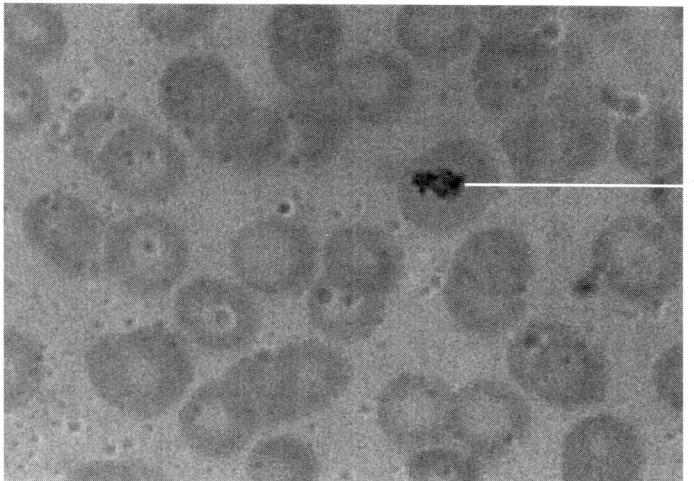

Figure 3.16 *Plasmodium vivax*, in the ring stage. (X1000)
1. Trophozoite in a red blood cell

DIVISION EUGLENOPHYTA (euglenoids)

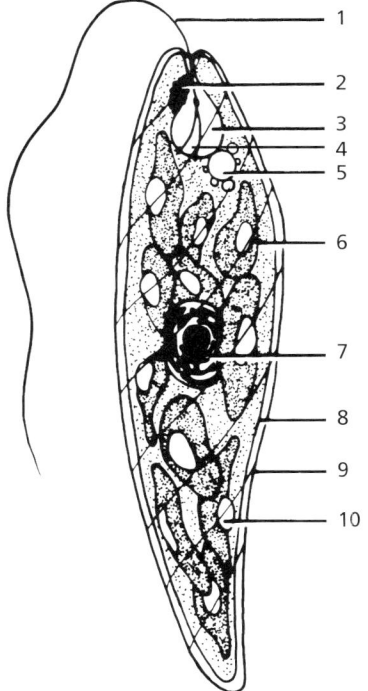

Figure 3.17 *Euglena* is a common flagellate that contains chloroplasts. They are fresh-water organisms that have a flexible pellicle rather than a rigid cell wall.

1. Flagellum
2. Photoreceptor
3. Reservoir
4. Basal body
5. Contractile vacuole
6. Chloroplast
7. Nucleus
8. Pellicle
9. Cell membrane
10. Paramylum granule (for food storage)

Figure 3.18 Light micrograph of *Euglena*. (X430)
1. Striated pellicle
2. Chloroplast
3. Paramylum granule
4. Flagellum
5. Reservoir
6. Nucleus
7. Paramylum granule

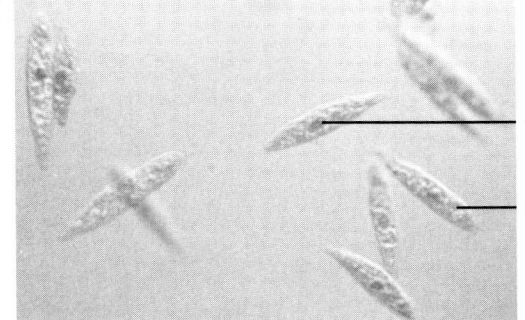

Figure 3.19 *Euglena*. (X100)
1. Nucleus
2. Chloroplast

DIVISION CILIOPHORA (ciliates, *Paramecium*)

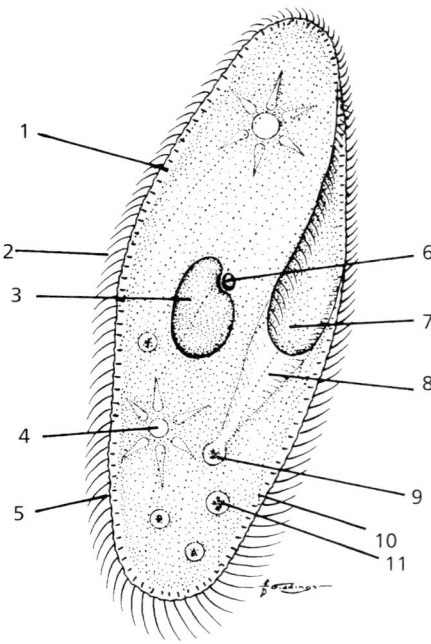

Figure 3.20 *Paramecium caudatum* is a ciliated protozoan. The poisonous trichocysts of these unicellular organisms are used for defense and capturing prey.

1. Trichocyst
2. Cilia
3. Macronucleus
4. Contractile vacuole
5. Pellicle
6. Micronucleus
7. Oral cavity
8. Gullet
9. Forming food vacuole
10. Anal pore
11. Food vacuole

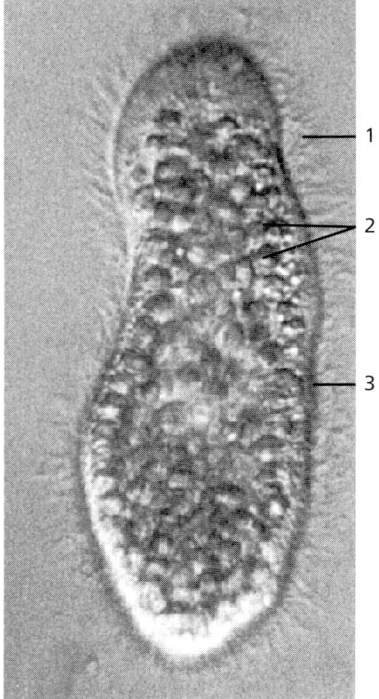

Figure 3.21 *Paramecium* is a unicellular, slipper-shaped organism. Paramecia are usually common in ponds containing decaying organic matter. (X400)

1. Cilia
2. Micronuclei
3. Pellicle

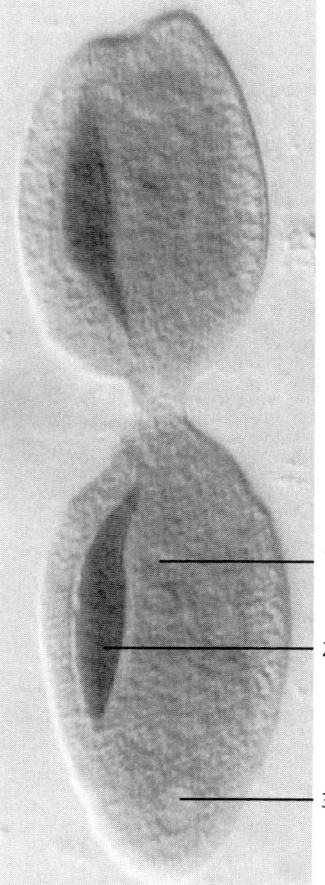

Figure 3.22 *Paramecium* in fission. (X400)

1. Micronucleus
2. Macronucleus
3. Contractile vacuole

Kingdom Protista: Primarily Multicellular Organisms

Included in the Kingdom Protista are the algae and protists that resemble fungi. Protists live in nearly all Earth's habitats. Protists can be aquatic or inhabitants of damp regions. Many symbiont protists inhabit the host's body cells, tissues, or fluids. Some parasitic protists are pathogens to plants and animals. Most protists are aerobic, using mitochondria for cellular respiration. Some protists have chloroplasts and are photoautrophs. Others are heterotrophs, absorbing or ingesting organic molecules.

Although protists vary considerably in their mode of reproduction, all can reproduce asexually. Some protists only reproduce asexually. Other undergo meiosis and syngamy and then reproduce asexually. Certain protists endure harsh conditions by forming protective cysts during a portion of their life cycle.

The three divisions of primarily multicellular algae are Phaeophyta (brown algae), Chlorophyta (green algae) and Rhodophyta (red algae). Most algae are multicellular, aquatic organisms. For example, seaweeds (phaeophytes) are multicellular, mostly marine, brown algae. Included in this division are giant kelp that may exceed 100 meters in length. Consisting of cellulose and algin, the cell walls of brown algae can withstand the movements of ocean currents and waves. These substances give seaweeds their characteristic slimy and rubbery feel.

Considerable evidence indicates that Chlorophyta are ancestral to plants. Most green algae live in fresh water, although marine planktonic and "attached forms" exist. Chlorophytes are photoautotrophic, manufacturing their own food. Certain chlorophytes live symbiotically with fungi, forming organisms known as lichens.

Most red algae are multicellular, marine forms. Colors other than red among Rhodophyta are not uncommon. Similar to brown algae, red algae are commonly called seaweeds. They reproduce sexually but lack flagellated stages. Alternation of generations is common. High in starch, some rhodophytes are harvested for food. Agar, used as a culture medium in bacteriology, is obtained from a species of red algae.

Organisms within the divisions Myxomycota (plasmodial slime molds), Acrasiomycota (cellular slime molds), and Oomycota (water mold, white rusts, downy mildews) resemble fungi; but the similarities are believed to be due to convergence. Many of these organisms are yellow, red, or orange. All are non-photosynthetic heterotrophs. The body of a Myxomycota is a multinucleated continuum of cytoplasm undivided by membranes or walls. During the feeding stage, an amoeboid mass called a *plasmodium* extends through moist organic soil, leaves, or decaying logs engulfing food particles by phagocytosis. Plasmodial slime molds are important decomposers in some habitats.

Acrasiomycota have a feeding stage consisting of solitary haploid cells, and an aggregate stage consisting of an amoeboid mass of cells. The aggregate stage is formed from thousands of individual cells that join at some signal to form a single body. Asexual fruiting bodies are produced by the aggregate stage.

Most Oomycota are colonial saprophytes on dead plants and animals, and many are important decomposers in freshwater ecosystems. Some Oomycota are parasitic on the skin and gills of fishes. Species of white rusts and downy mildews live on land as plant parasites. Distributed by windblown spores, these organisms are of concern in the potato industry (potato blight) and the grape and wine industry (downy mildew).

TABLE 4.1.
Some Representatives of the Kingdom Protista: Primarily Multicellular Organisms

Division and Representative Kinds	Characteristics
ALGAE	
Division Chlorophyta (green algae)	Unicellular, colonial, and multicellular forms; mostly fresh water; reproduce asexually and sexually; gametes often biflagellated, with cup-shaped chloroplasts
Division Phaeophyta (brown algae, giant kelp)	Multicellular, mostly marine in the intertidal zone; most with alternation of generations
Division Rhodophyta (red algae)	Multicellular, mostly marine; sexual reproduction but with no flagellated cells; alternation of generations common
PROTISTS RESEMBLING FUNGI	
Division Myxomycota (plasmodial slime molds)	Multinucleated continuum of cytoplasm without internal membranes; amoeboid plasmodium during feeding stage; produce asexual, fruiting bodies
Division Acrasiomycota (cellular slime molds)	Solitary cells during feeding stage; aggregate of cells when food is scarce; produce asexual, fruiting bodies
Division Oomycota (water molds, white rusts, downy mildews)	Decomposers or parasitic forms; walls of cellulose, dispersal by spores or flagellated zoospores

DIVISION CHLOROPHYTA (green algae)

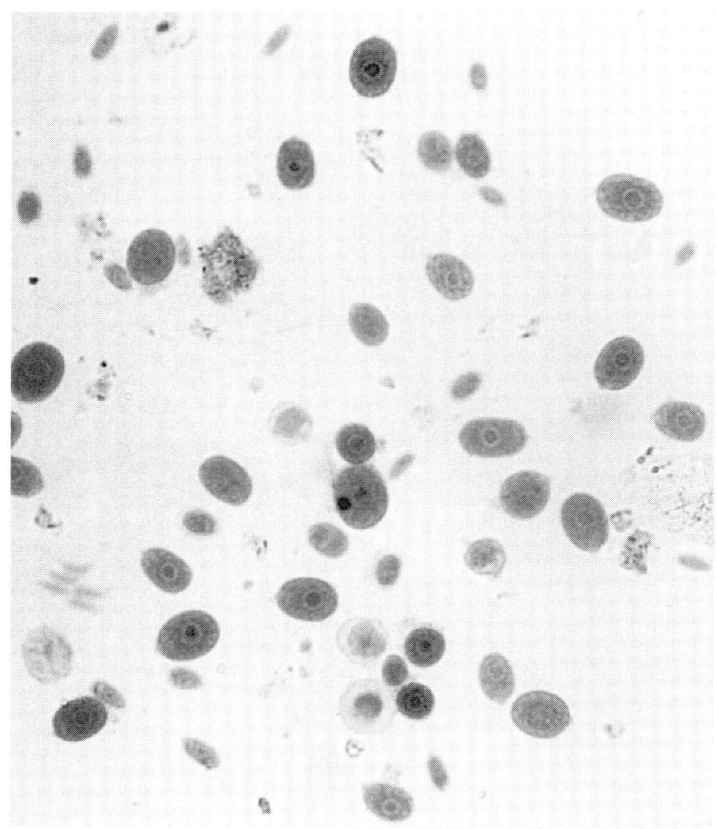

Figure 4.1 *Chlamydamonas*, a common unicellular green alga. (X800)

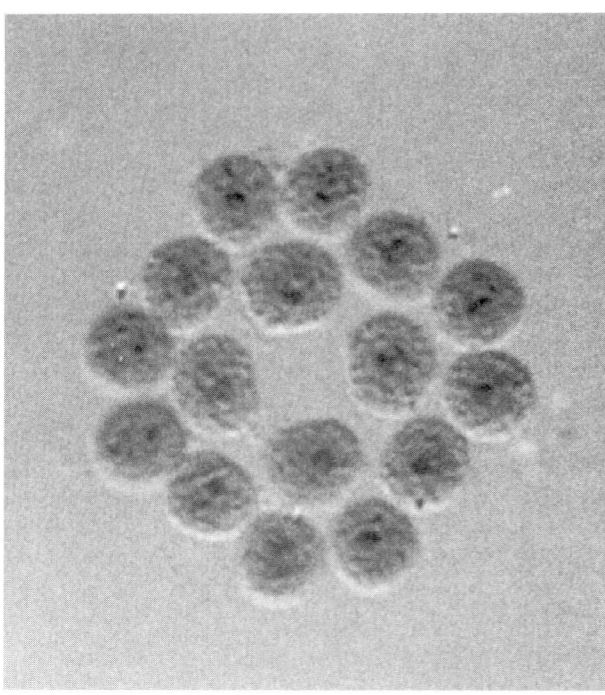

Figure 4.2 *Gonium* colony. *Gonium* is a 16-celled flat colony of *Chlamydamonas*-like cells. (X450)

DIVISION CHLOROPHYTA (green algae)

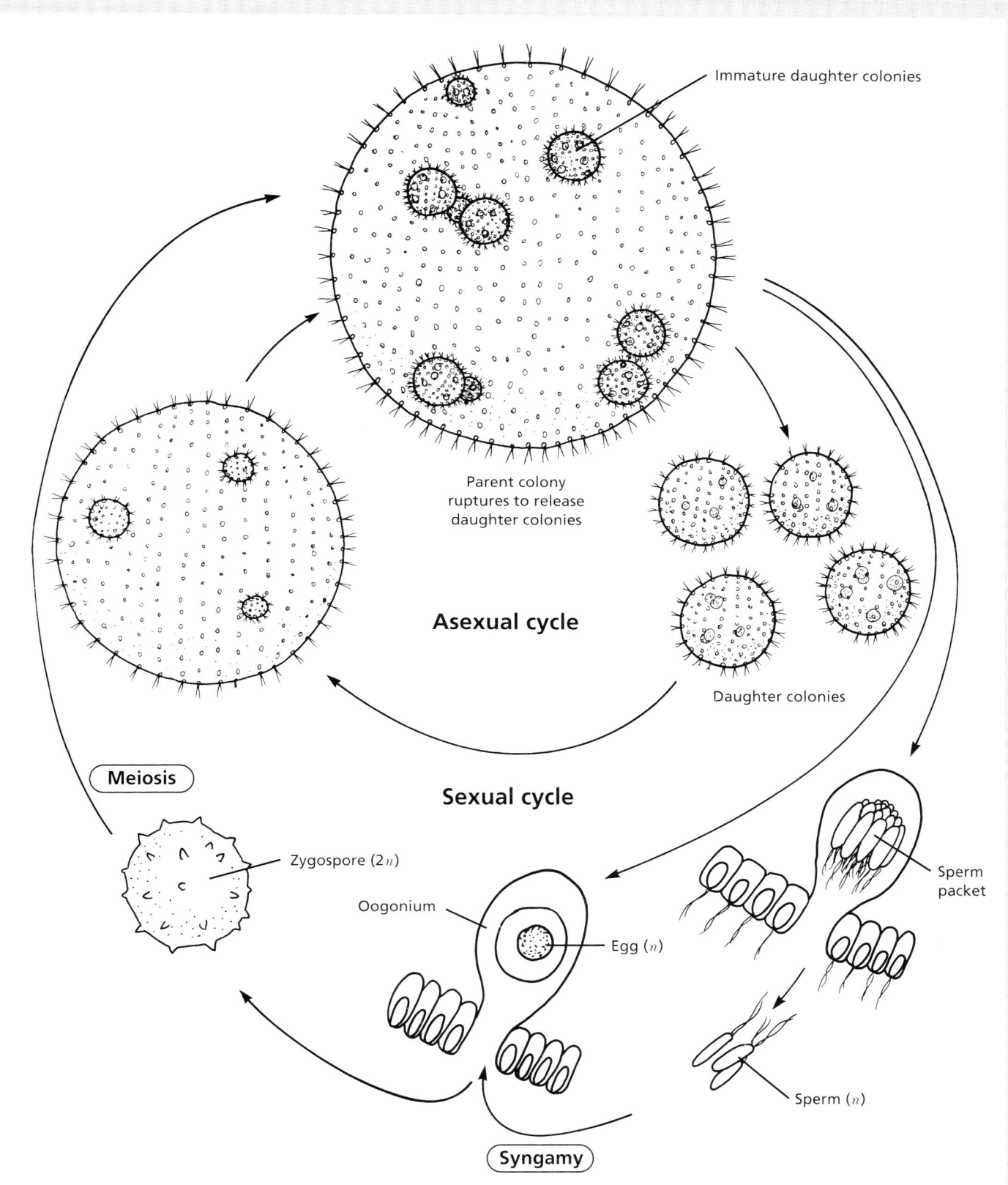

Figure 4.3 Life cycle of *Volvox*, a common freshwater chlorophyte. *Volvox* is considered by some to be a colony and by others to be a single, integrated plant.

DIVISION CHLOROPHYTA (green algae)

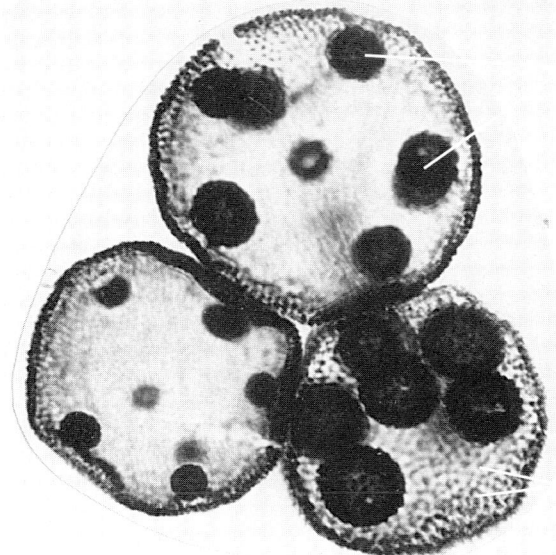

Figure 4.4 *Volvox*. Three separate organisms are shown in this photomicrograph, each containing daughter colonies. (X100)
1. Daughter colonies
2. Vegetative cells

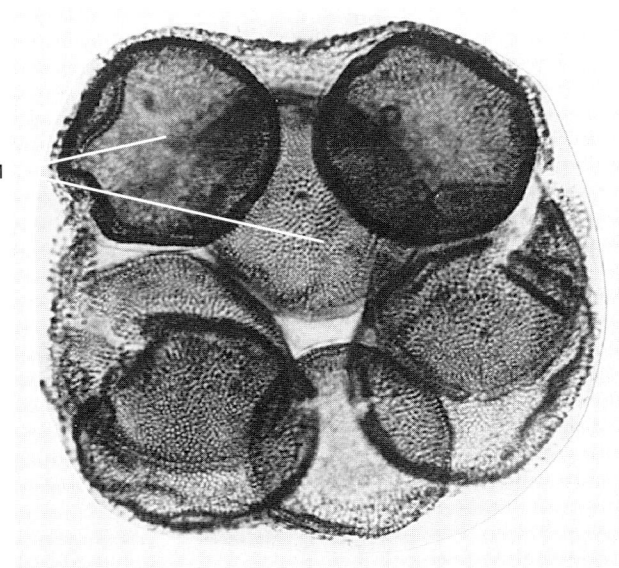

Figure 4.5 *Volvox*, a single organism with several large daughter colonies. (X100)
1. Daughter colonies

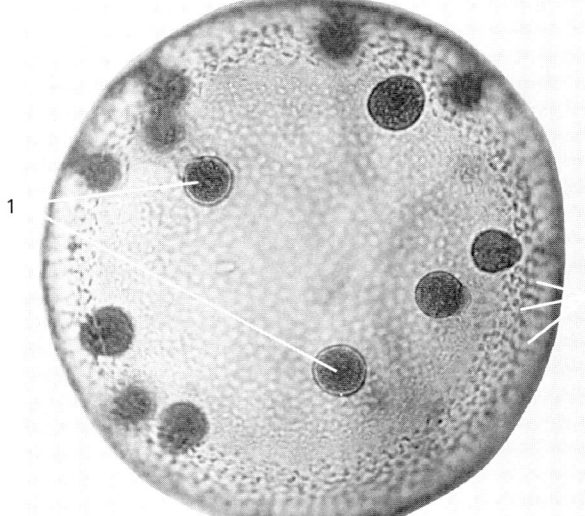

Figure 4.6 *Volvox*, a single mature specimen with several eggs and zygotes. (X100)
1. Zygotes
2. Vegetative cells

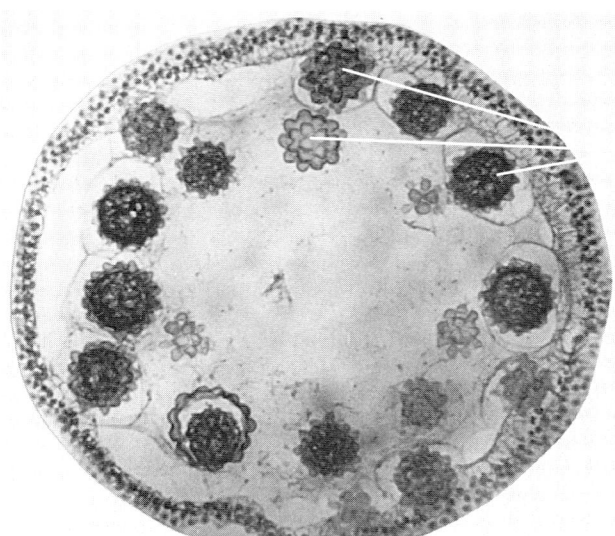

Figure 4.7 *Volvox*, a single mature organism with zygospores. (X100)
1. Zygospores

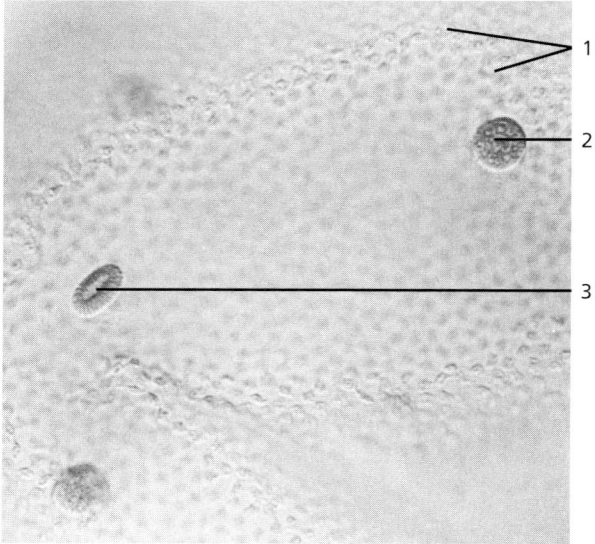

Figure 4.8 *Volvox*, a motile green alga. This photograph is a highly magnified view of a single organism. (X430)
1. Vegetative cells   2. Egg   3. Sperm packet

DIVISION CHLOROPHYTA (green algae)

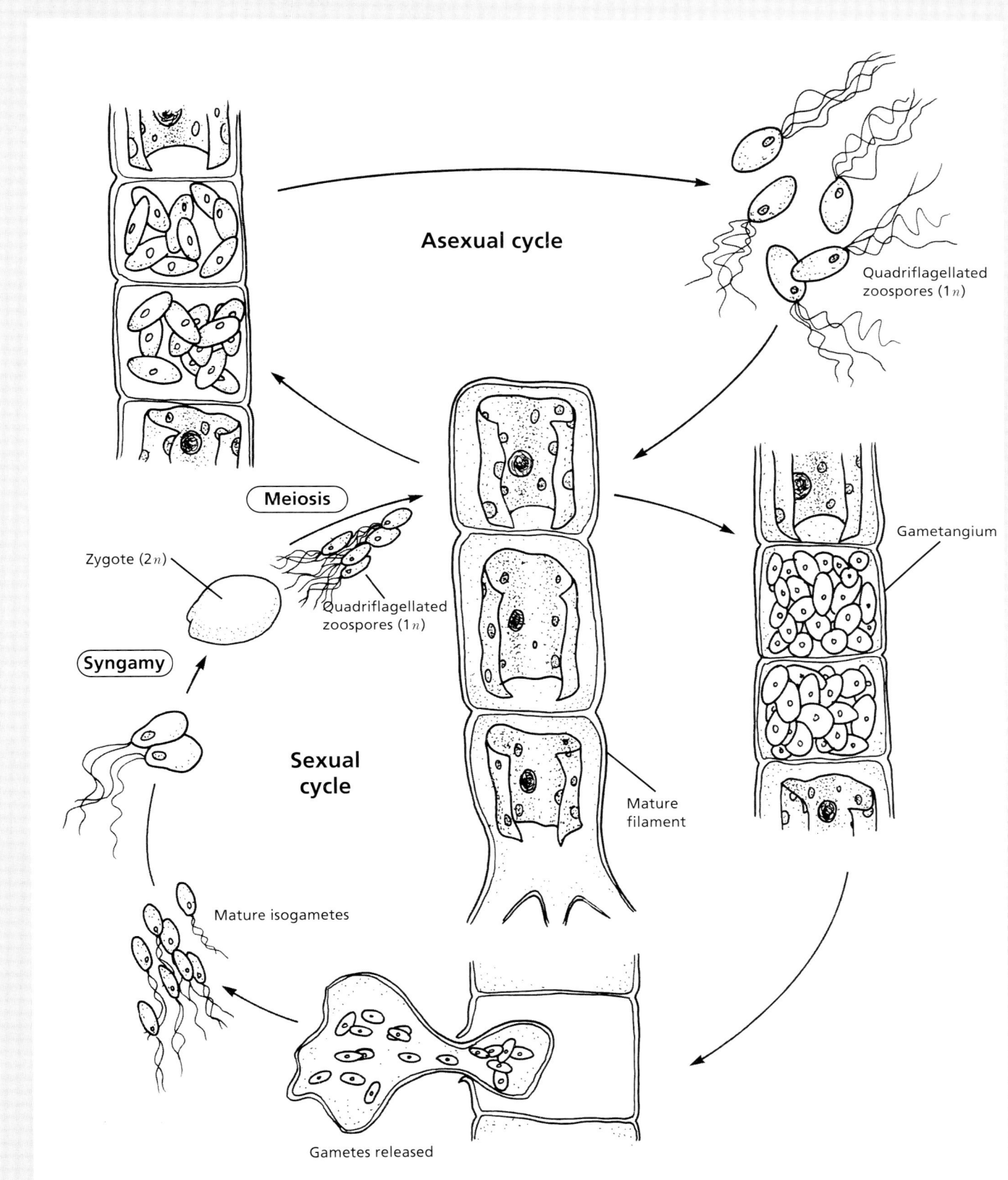

Figure 4.9 Life cycle of *Ulothrix*, a green alga within the class Ulvophyceae.

DIVISION CHLOROPHYTA (green algae)

Figure 4.10 *Ulothrix*, a filamentous, unbranched, green alga. (X100)
1. Chloroplasts
2. Cell wall

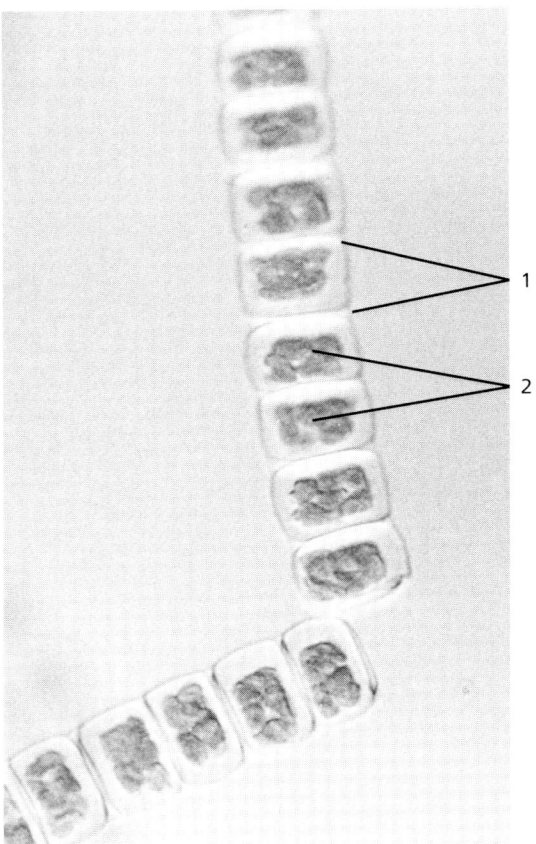

Figure 4.11 *Ulothrix*, an unbranched and filamentous green alga. (X100)
1. Individual cells (known as sporangia when they produce spores)
2. Zoospores

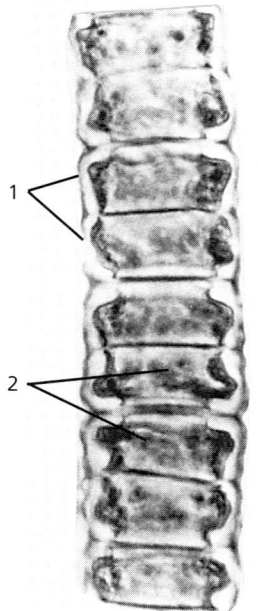

A vegetative filament

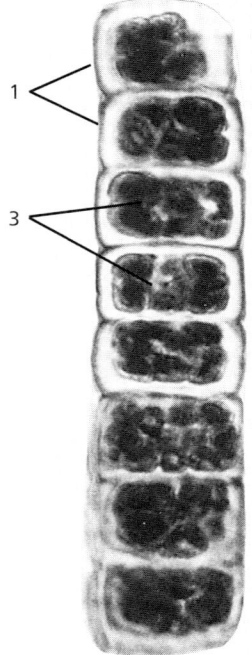

Filament with zoospores

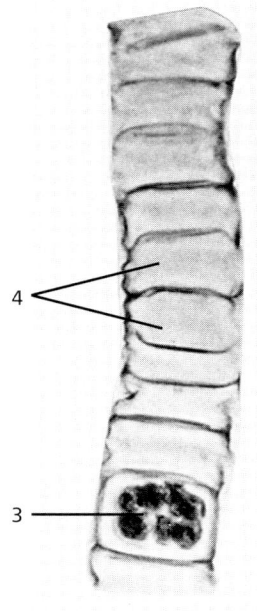

Empty filament, after zoospores have been released

Figure 4.12 The production and release of zoospores in the green alga *Ulothrix*. (all X200)
1. Filament
2. Chloroplasts
3. Zoospores
4. Empty cells

DIVISION CHLOROPHYTA (green algae)

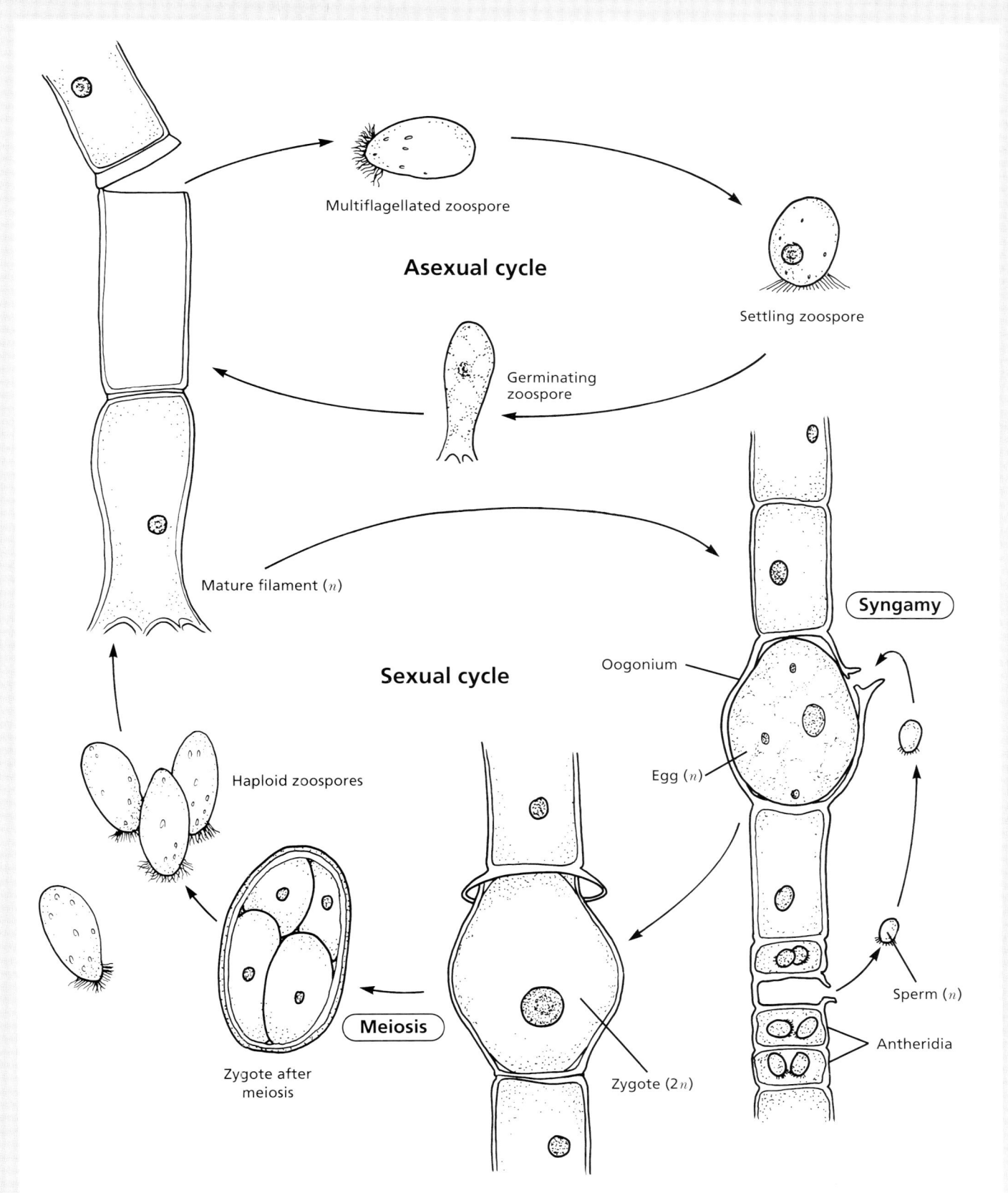

Figure 4.13 Life cycle of *Oedogonium*, an unbranched, filamentous green alga.

DIVISION CHLOROPHYTA (green algae)

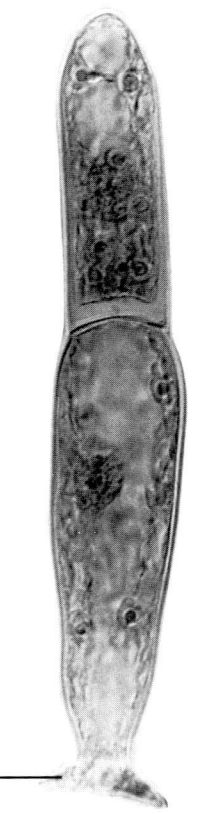

Figure 4.14 New filament of *Oedogonium*. (X430)
1. Holdfast

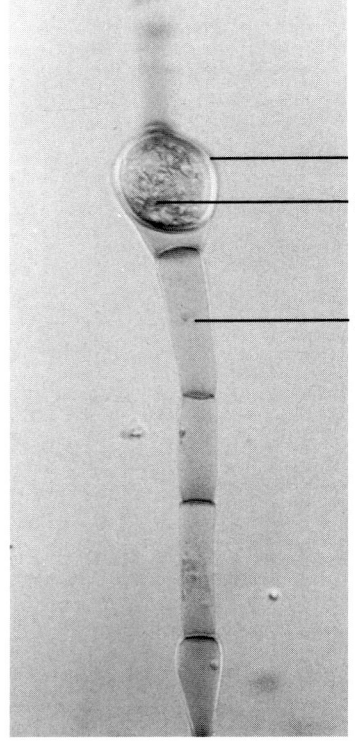

Figure 4.15 The oogonium of the unbranched, green alga, *Oedogonium*. (X200)
1. Oogonium
2. Egg
3. Vegetative cell

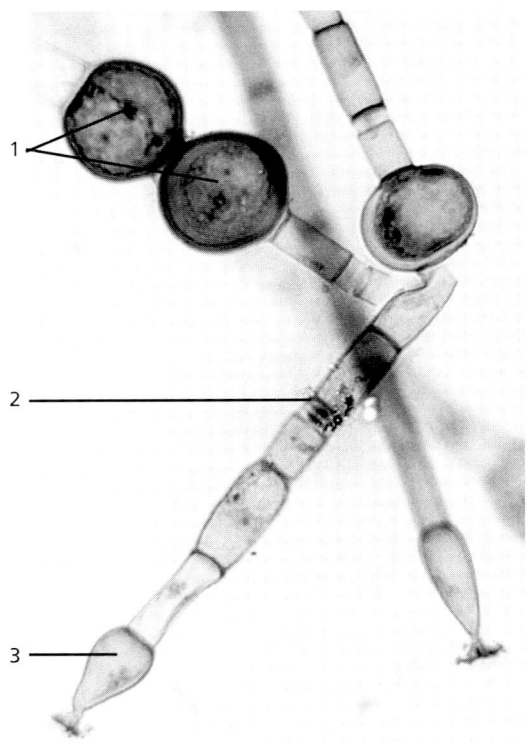

Figure 4.16 *Oedogonium*, a filamentous, unbranched, green alga. (X430)
1. Oogonia
2. Antheridium
3. Basal, or holdfast, cell

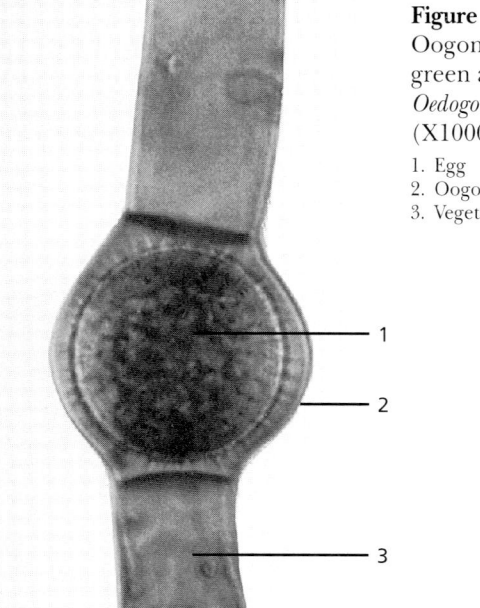

Figure 4.17 Oogonium of the green alga *Oedogonium*. (X1000)
1. Egg
2. Oogonium
3. Vegetative cell

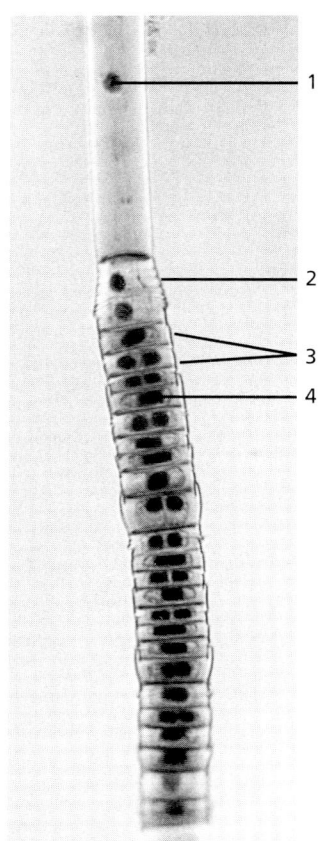

Figure 4.18 A filament of the unbranched green alga, *Oedogonium*. (X430)
1. Cell nucleus
2. Annular scars from cell division
3. Antheridia
4. Sperm

DIVISION CHLOROPHYTA (green algae)

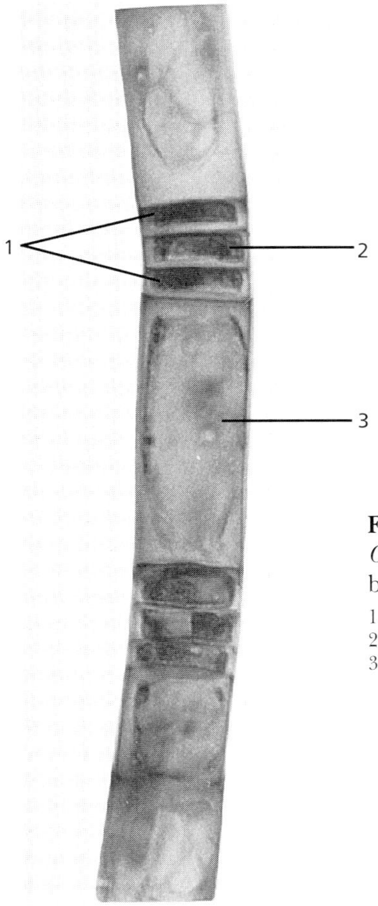

Figure 4.19 The green alga, *Oedogonium*, showing antheridia between vegetative cells. (X600)

1. Antheridia
2. Sperm
3. Vegetative cell

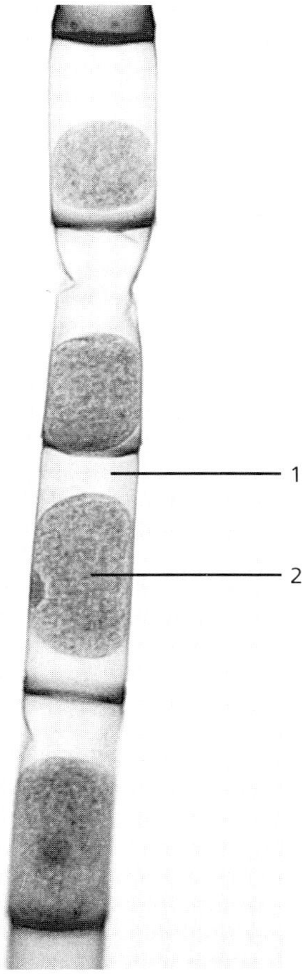

Figure 4.20 The zoosporangium of the unbranched green alga, *Oedogonium*. (X600)

1. Zoosporangium
2. Zoospore

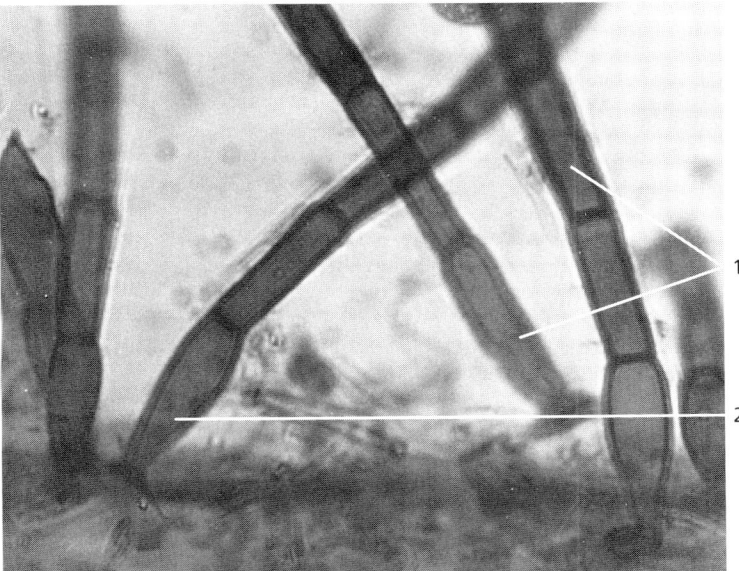

Figure 4.21 Several filaments of *Oedogonium*, growing as epiphytes on another alga. An epiphyte is a non-parasitic organism that grows upon another organism. (X430)

1. Filaments of *Oedogonium*
2. Basal, or holdfast, cell

DIVISION CHLOROPHYTA (green algae)

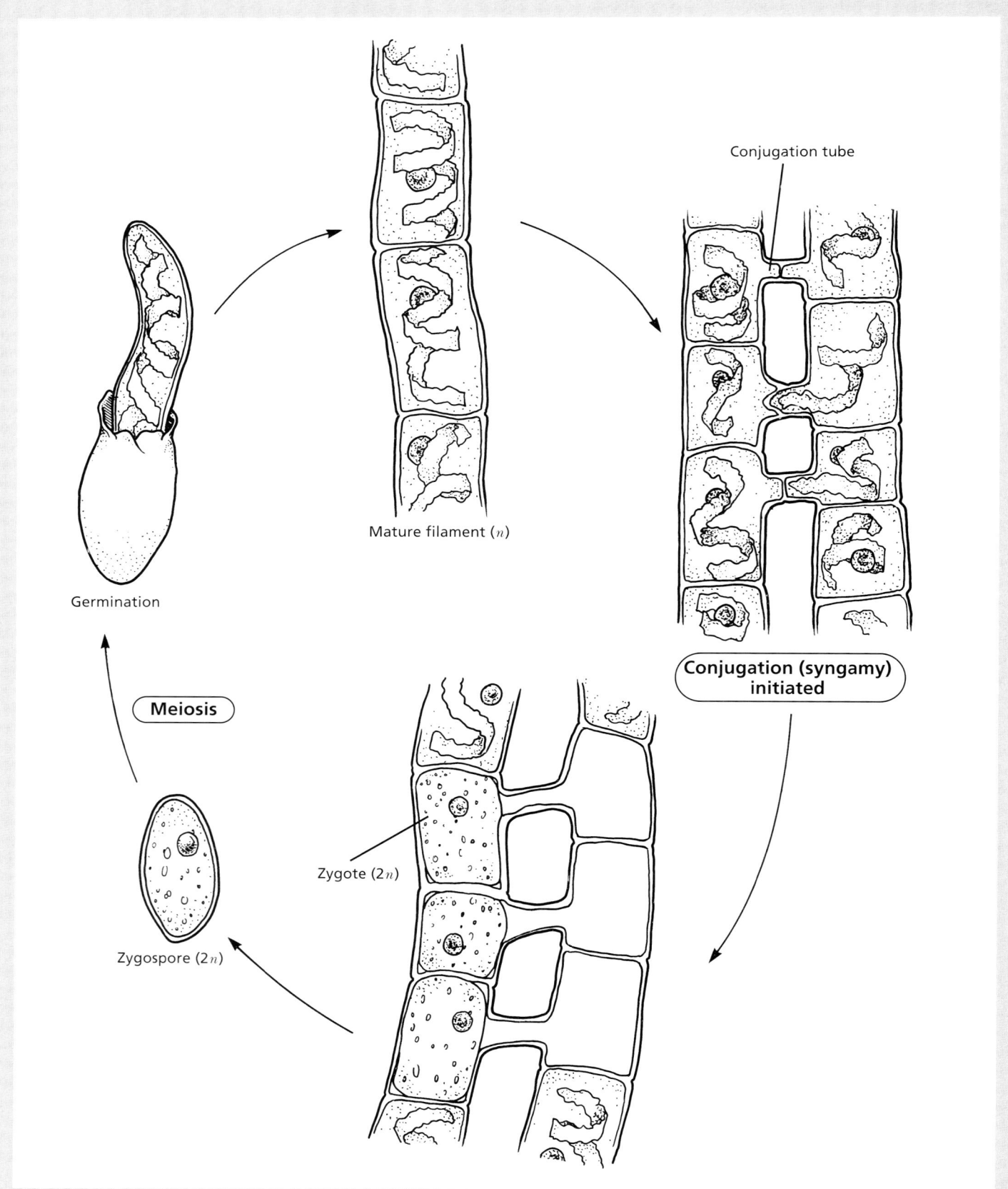

Figure 4.22 Life cycle of *Spirogyra*, a common fresh-water green alga.

DIVISION CHLOROPHYTA (green algae)

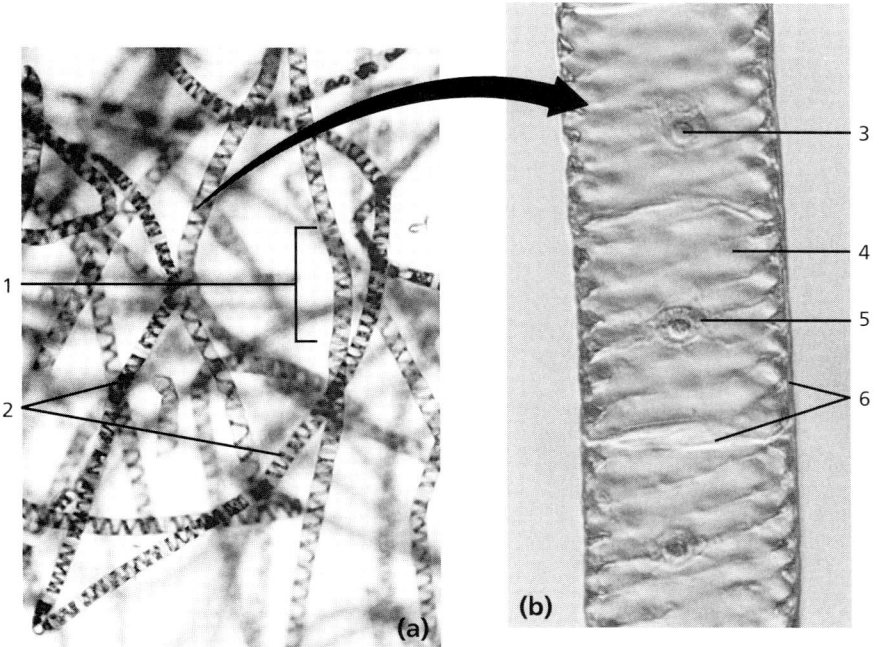
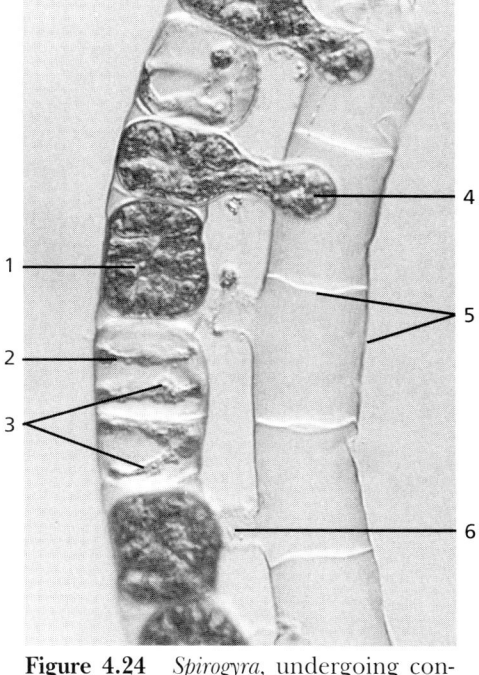

Figure 4.23 Species of *Spirogyra* are filamentous, green algae commonly found in green masses on the surfaces of ponds and streams. Their chloroplasts are arranged as a spiral within the cell. (a) Several cells comprise a filament (X65); and (b) a magnified view of a single filament composed of several cells. (X430)

1. Single cell
2. Filaments
3. Nucleolus
4. Chloroplast
5. Nucleus
6. Cell wall

Figure 4.24 *Spirogyra*, undergoing conjugation. (X200)

1. Zygote (zygospore)
2. Chloroplast
3. Pyrenoid
4. Donor gamete
5. Cell wall
6. Conjugation tube

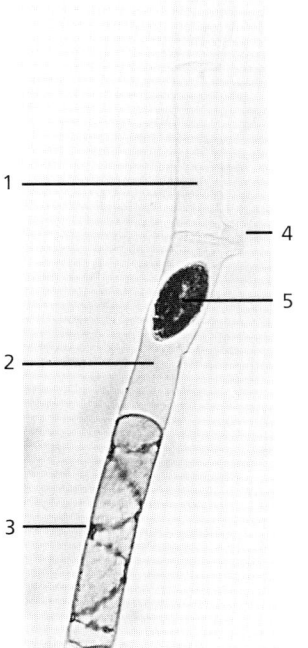
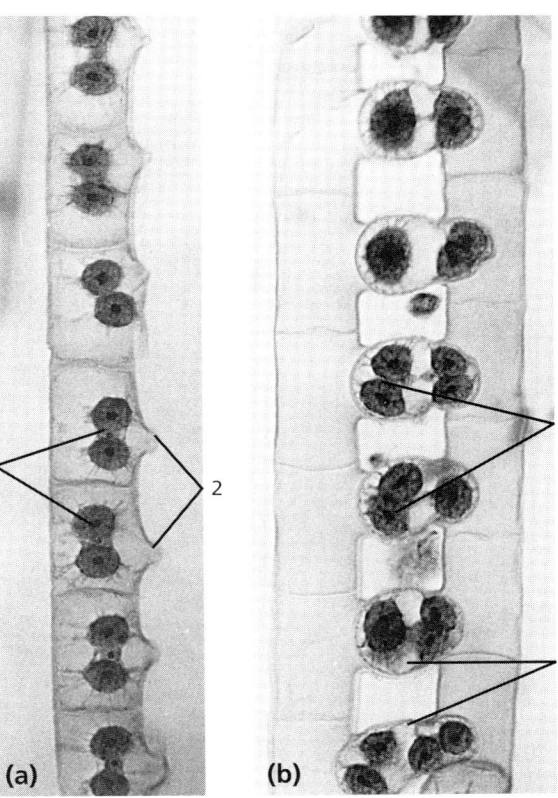

Figure 4.25 A self-fertile species of *Spirogyra*. A gamete has migrated from the upper cell to form a zygote in the lower cell. (X100)

1. Upper cell
2. Lower cell
3. Chloroplast
4. Conjugation tube
5. Zygote

Figure 4.26 *Zygnema*, undergoing conjugation. (a) The filament is just forming conjugation tubes; and (b) two conjugated filaments. Zygotes in this species are produced in the conjugation tubes. (X100)

1. Chloroplasts
2. Developing conjugation tubes
3. Developing zygotes
4. Conjugation tubes

DIVISION CHLOROPHYTA (green algae)

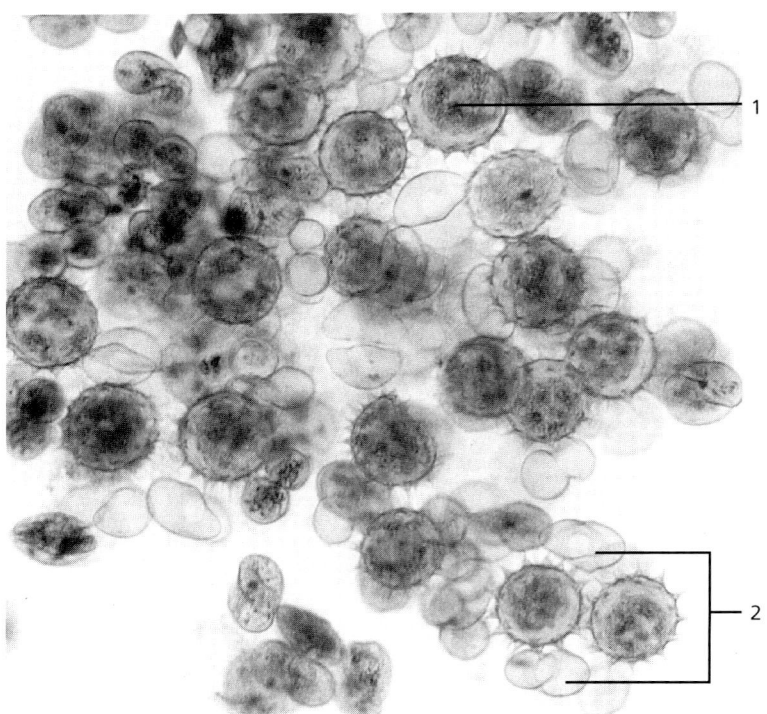

Figure 4.27 Cells of the desmid *Cosmarium* in the process of conjugation. Desmids are unicellular, green algae. (X450)
1. Zygote produced from conjugation
2. Parent cells that have produced gametes

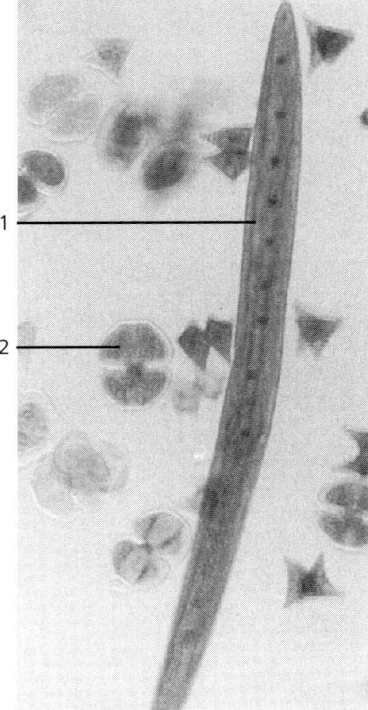

Figure 4.28 Representative desmids. Desmids are unicellular, fresh-water green algae. (X200)
1. Closterium
2. Cosmarium

Figure 4.29 Several green algae are common marine organisms. *Codium* occurs in intertidal zones and is commonly known as dead man's fingers. (X0.5)

Figure 4.30 Sea lettuce, *Ulva*, lives as a flat membranous form in marine environments. (X0.5)

DIVISION CHLOROPHYTA (green algae)

Figure 4.31 Branching filaments of *Cladophora glomerata*. This member of class Ulvophyceae is found in both fresh-water and marine habitats. (X10)

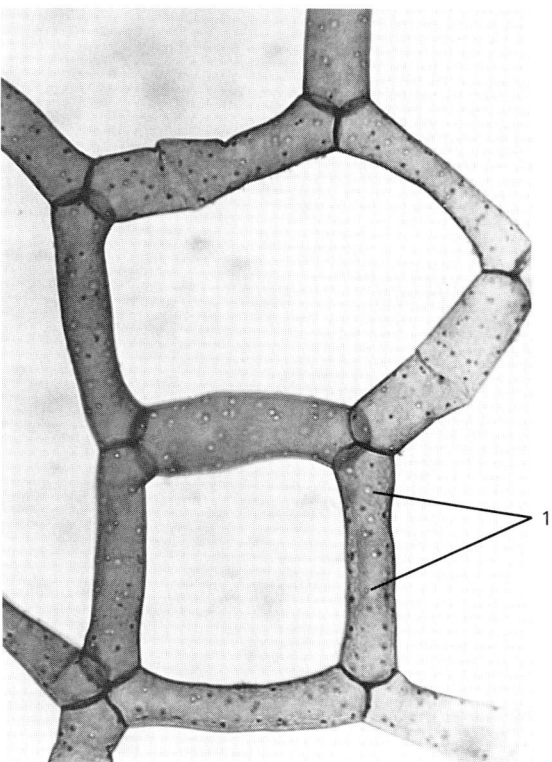

Figure 4.32 *Hydrodictyon*. The large, multinucleated cells form net-shaped colonies. (X430)

1. Nuclei

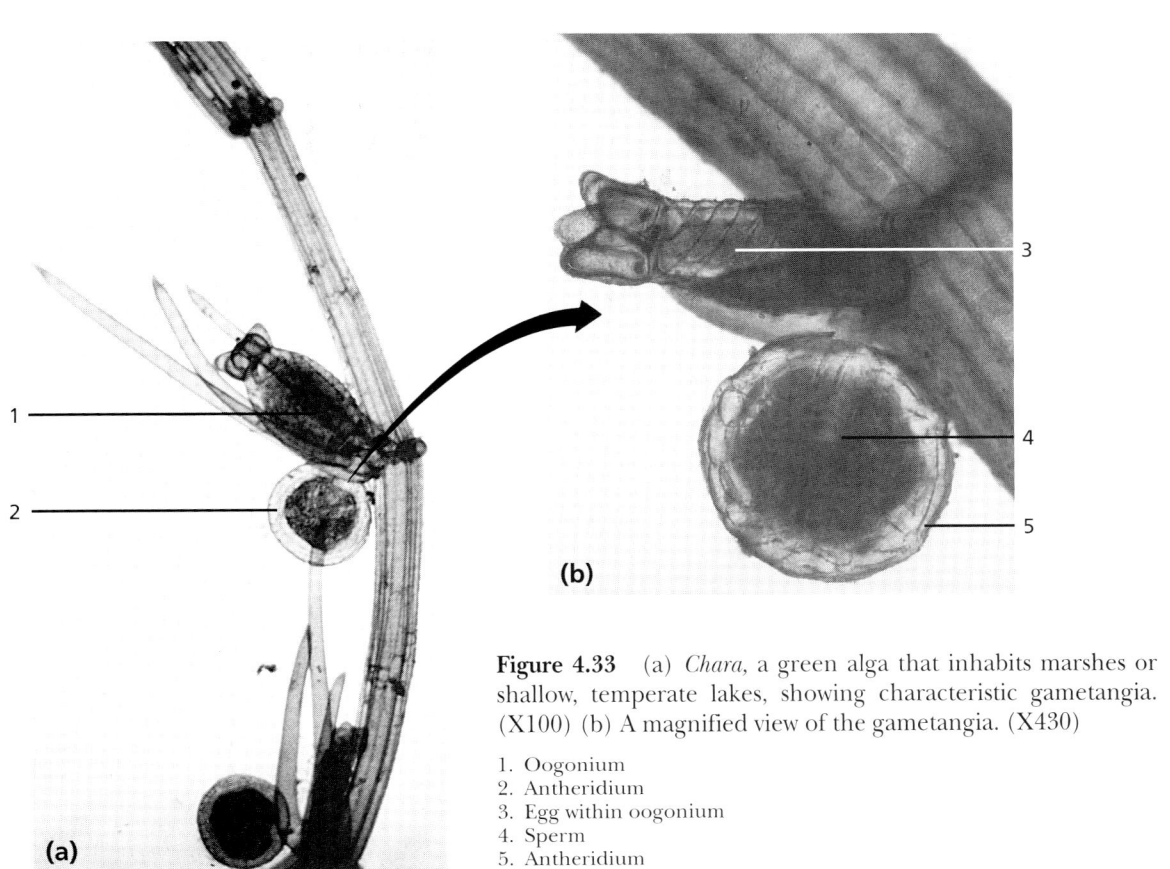

Figure 4.33 (a) *Chara*, a green alga that inhabits marshes or shallow, temperate lakes, showing characteristic gametangia. (X100) (b) A magnified view of the gametangia. (X430)

1. Oogonium
2. Antheridium
3. Egg within oogonium
4. Sperm
5. Antheridium

DIVISION PHAEOPHYTA (brown algae, giant kelp)

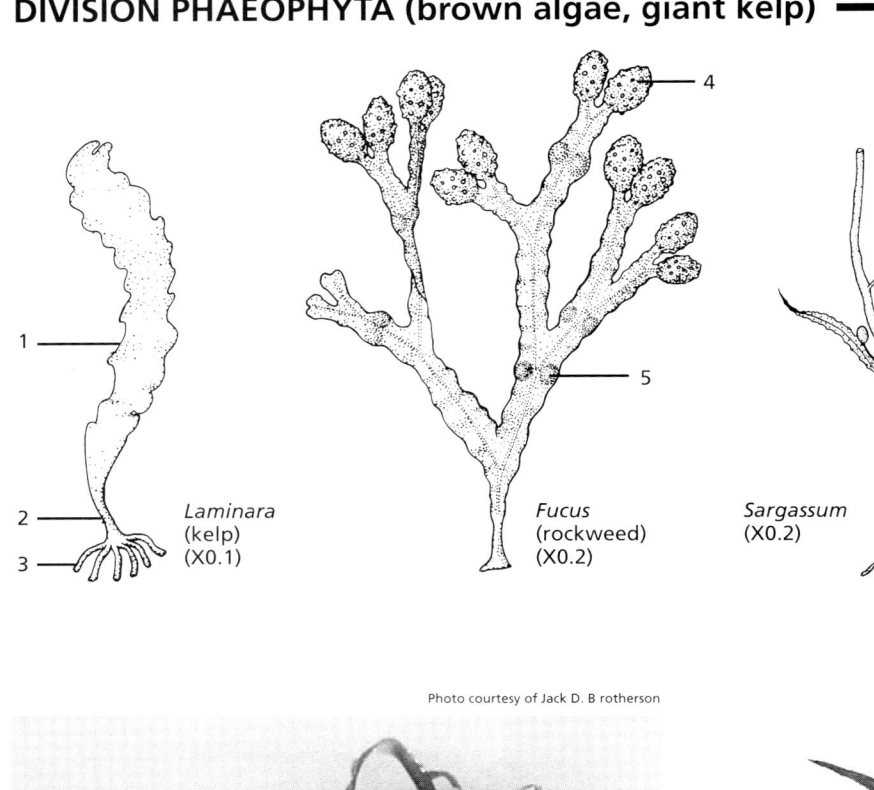

Figure 4.34 Brown algae are marine organisms commonly referred to as seaweeds. They range in size from small filamentous organisms a few millimeters in length to the giant kelp between 50 and 100 meters long.

1. Blade
2. Stipe
3. Holdfast
4. Receptacle
5. Air bladder
6. Air bladder
7. Holdfast

Figure 4.35 "Sea palm," *Postelsia palmaeformis*, a common brown alga found on the west coast of North America. (X0.25)

Figure 4.36 Herbarium specimen of the "sea palm," *Postelsia palmaeformis*. This brown alga occurs on rocks where the surf begins to break. It withstands water movement because of its large holdfast and thick stipe. "Sea palms" occur along the Pacific coastline of North America.

DIVISION PHAEOPHYTA (brown algae, giant kelp)

Figure 4.37 The kelp, *Laminaria*, is one of the common "sea weeds" found along many rocky coasts. Note that the holdfast of this specimen is attached to a mussel shell. (X0.3)

1. Blade 2. Stipe 3. Holdfast 4. Mussel shell

Figure 4.38 A tidal pool at low tide. Several different species of brown algae can be seen attached to the rocks.

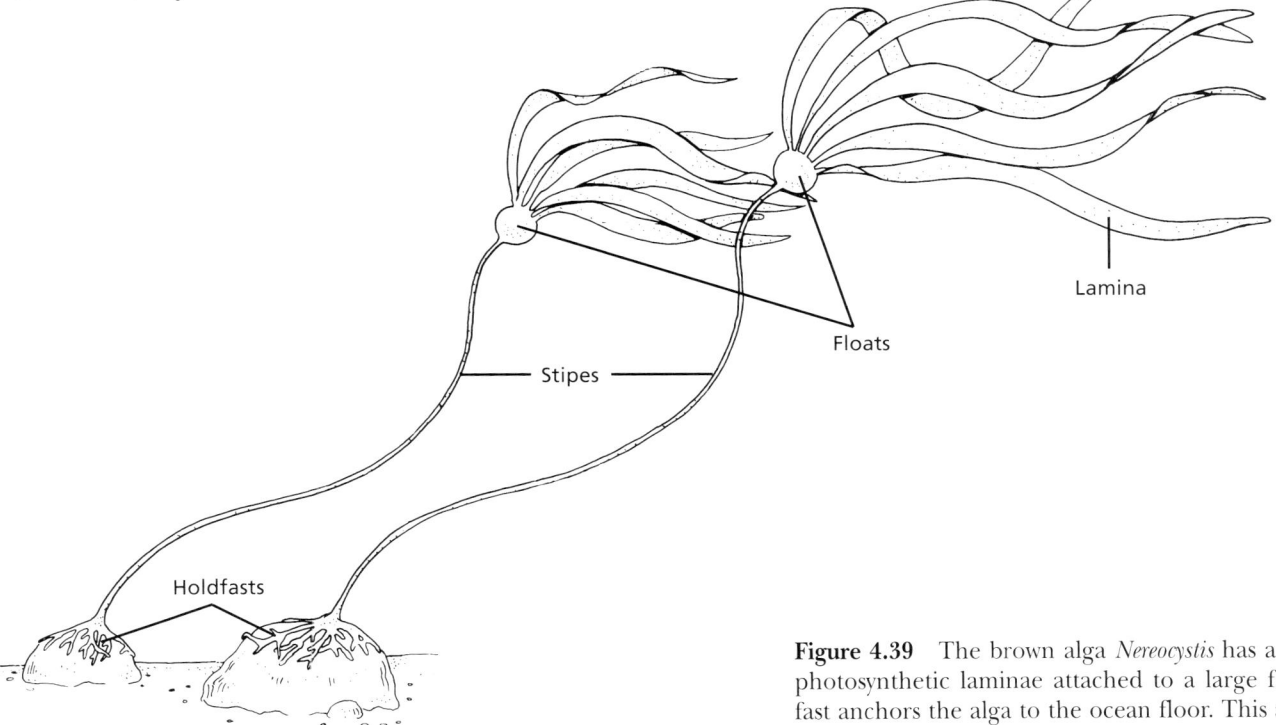

Figure 4.39 The brown alga *Nereocystis* has a long stipe and photosynthetic laminae attached to a large float. The holdfast anchors the alga to the ocean floor. This alga and others can grow to lengths of several meters. (X0.03)

DIVISION PHAEOPHYTA (brown algae, giant kelp)

Figure 4.40 An algal hummock, formed by detached brown algae washing ashore and becoming entangled. (X0.05)

Figure 4.41 *Sargassum*, a brown alga. (X0.25)

1. Float (air-filled bladder)
2. Blade
3. Stipe

CHAPTER 4 Kingdom Protista: Primarily Multicellular Organisms 41

DIVISION PHAEOPHYTA (brown algae, giant kelp)

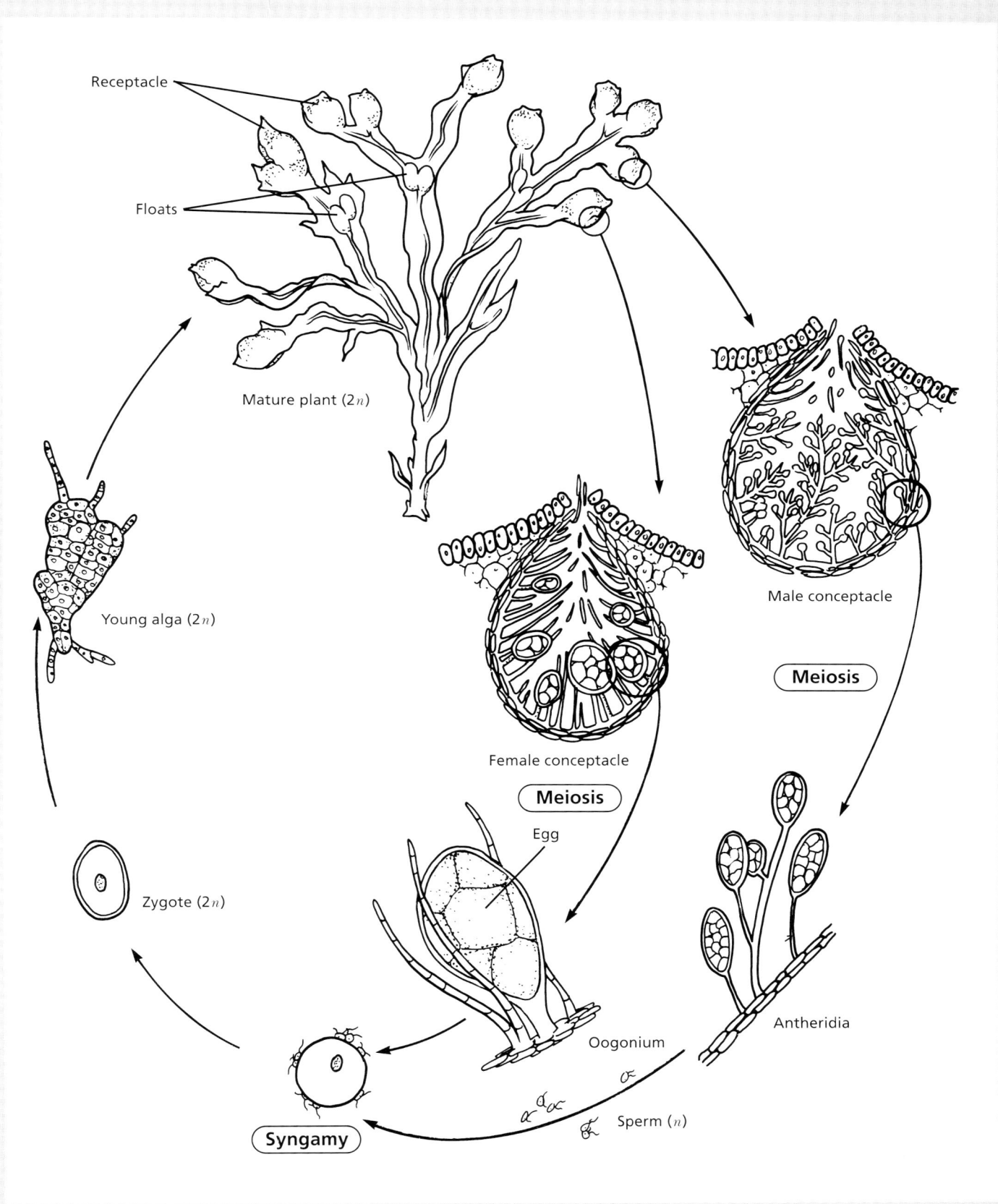

Figure 4.42 Life cycle of *Fucus*, a common brown alga.

DIVISION PHAEOPHYTA (brown algae, giant kelp)

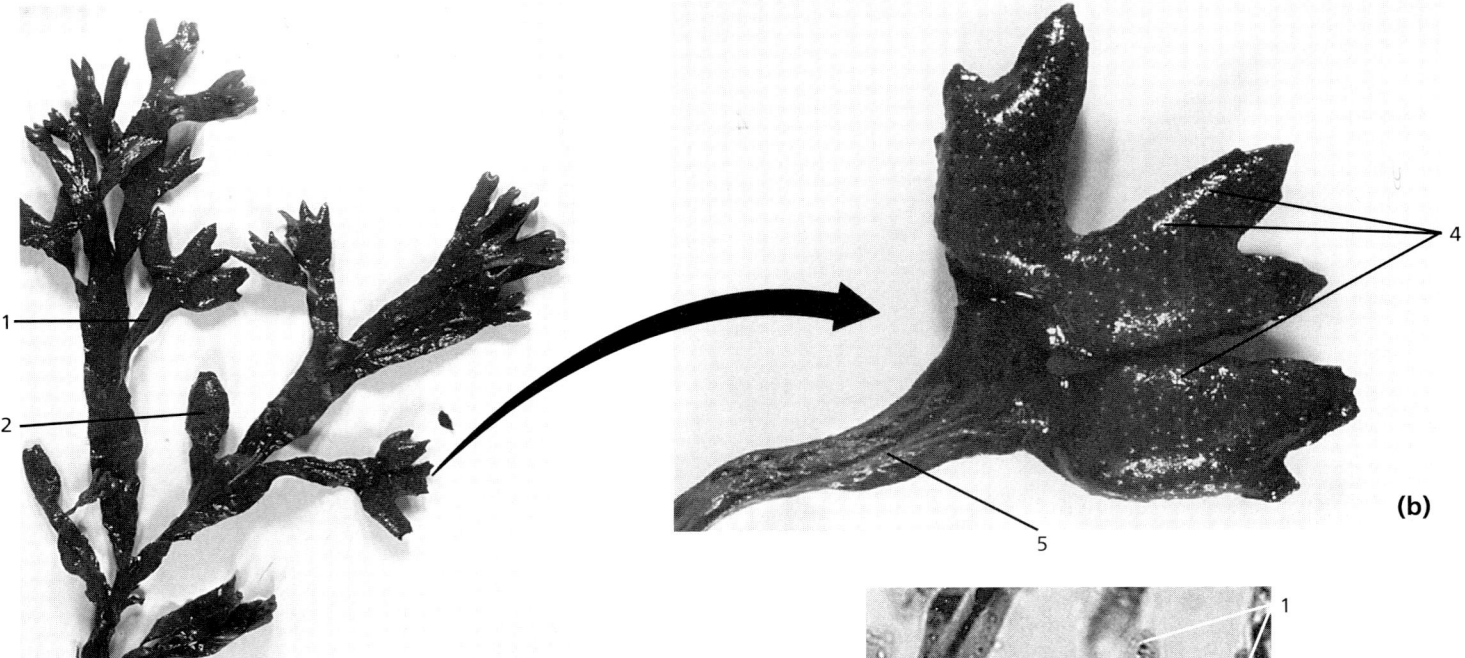

Figure 4.43 (a) *Fucus*, a brown alga, commonly called rockweed. (about X0.5) (b) An enlargement of a blade supporting the receptacles. (about X2)

1. Blade
2. Receptacle
3. Stipe
4. Conceptacles (light-colored spots) are chambers imbedded in the receptacles
5. Blade

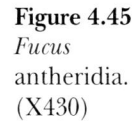

Figure 4.45 *Fucus* antheridia. (X430)

1. Paraphyses
2. Antheridium
3. Sperm within antheridium

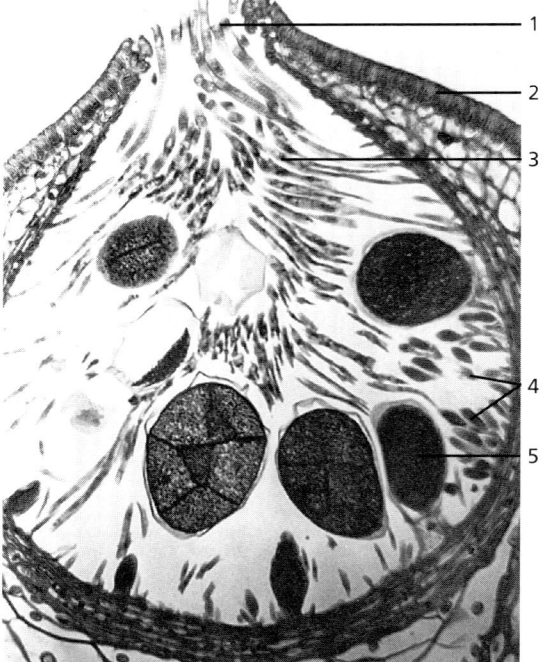

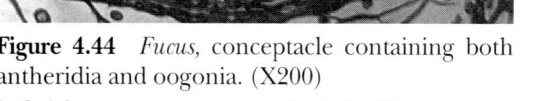

Figure 4.44 *Fucus*, conceptacle containing both antheridia and oogonia. (X200)

1. Ostiole
2. Surface of receptacle
3. Paraphyses (sterile hairs)
4. Antheridia
5. Oogonium

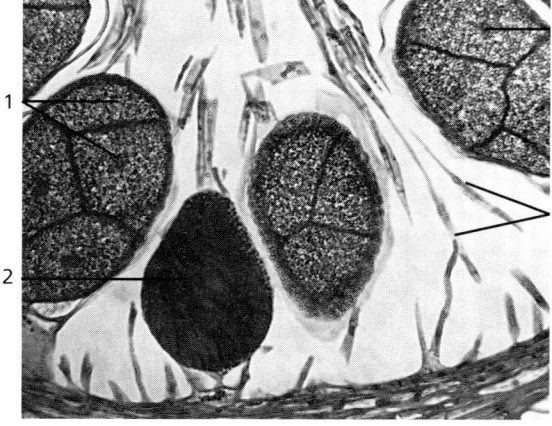

Figure 4.46 *Fucus*, closeup of female conceptacle. (X200)

1. Eggs
2. Oogonium
3. Nucleus of egg
4. Paraphyses

DIVISION RHODOPHYTA (red algae)

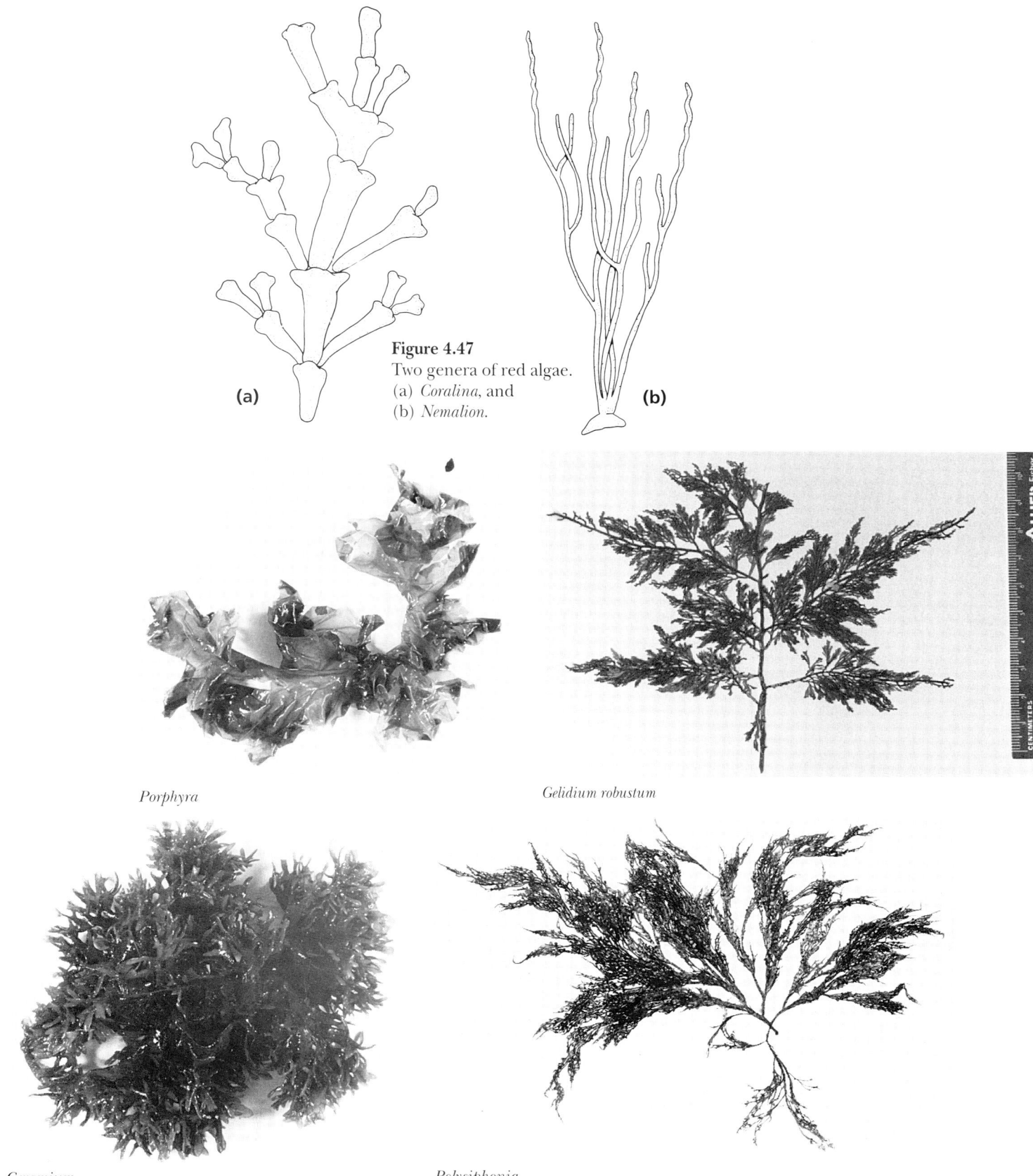

Figure 4.47
Two genera of red algae.
(a) *Coralina*, and
(b) *Nemalion*.

Porphyra

Gelidium robustum

Ceramium

Polysiphonia

Figure 4.48 Examples of common marine red algae.

DIVISION RHODOPHYTA (red algae)

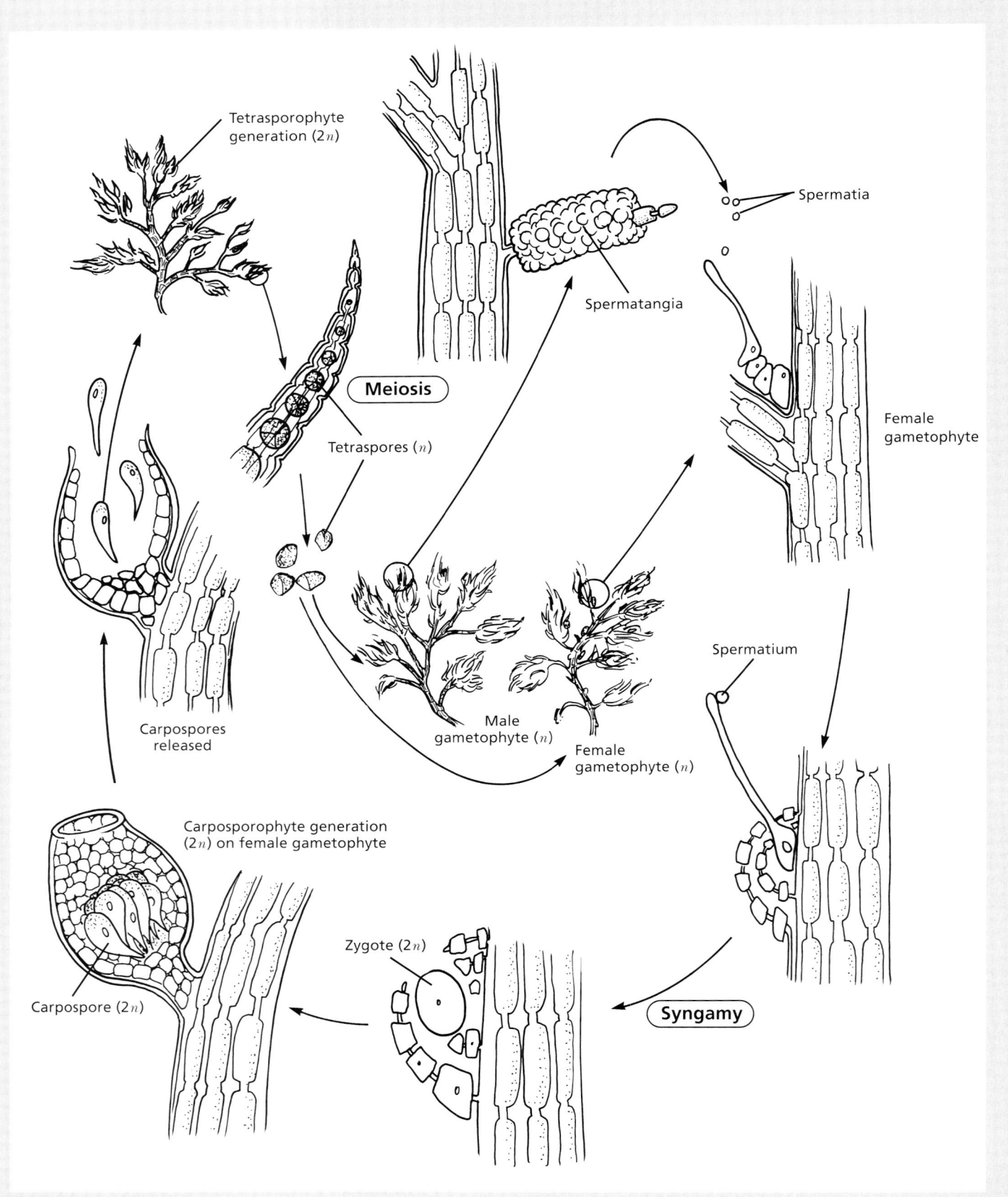

Figure 4.49 Life cycle of the red alga *Polysiphonia*.

DIVISION RHODOPHYTA (red algae)

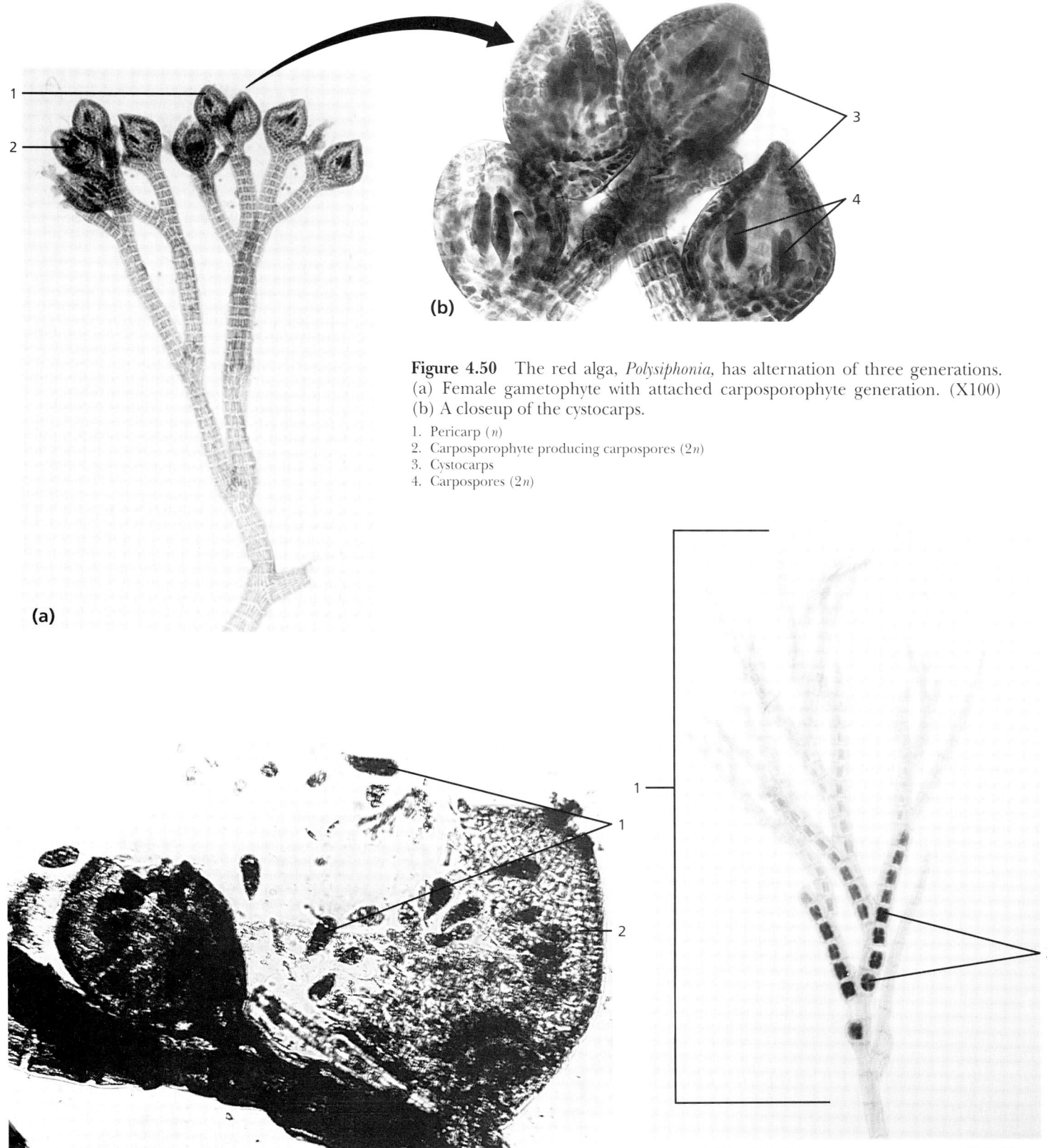

Figure 4.50 The red alga, *Polysiphonia*, has alternation of three generations. (a) Female gametophyte with attached carposporophyte generation. (X100) (b) A closeup of the cystocarps.

1. Pericarp (*n*)
2. Carposporophyte producing carpospores (2*n*)
3. Cystocarps
4. Carpospores (2*n*)

Figure 4.51 *Polysiphonia*, showing the release of carpospores. (X430)

1. Carpospores (2n)   2. Ruptured cystocarp

Figure 4.52 *Polysiphonia*, tetrasporophyte. (X100)

1. Tetrasporophyte (2*n*)   2. Tetraspores (*n*)

DIVISION RHODOPHYTA (red algae)

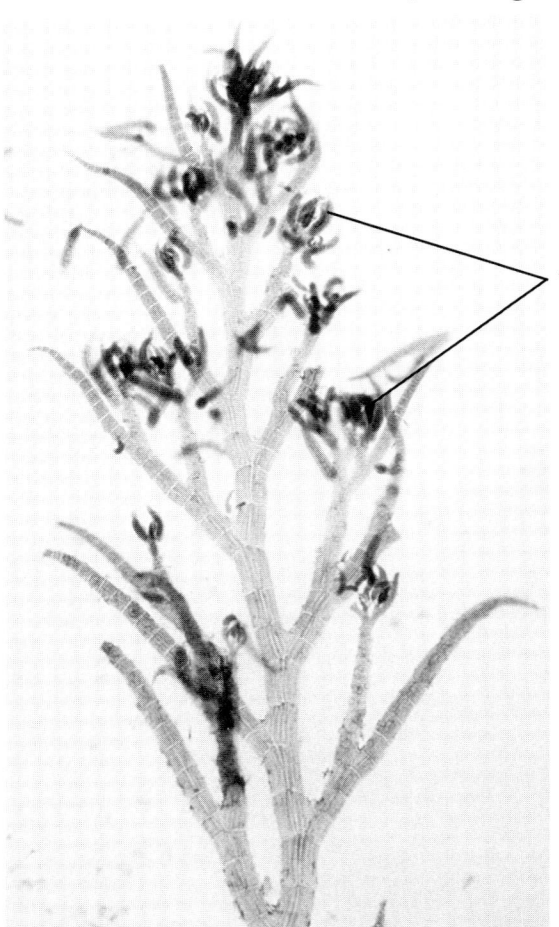

Figure 4.53 *Polysiphonia*, male gametophyte. Male reproductive structures, known as spermatangia, produce non-motile spermatia. (X40)
1. Spermatangia

Figure 4.54 Herbarium specimen of the red alga, *Phycodrys rubens*. Because they require little sunlight, red algae can survive at considerable depths in ocean waters.

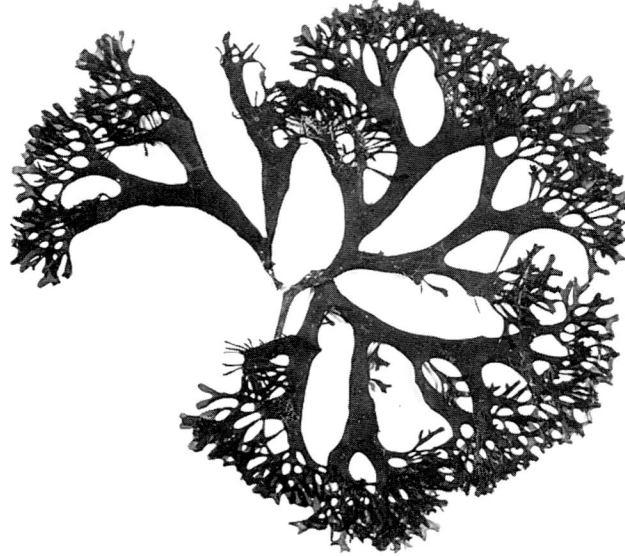

Figure 4.55 *Chondrus crispus*, Irish moss, is an industrially important red alga in Europe.

Figure 4.56 A photograph of small encrusting colonies of a species of red alga on a stone. The colonies shown are bright red and are only a few millimeters in size.

DIVISION MYXOMYCOTA (plasmodial slime molds)

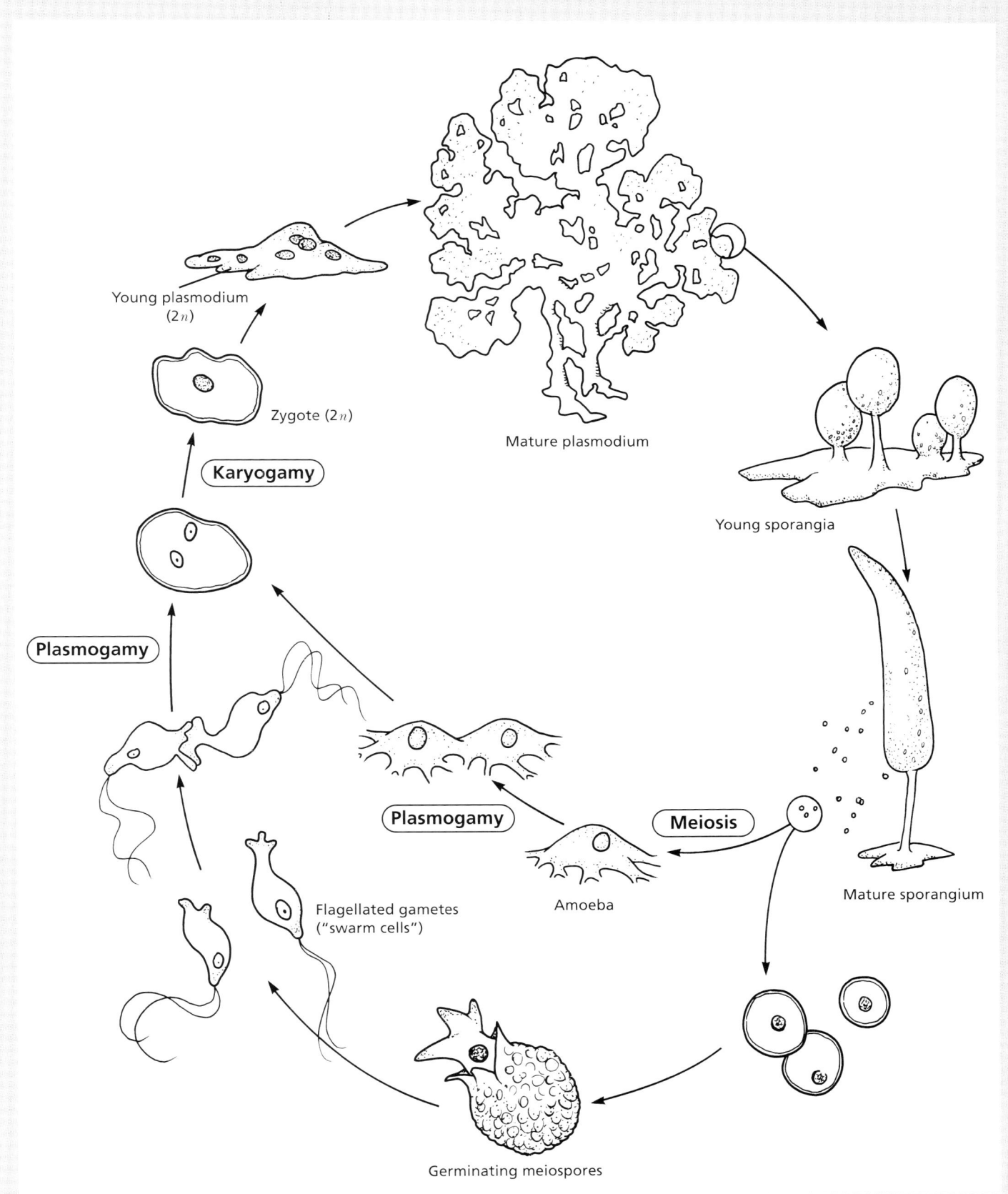

Figure 4.57 Life cycle of a plasmodial slime mold.

DIVISION MYXOMYCOTA (plasmodial slime molds)

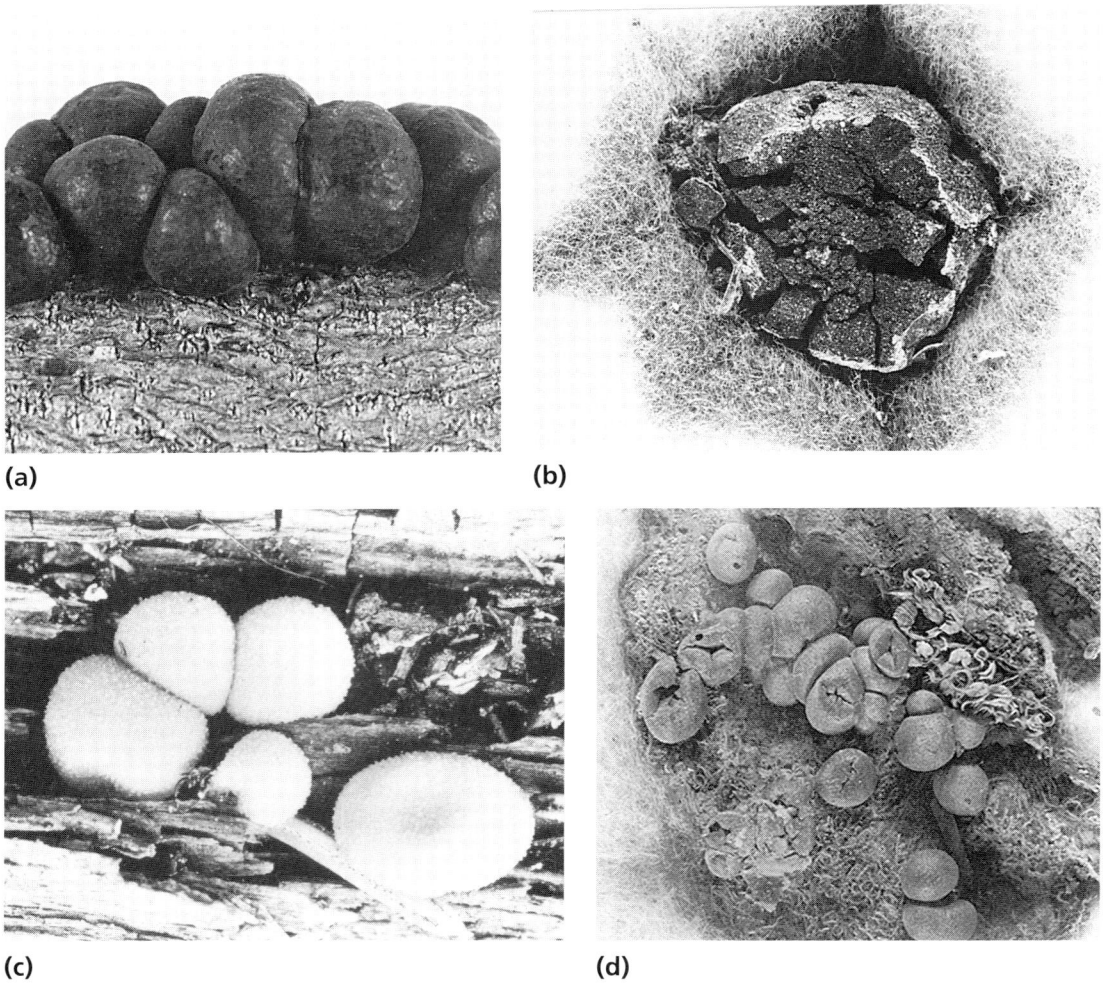

Figure 4.58 Slime mold sporangia vary considerably in size and shape. (a) and (b) are species of *Fuligo*; (c) and (d) are species of *Lycogala*.

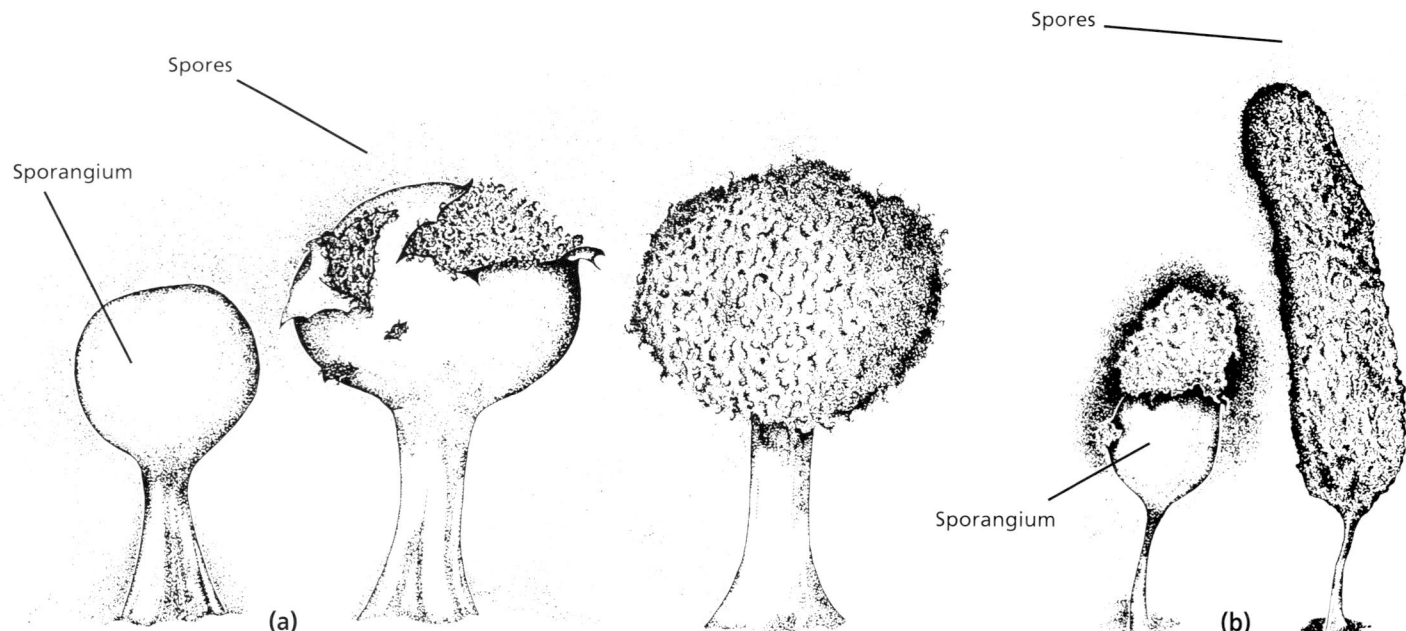

Figure 4.59 Diagrams of slime mold sporangia. (a) Stages in spore releasal into the wind from (a) *Hemitrichia*, and (b) *Arcyria*.

DIVISION MYXOMYCOTA (plasmodial slime molds)

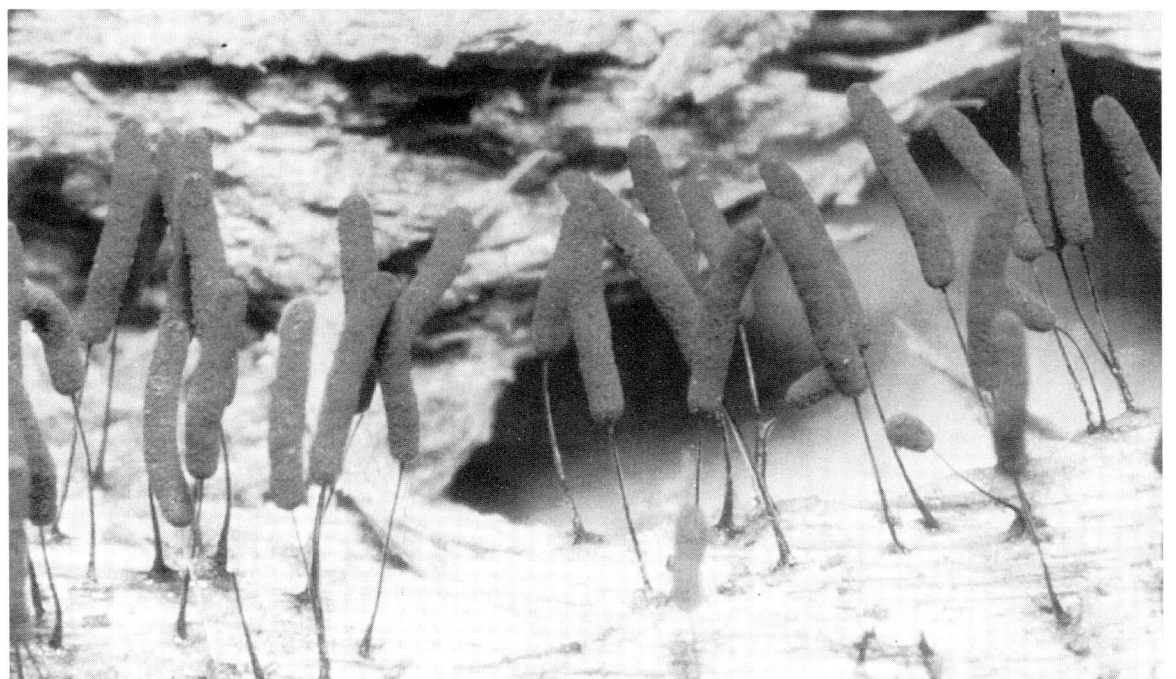

Figure 4.60 Sporangia of the slime mold, *Comatricha typhoides*.

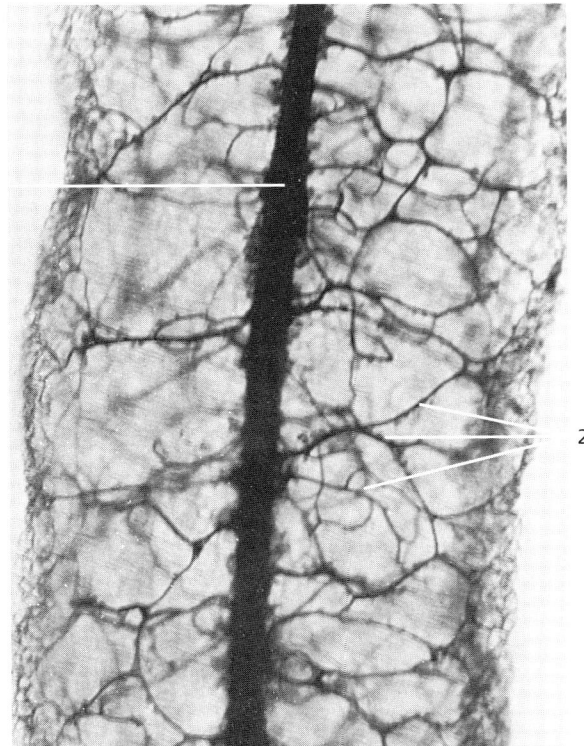

Figure 4.61 A longitudinal section through the sporangium of *Stemonitis*. (X25)
1. Columella
2. Cellular filaments (capillitum)

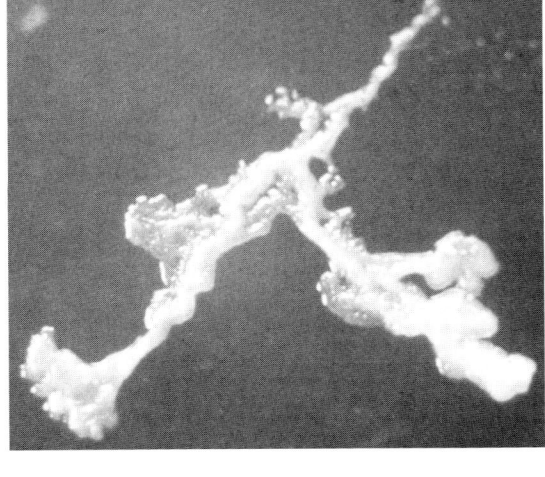

Figure 4.62 Slime mold, *Physarum*, growing on an agar culture medium. (X40)

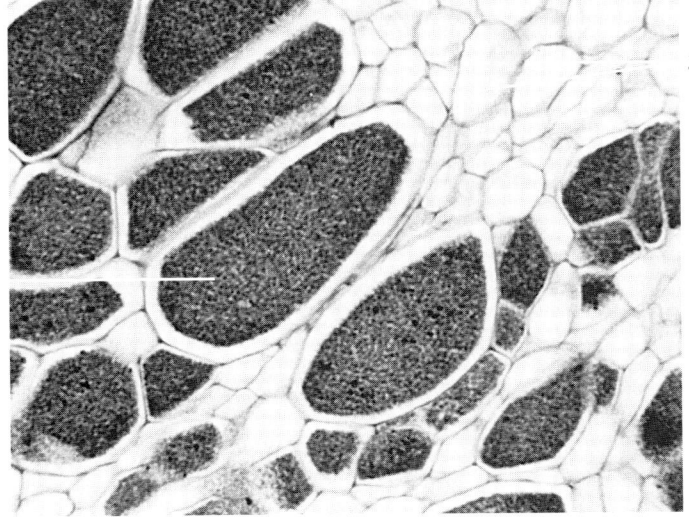

Figure 4.63 *Plasmodiophora brassicae*, a myxomycete responsible for clubroot in cabbage. Depicted is a section through a cabbage root, showing spores of *Plasmodiophora* in the cabbage root cells.
1. Spores
2. Cabbage root cells

DIVISION OOMYCOTA (water molds, white rusts, downy mildews)

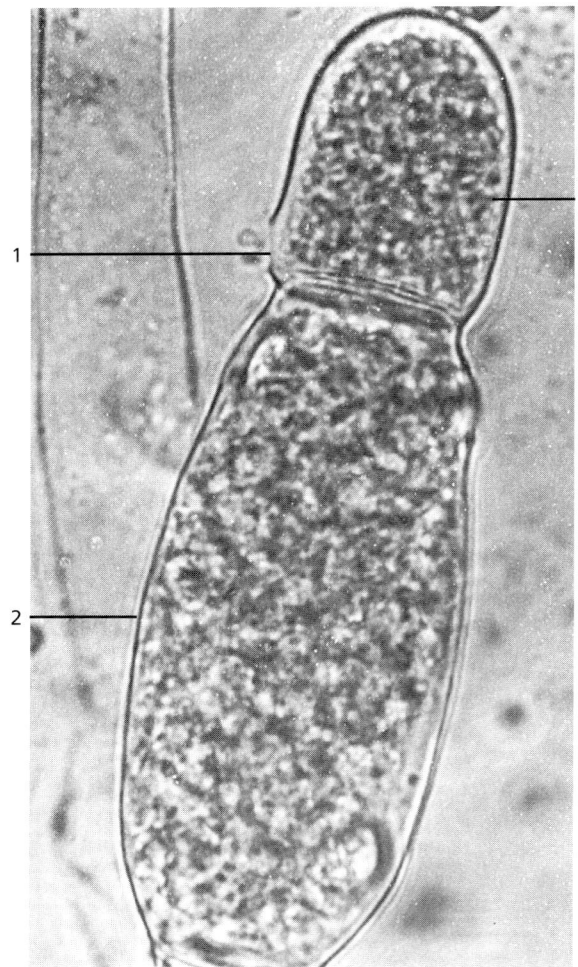

Figure 4.64 The gametangia of the water mold, *Allomyces*. Male gametes escape through exit pores. (X430)

1. Exit pore
2. Female gametangium
3. Male gametangium

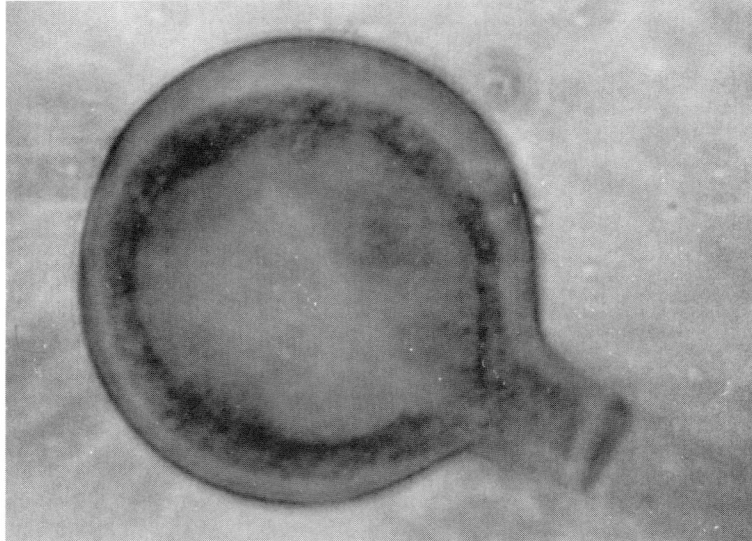

Figure 4.65 A water mold, *Saprolegnia*, showing a young oogonium before eggs have been formed. (X430)

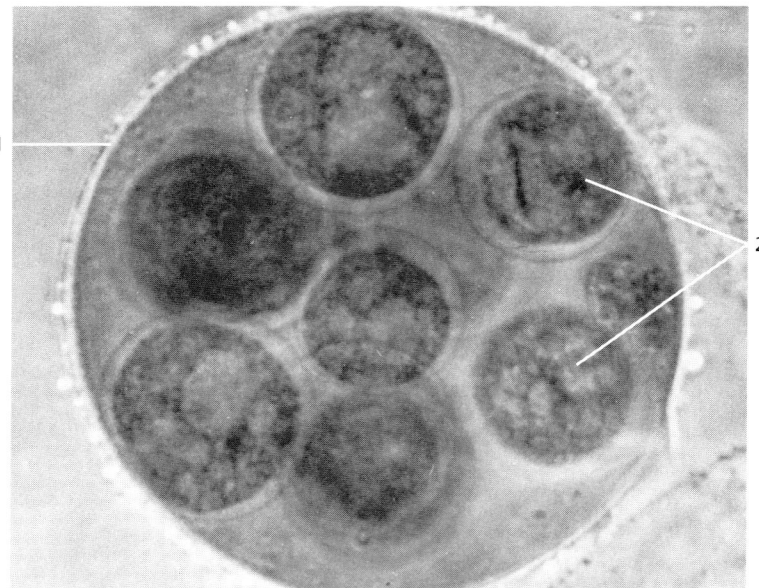

Figure 4.66 A mature oogonium of the water mold, *Saprolegnia*. (X430)

1. Oogonium   2. Eggs

Figure 4.67 The skin of this brown trout has been infected by the common water mold, *Saprolegnia*.

Kingdom Fungi

About 250,000 fungi species exist. All fungi are heterotrophs because they absorb nutrients through their cell walls and cell membranes. The kingdom Fungi includes the typical conjugation fungi, yeasts, mushrooms, toadstools, rusts, and lichens. Most are saprobes, absorbing nutrients from dead organic material, while a few are parasitic, absorbing nutrients from living hosts.

Except for the unicellular yeasts, fungi consist of elongated filaments called *hyphae*. Hyphae begin as tubular extensions of spores that branch as they grow to form a network of hyphae called a *mycelium*. Even the body of a mushroom consists of a mass of tightly packed hyphae, attached to an underground mycelium.

Fungi are nonmotile and produce no motile cells. Fungi reproduce by means of spores, which are produced sexually or asexually.

Fungi help decompose organic material, helping to recycle the inorganic nutrients essential for plant growth.

Many species of fungi are commercially important. Some are used as food, such as mushrooms; or in the production of foods, such as the use of yeasts in making bread, cheese, beer, and wine. Other species are important in medicine, for example, in the production of antibiotic penicillin. Many other species of fungi are of medical and economic concern because they cause plant and animal diseases and destroy crops and stored goods.

TABLE 5.1
Some Representatives of the Kingdom Fungi

Divisions and Representative Kinds	Characteristics
Zygomycota (conjugation fungi)	Hyphae lack cross walls between nuclei
Ascomycota (yeasts, molds, morels, truffles)	Septate hyphae, reproductive structures usually contain eight ascospores within asci; asexual reproduction by budding or conidia
Basidiomycota (mushrooms, toadstools, rusts, smuts)	Septate hyphae; 4 spores produced externally on cells called basidia contained in basidiocarp (basidioma)
Lichens (*not a division*, but rather a group of symbiotic organisms comprised of an alga and a fungus)	Algal component (usually a green alga) provides food from photosynthesis; fungal component (usually an ascomycete) may provide anchorage, water retention, and/or nutrient absorbance

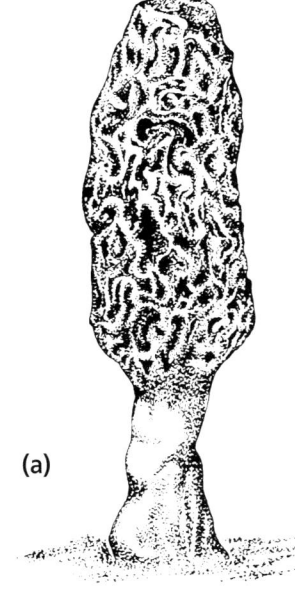
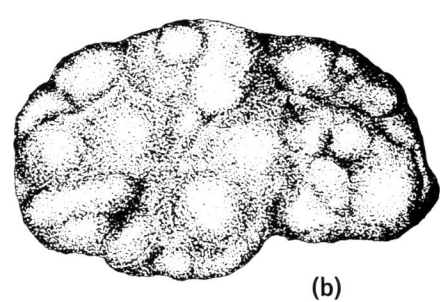

Figure 5.1 Diagrams of two commercially important Ascomycetes. (a) An ascocarp of *Morchella esculenta*, the common morel prized as a gourmet food; (b) an ascocarp of the truffle *Tuber* species. Truffles develop their ascocarps underground, where they are difficult to find and harvest.

DIVISION ZYGOMYCOTA (conjugation fungi)

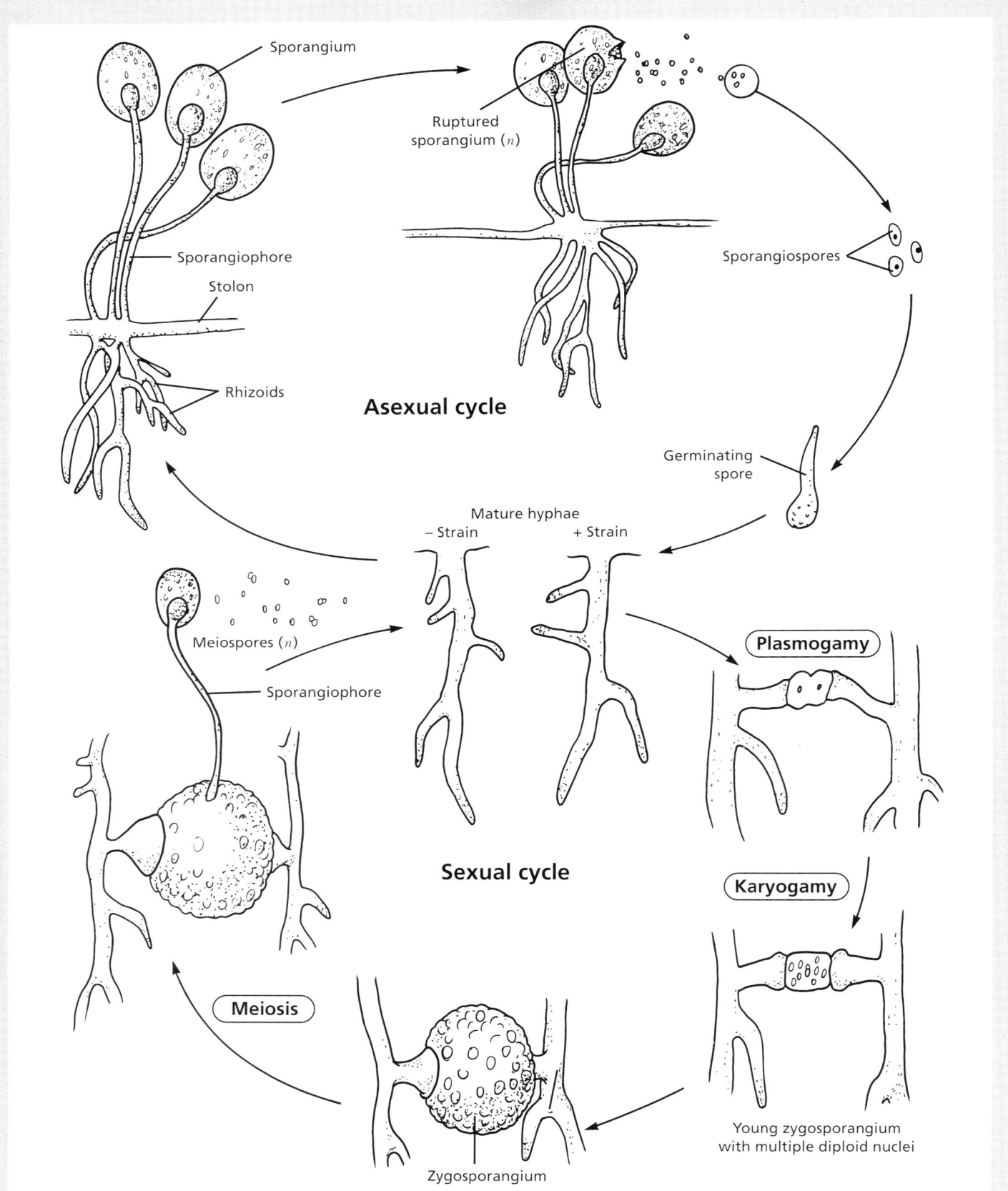

Figure 5.2 Life cycle of *Rhizopus*, the common bread mold.

CHAPTER 5 Kingdom Fungi

DIVISION ZYGOMYCOTA (conjugation fungi)

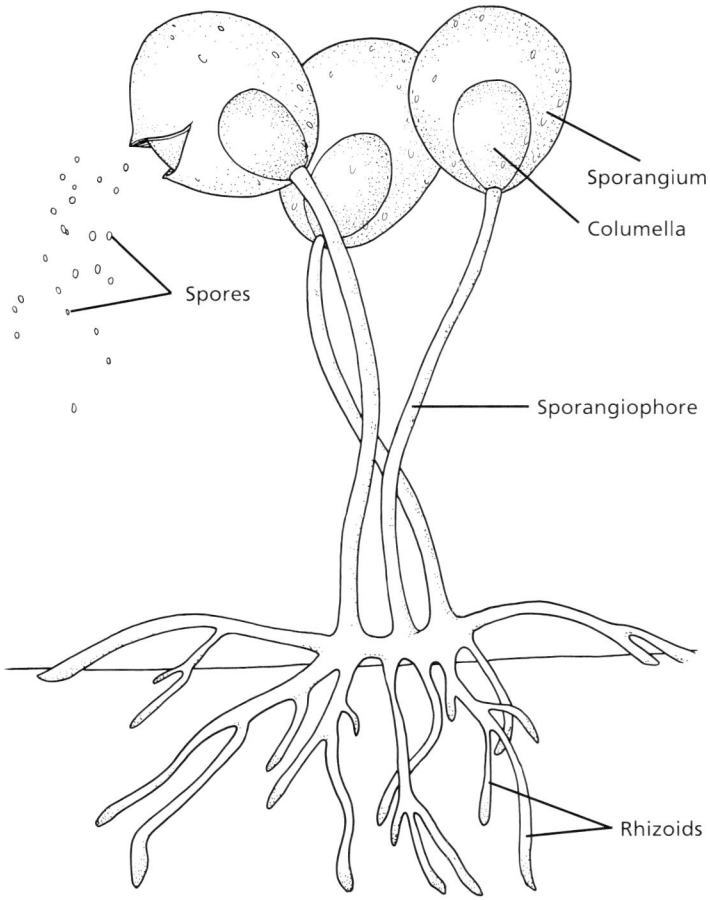

Figure 5.3 Diagram of the sporangia of the bread mold, *Rhizopus*. Spores are released when the wall of the sporangium ruptures.

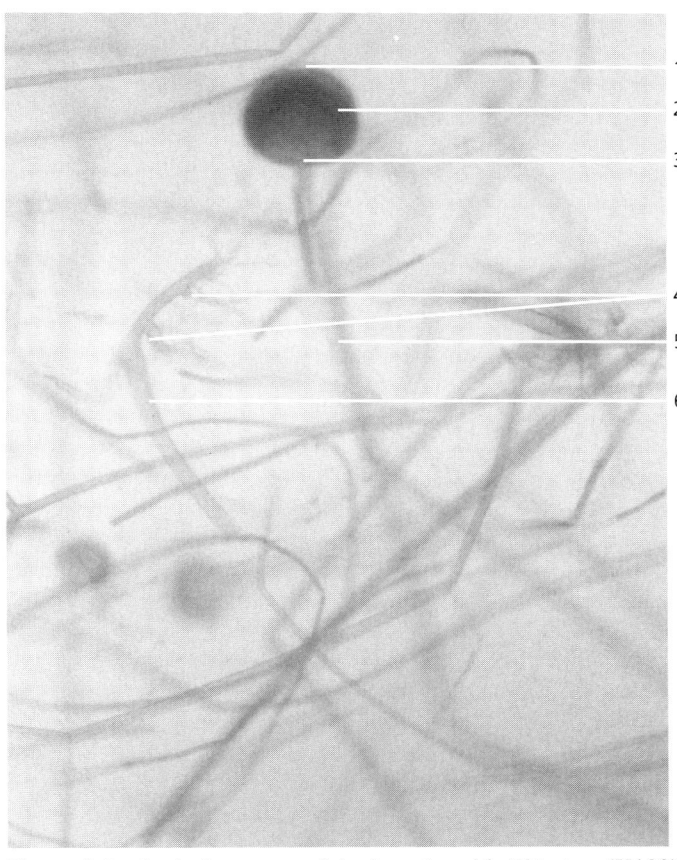

Figure 5.4 A whole mount of the bread mold, *Rhizopus*. (X100)

1. Sporangium
2. Spores
3. Columella
4. Rhizoids
5. Sporangiophore
6. Stolon (hyphae)

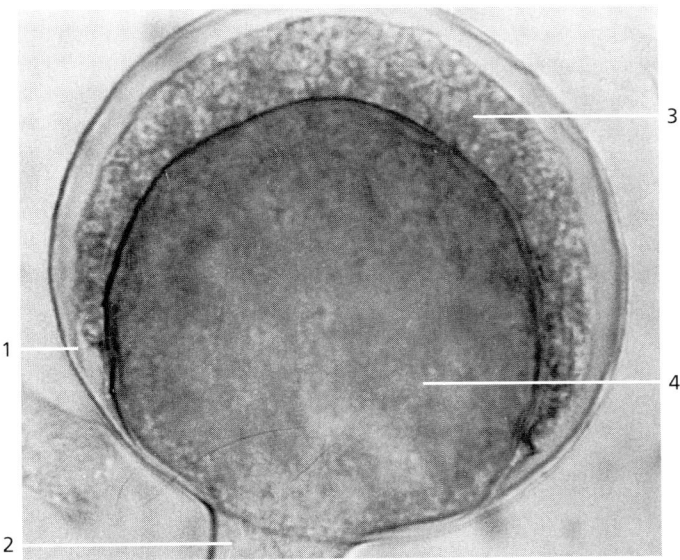

Figure 5.5 A mature sporangium in the asexual reproductive cycle of the bread mold, *Rhizopus*. (X430)

1. Sporangium
2. Sporangiophore
3. Spores
4. Columella

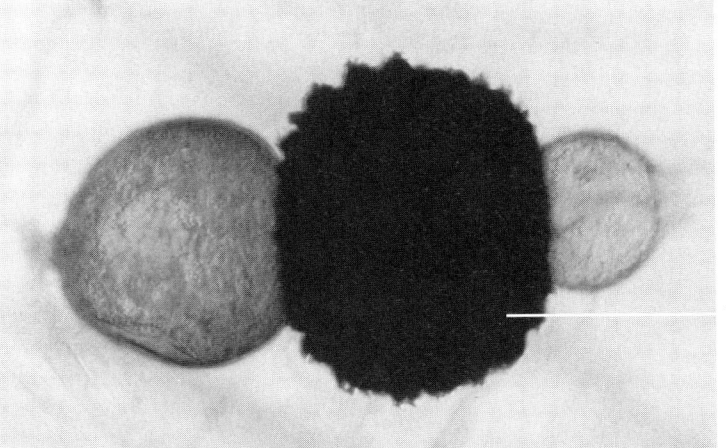

Figure 5.6 Conjugation of two hyphae in the bread mold, *Rhizopus*. (X430)

1. Zygospore (zygote)

DIVISION ASCOMYCOTA (yeasts, molds, morels, truffles)

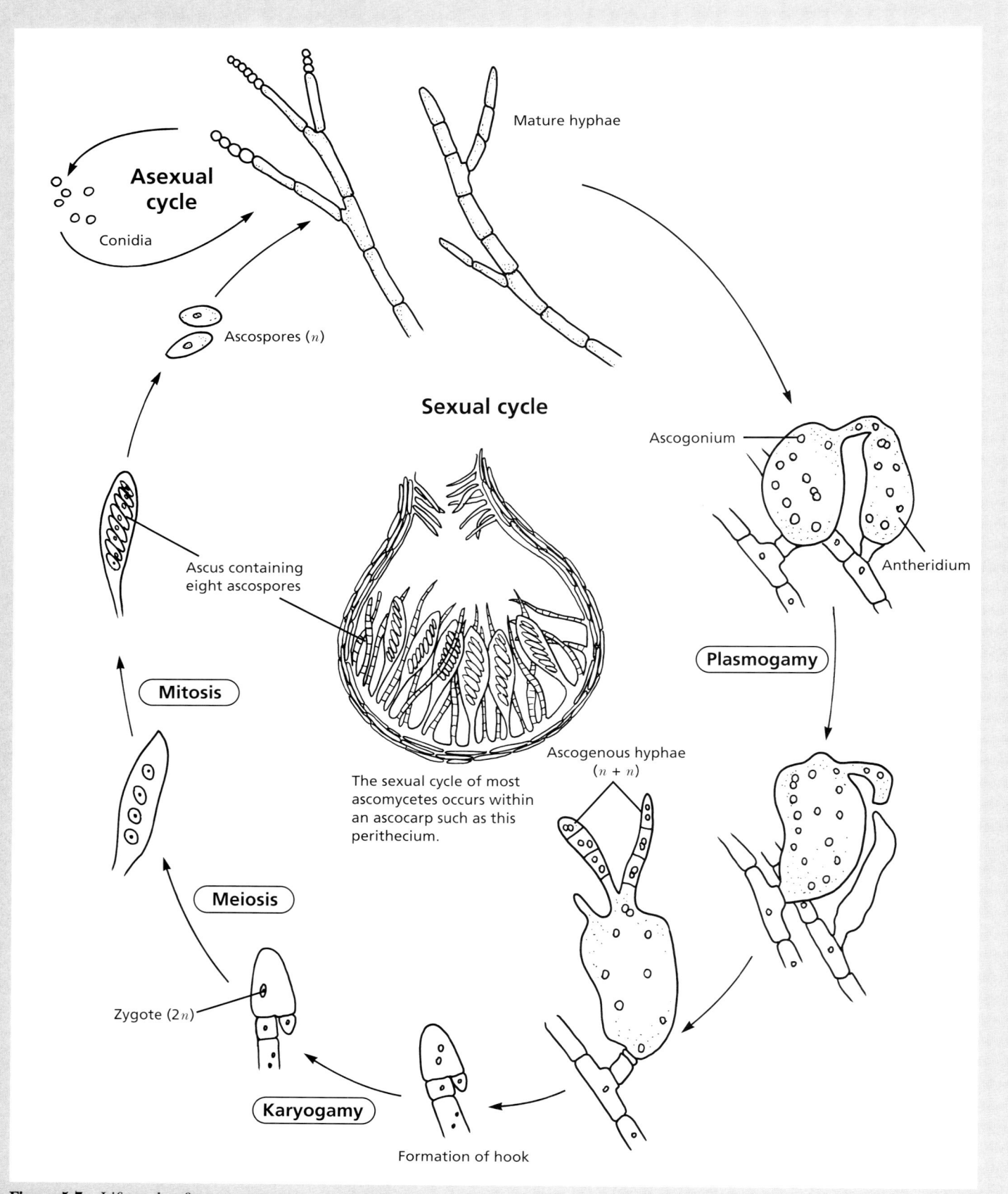

Figure 5.7 Life cycle of an ascomycete.

DIVISION ASCOMYCOTA (yeasts, molds, morels, truffles)

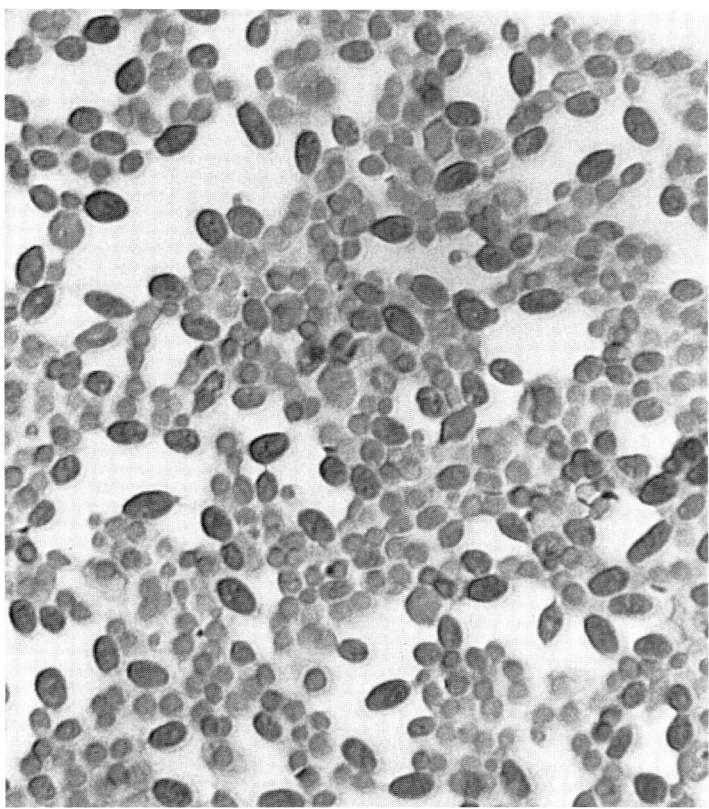

Figure 5.8 Baker's yeast, *Saccharomyces cerevisiae*. The ascospores of this unicellular ascomycete are characteristically spheroidal or ellipsoidal in shape. (X1000)

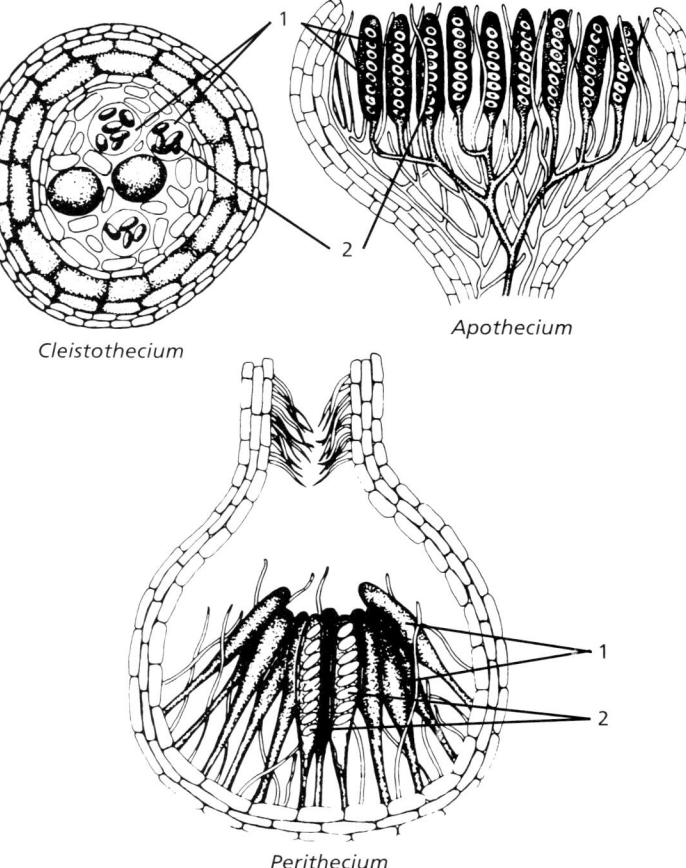

Figure 5.9 Examples of various types of ascocarps, or fruiting bodies, of ascomycetes.

1. Asci   2. Ascospores

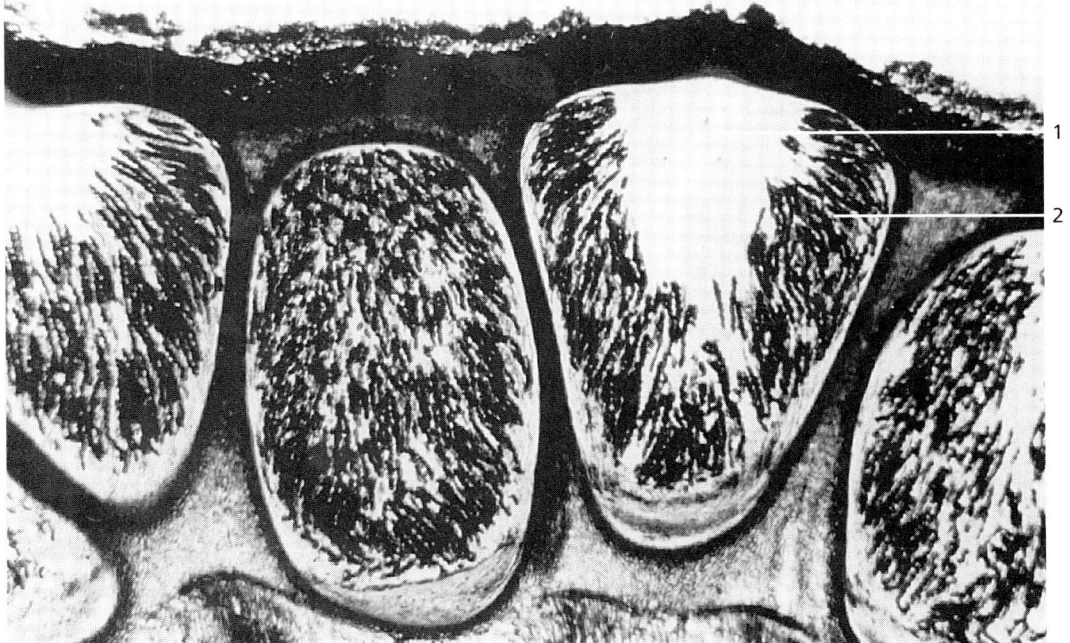

Figure 5.10 Closeup of the parasitic ascomycete, *Hypoxylon*, showing imbedded perithecia. (X200)

1. Perithecium   2. Hymenium

DIVISION ASCOMYCOTA (yeasts, molds, morels, truffles)

Morchella

Helvella

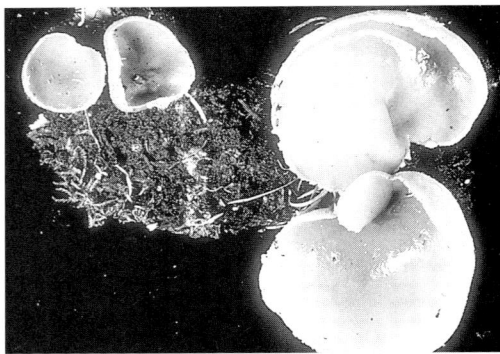
Peziza repanda

Monolina fructicola

Figure 5.11 Fruiting bodies (ascocarps) of common ascomycetes. *Morchella* is a common edible morel. *Helvella* is sometimes known as a saddle fungus since the fruiting body is thought by some to resemble a saddle. *Peziza repanda* is a common woodland cup fungus. *Monolina fructicola* is an important plant pathogen causing brown rot of fruit.

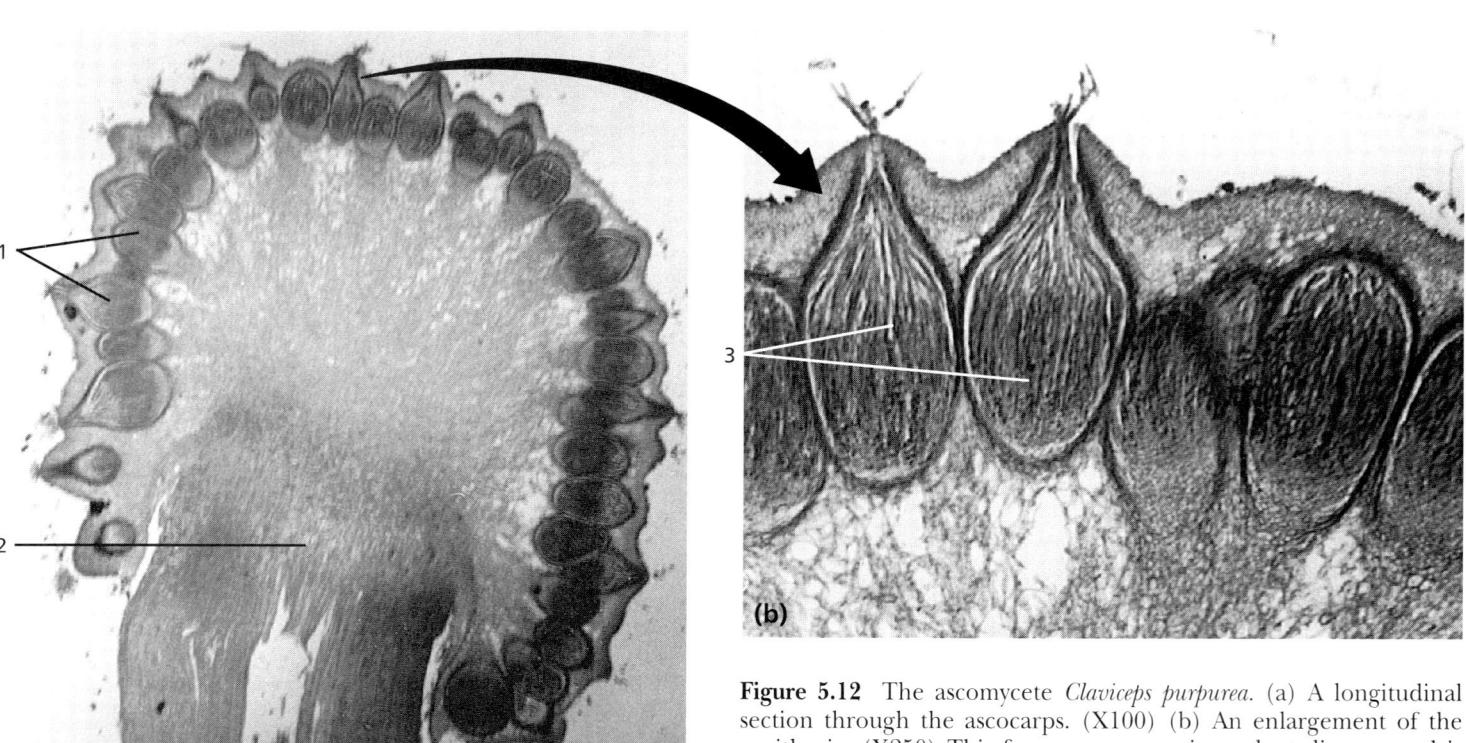

Figure 5.12 The ascomycete *Claviceps purpurea*. (a) A longitudinal section through the ascocarps. (X100) (b) An enlargement of the perithecia. (X250) This fungus causes serious plant diseases and is toxic to humans.

1. Perithecia   2. Stroma   3. Perithecia containing asci with ascospores

CHAPTER 5 — Kingdom Fungi

DIVISION ASCOMYCOTA (yeasts, molds, morels, truffles)

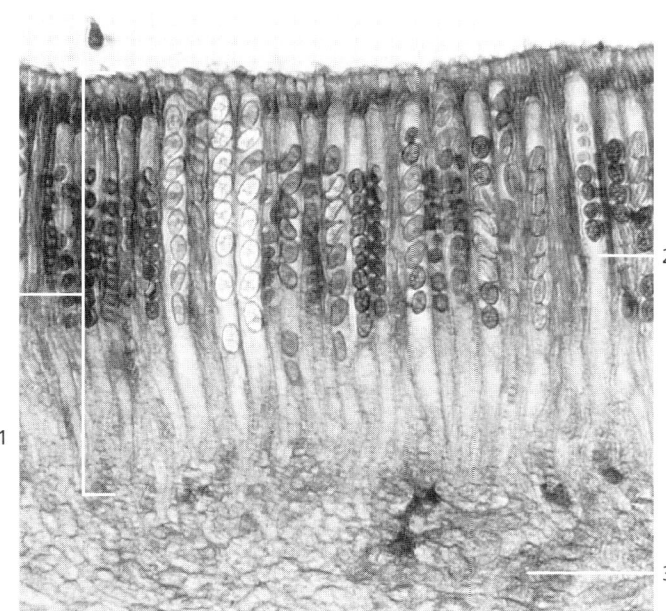

Figure 5.13 A section through the hymenial layer of the apothecium of *Peziza*, showing asci with ascospores. (X100)

1. Hymenial layer 2. Ascus with ascospores 3. Ascocarp mycelium

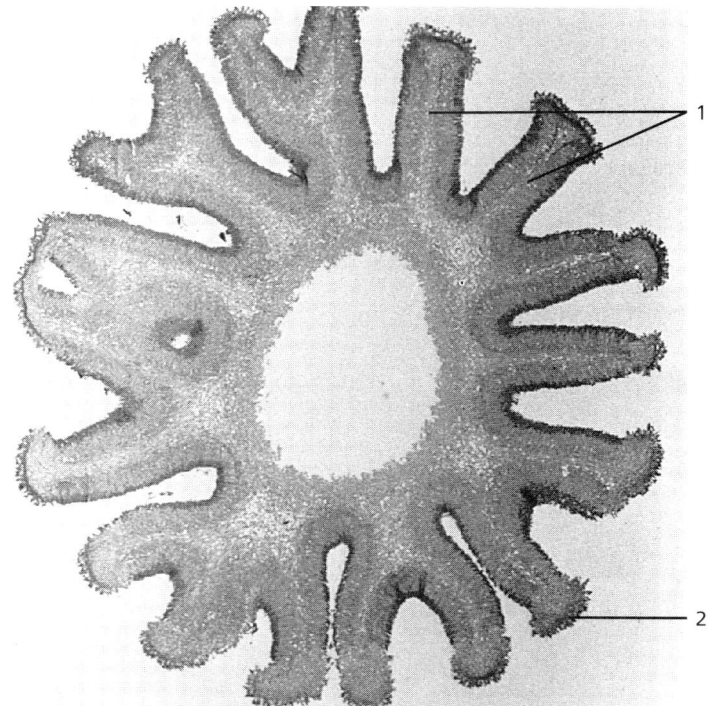

Figure 5.14 A section through an ascocarp of a morel, *Morchella*. True morels are prized for their excellent flavor. (X100)

1. Convoluted fruiting body 2. Hymenium

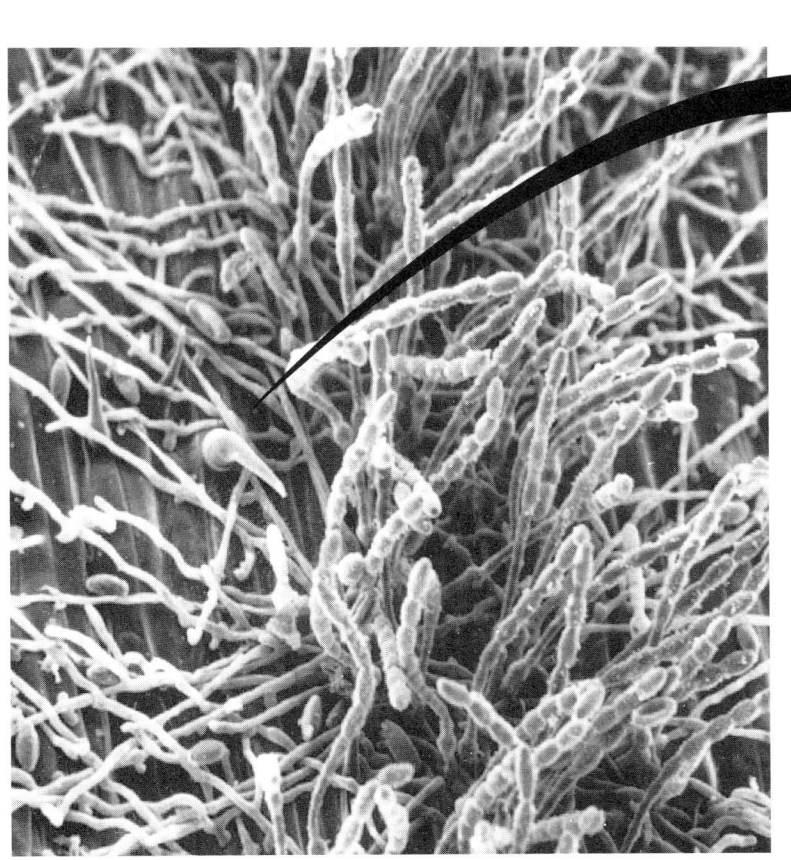

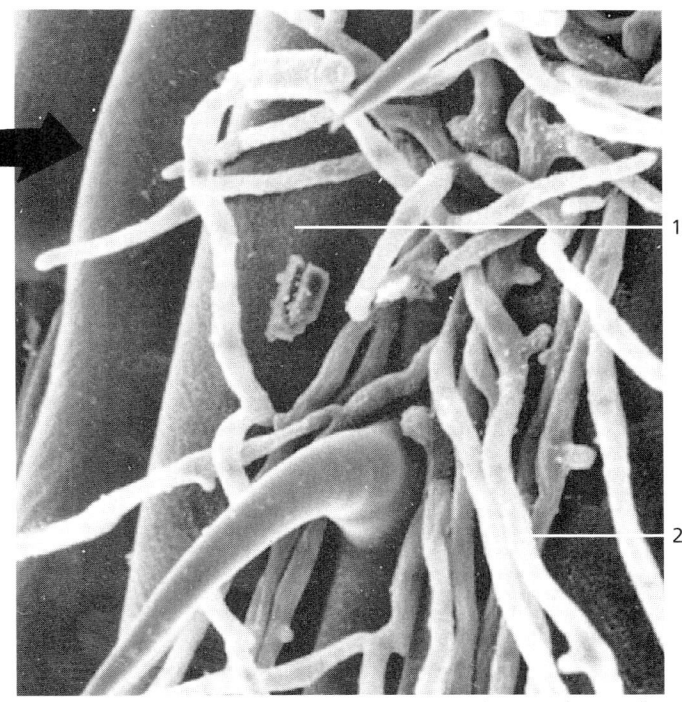

Photographs courtesy of James V. Allen

Figure 5.15 Scanning electron micrographs of the powdery mildew, *Erysiphe graminis*, on the surface of wheat. As the mycelium develops, it produces spores (conidia) that give a powdery appearance to the wheat.

1. Wheat host 2. Mycelium

DIVISION ASCOMYCOTA (yeasts, molds, morels, truffles)

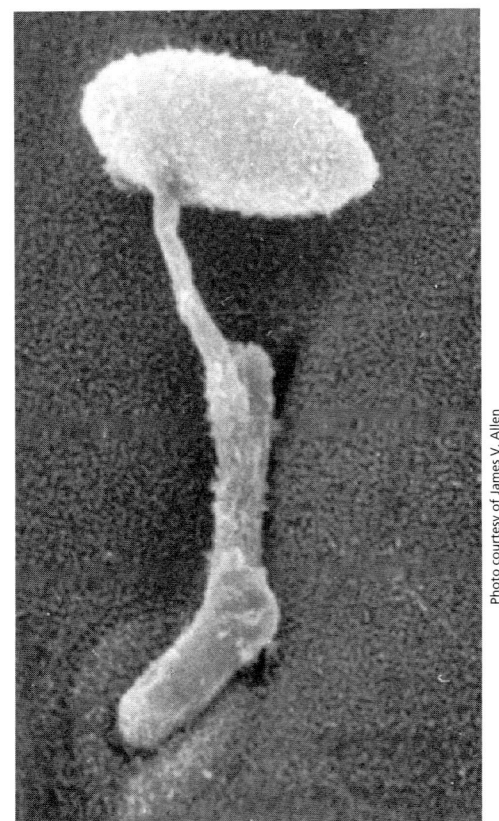

Figure 5.16 Scanning electron micrograph of a germinating spore (conidium) of the powdery mildew, *Erysiphe graminis*. The spore develops into a mycelium that spreads over the host plant.

Figure 5.17 The fungus *Penicillium* causes economic damage as a mold but is also the source of important antibiotics. (X100)

1. Conidia (spores)   2. Conidiophores

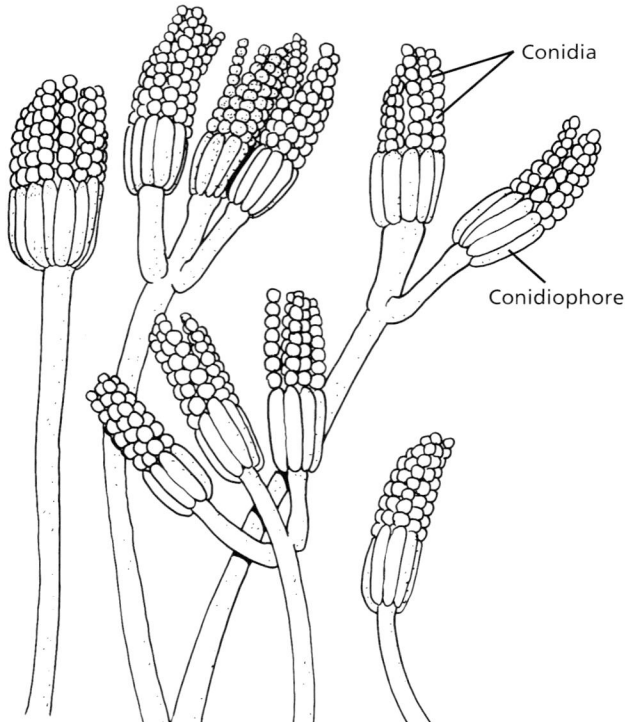

Figure 5.18 Conidiophores and conidia of *Penicillium*.

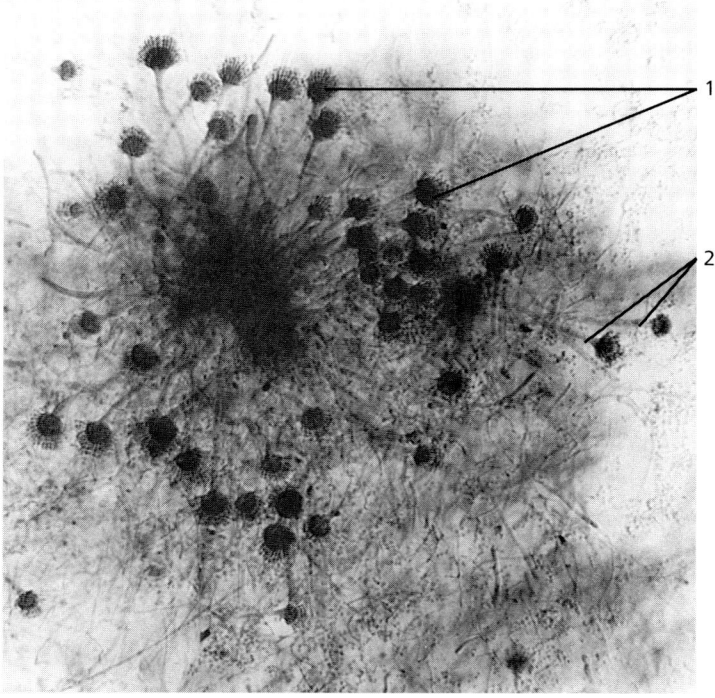

Figure 5.19 The common mold, *Aspergillus*. (X100)

1. Conidia   2. Conidiophores

DIVISION ASCOMYCOTA (yeasts, molds, morels, truffles)

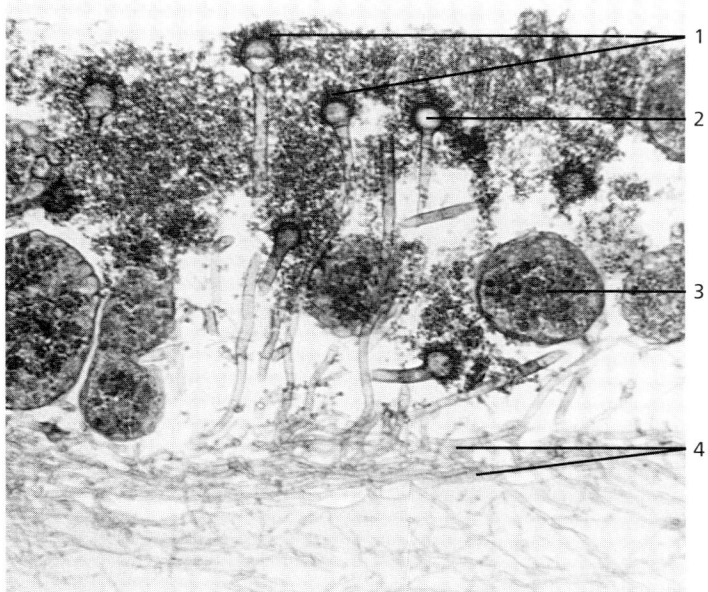

Figure 5.20 The common mold, *Aspergillus*. (X200)
1. Conidia (spores)
2. Conidiophore
3. Cleistothecium
4. Hyphae

Photo courtesy of Wilford M. Hess

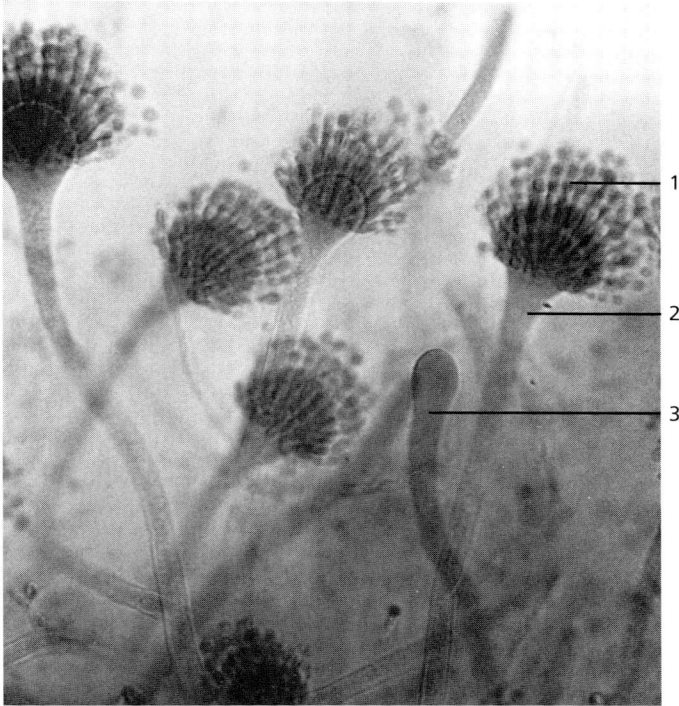

Figure 5.21 Closeup of sporangia of the mold, *Aspergillus*. The conidia, or spores, of this genus are produced in a characteristic radiate pattern. (X430)
1. Conidia (spores)   2. Conidiophore   3. Developing conidiophore

Figure 5.22 An electron micrograph of an *Aspergillus* spore. Note the rodlet pattern on the spore wall. (X1500)

Figure 5.23 A parasitic ascomycete, *Dibotryon morbosum*, on a branch of a chokecherry.

DIVISION BASIDIOMYCOTA (mushrooms, toadstools, rusts, smuts)

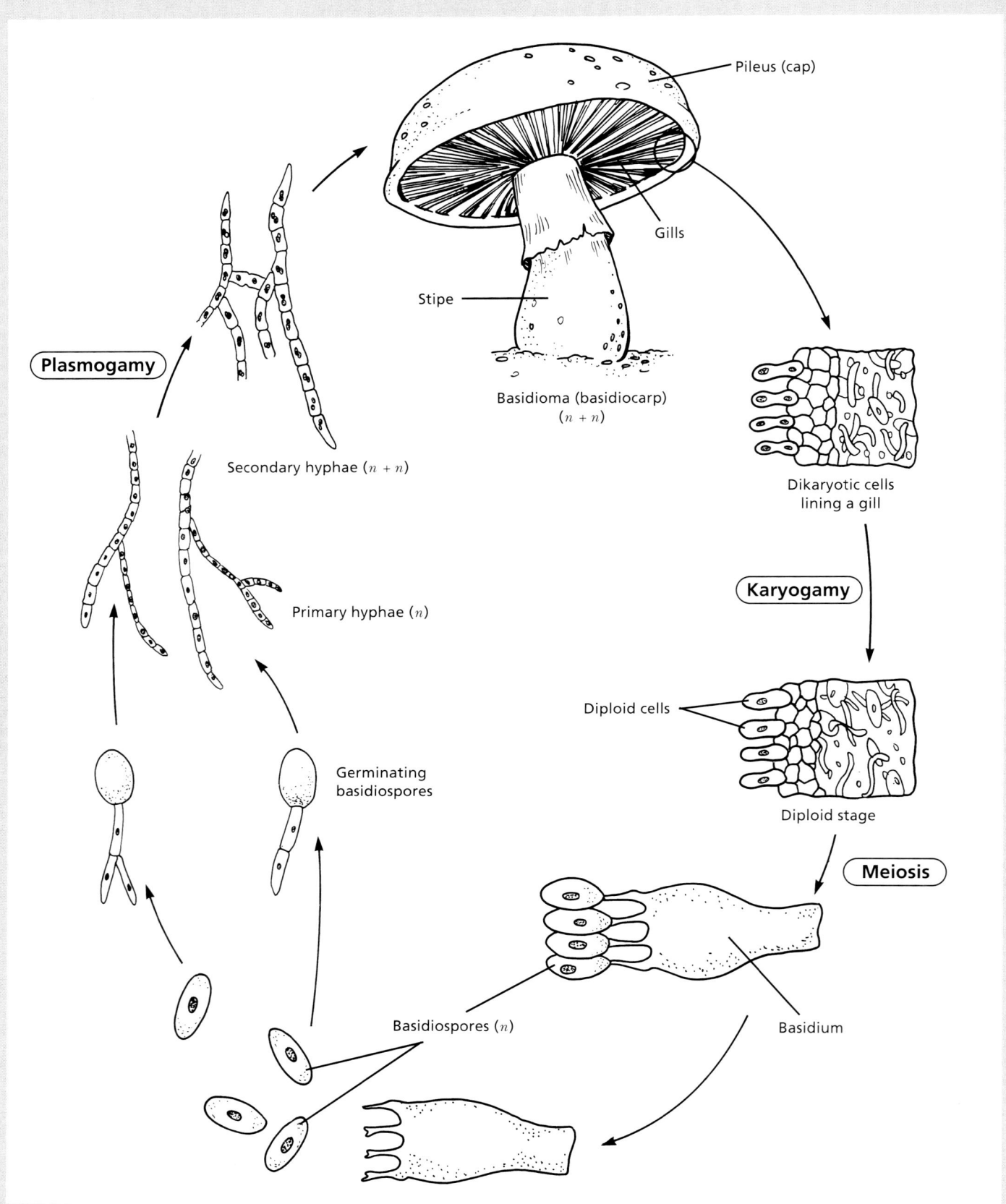

Figure 5.24 Life cycle of a mushroom.

DIVISION BASIDIOMYCOTA (mushrooms, toadstools, rusts, smuts)

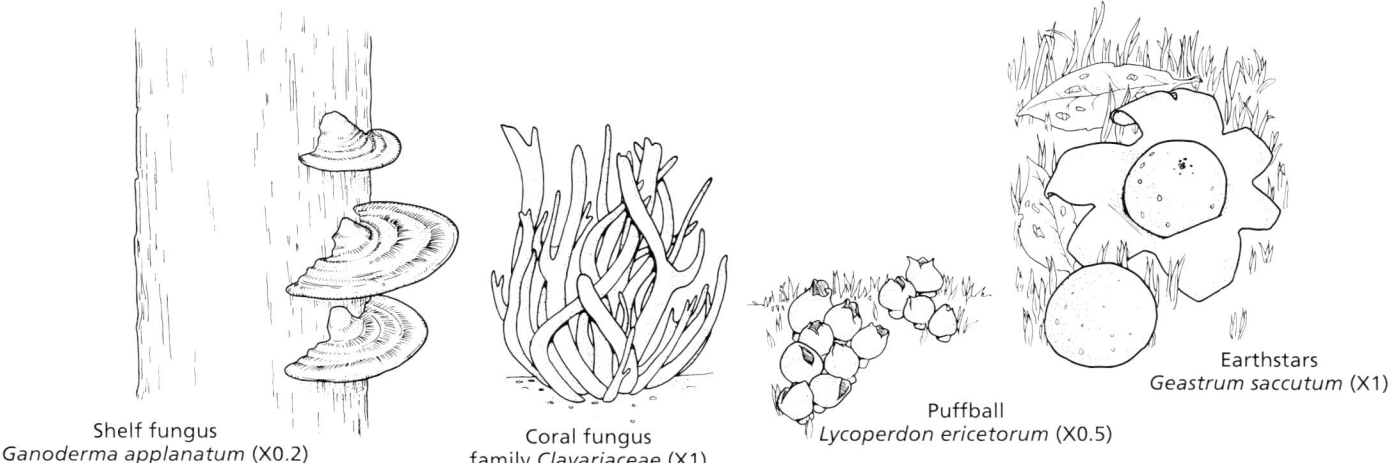

Shelf fungus
Ganoderma applanatum (X0.2)

Coral fungus
family *Clavariaceae* (X1)

Puffball
Lycoperdon ericetorum (X0.5)

Earthstars
Geastrum saccutum (X1)

Figure 5.25 Drawings of various basidiomycetes.

Agaricus rodmani

Cortinarius species

Amanita pantherina

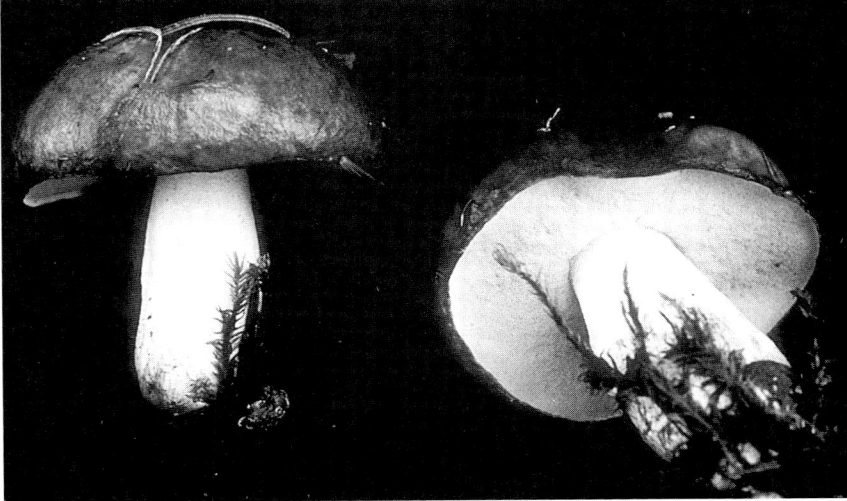

Boletus species

Figure 5.26 Fruiting bodies (basidiocarps) of common mushrooms. *Agaricus rodmani* is a prized edible mushroom. The genus *Cortinarius* contains more species than any other in North America. *Amanita pantherina* is a deadly toxic mushroom. *Boletus* is a mushroom with pores on the undersurface of the cap rather than gills.

DIVISION BASIDIOMYCOTA (mushrooms, toadstools, rusts, smuts)

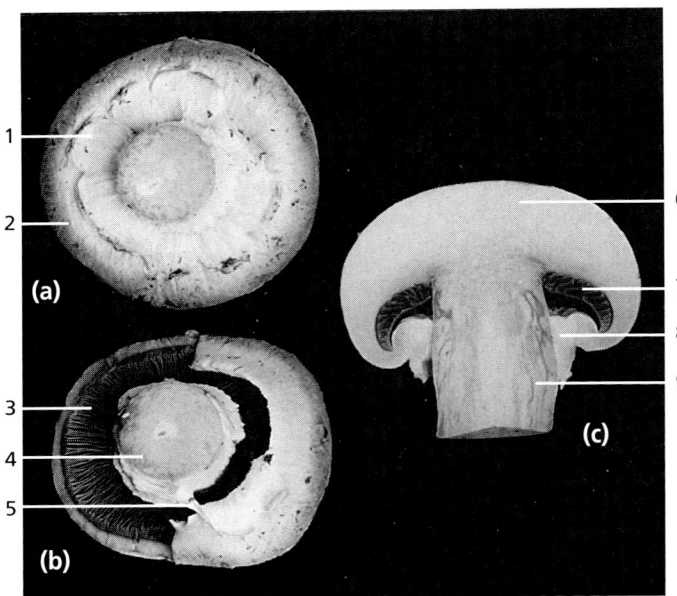

Figure 5.27 Mushroom. (a) Inferior view with the annulus intact; (b) inferior view with a portion of the annulus removed to show the gills; (c) longitudinal section.

1. Veil
2. Pileus (cap)
3. Gills
4. Stipe (stalk)
5. Annulus
6. Pileus (cap)
7. Gills
8. Veil
9. Stipe (stalk)

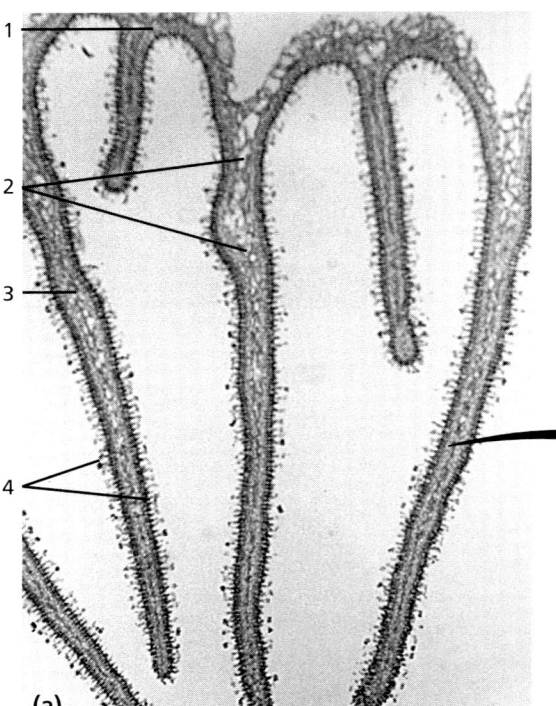

Figure 5.30 Gills of the mushroom, *Coprinus*. (a) Closeup of several gills, (X40); and (b) closeup of a portion of a single gill. (X100)

1. Pileus (cap) comprised of gills
2. Hyphae comprising the gills
3. Gill
3. Basidiospores
5. Sterigma
6. Gill (comprised of hyphae)
7. Basidia
8. Basidiospore

Figure 5.28 Basidiomycete puffballs growing on a decaying log. (X0.25)

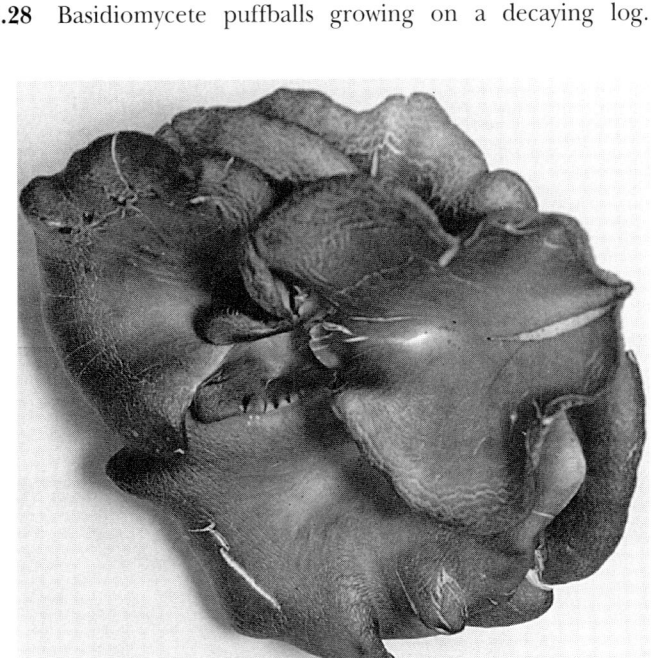

Figure 5.29 Herbarium specimen of the wood fungus, *Stropharia semiglobata*. Growing on decaying wood and other organic matter, basidiomycetes are important decomposers in forest communities.

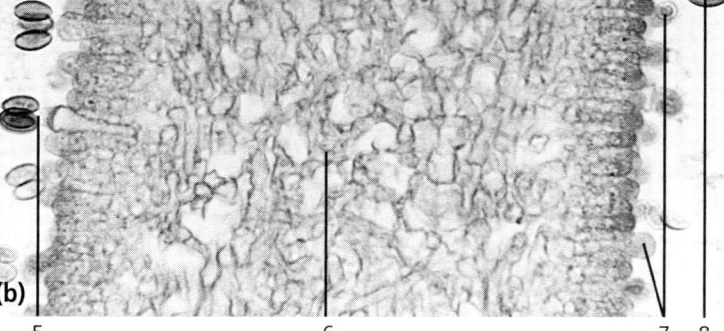

DIVISION BASIDIOMYCOTA (mushrooms, toadstools, rusts, smuts)

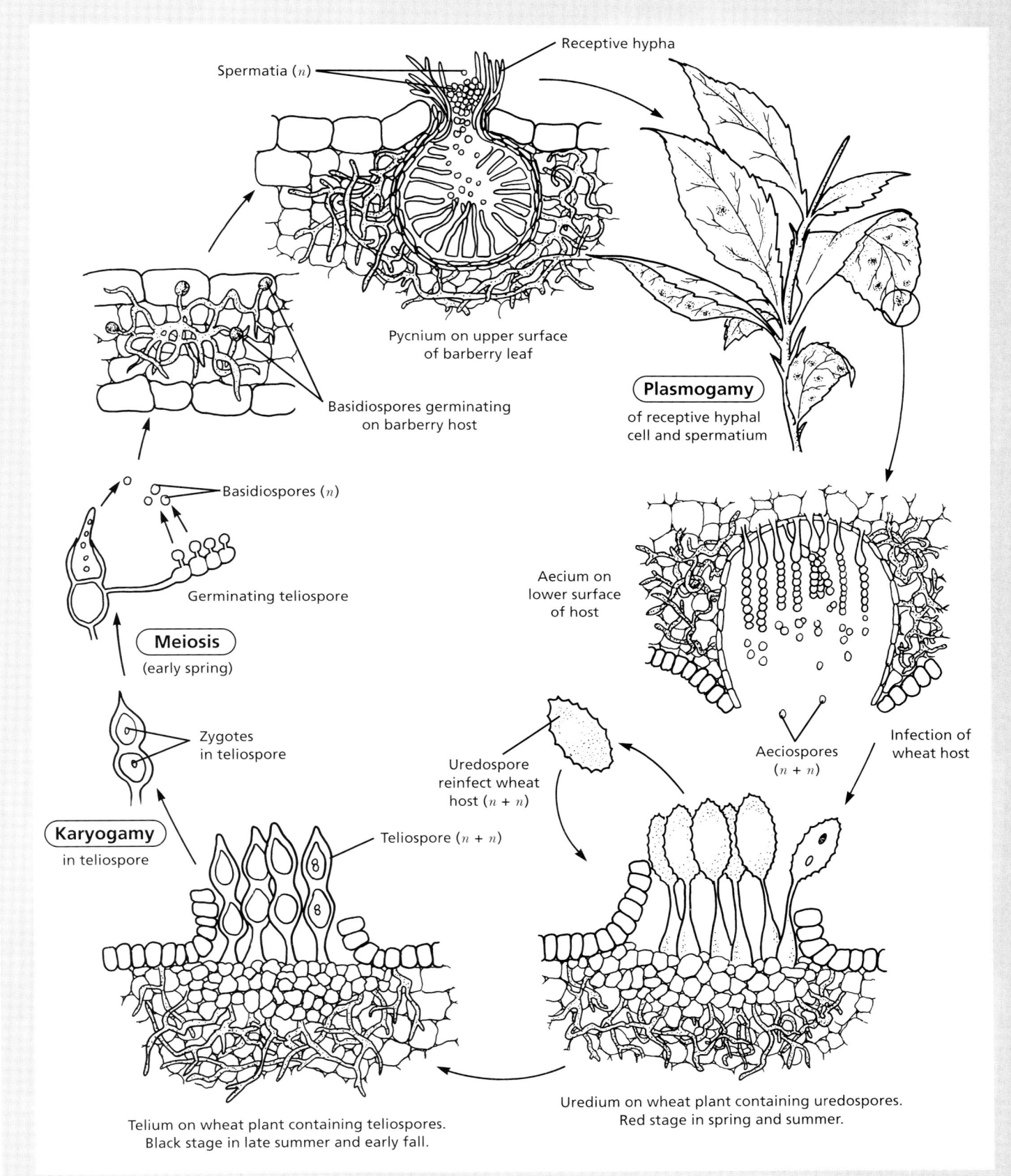

Figure 5.31 Life cycle of wheat rust, *Puccinia graminis*.

DIVISION BASIDIOMYCOTA (mushrooms, toadstools, rusts, smuts)

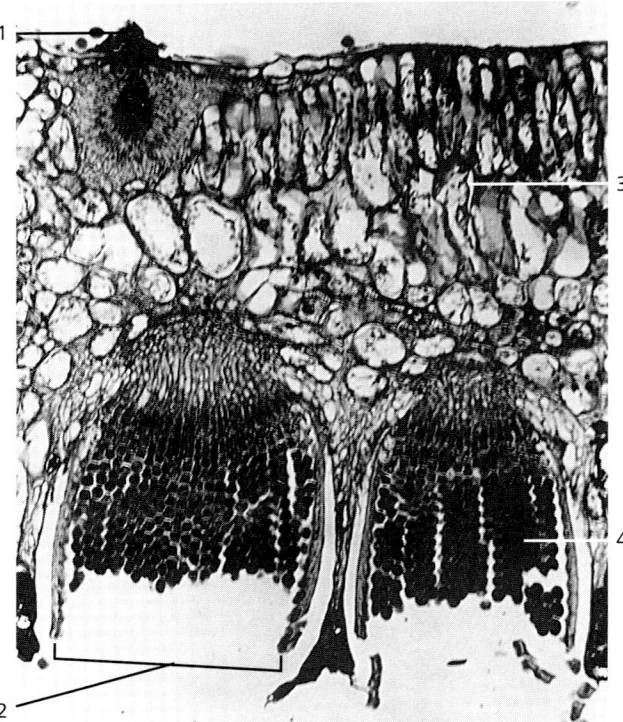

Figure 5.32 Wheat rust, *Puccinia graminis*, pycnium and aecia on barberry leaf. (X100)

1. Pycnium
2. Aecium
3. Barberry leaf
4. Aeciospores

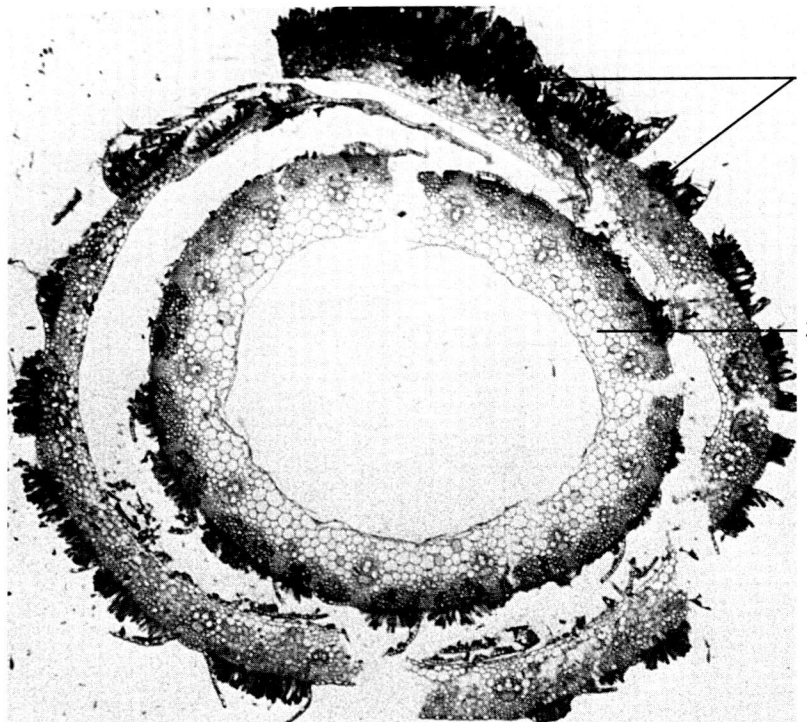

Figure 5.33 Section of a wheat stem showing the wheat rust *Puccinia graminis*. (X40)

1. Telia
2. Wheat stem

Figure 5.34 Black stem wheat rust, *Puccinia graminis*, on the lower surface of barberry leaves.

1. Clusters of aecia

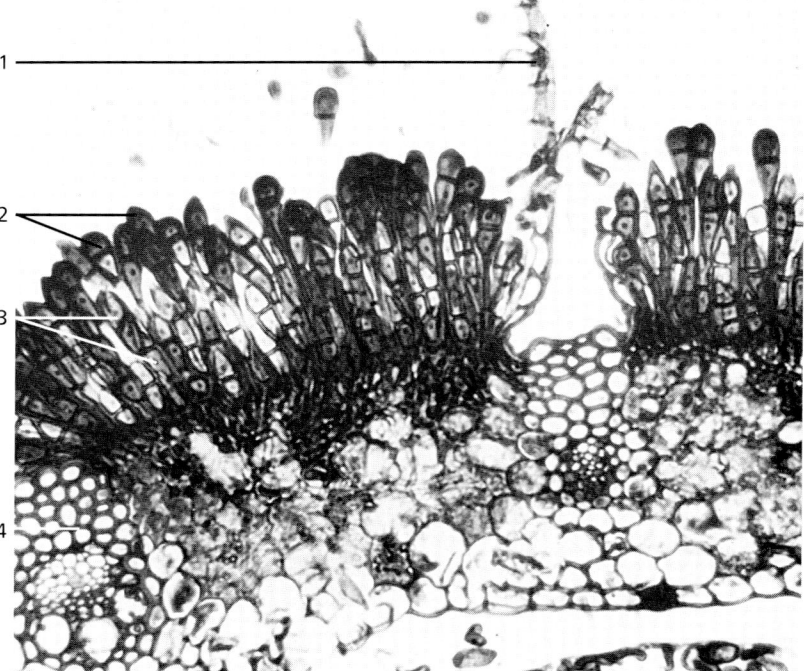

Figure 5.35 Closeup of a wheat leaf sheath showing the wheat rust *Puccinia graminis*. (X100)

1. Epidermis of leaf
2. Teliospores
3. Nuclei of teliospores
4. Wheat leaf sheath

DIVISION BASIDIOMYCOTA (mushrooms, toadstools, rusts, smuts)

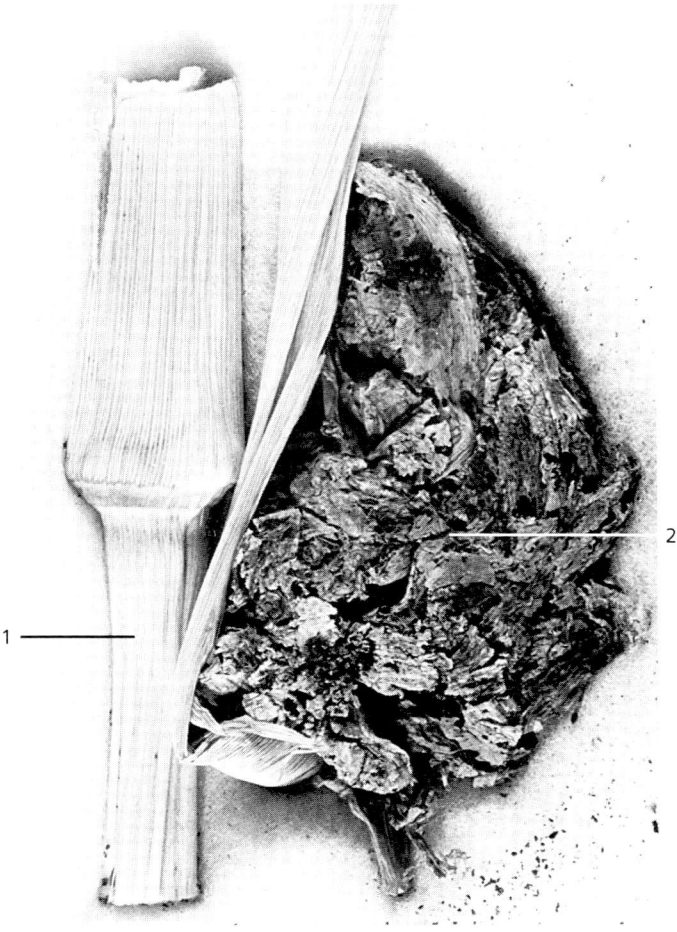

Figure 5.36. A photograph of a corn plant infected by the smut *Ustilago maydis*.
1. Corn stalk
2. A corn ear completely destroyed by the fungus

Figure 5.37 A photograph of a smut-infected brome grass. The grains have been destroyed by the fungus.

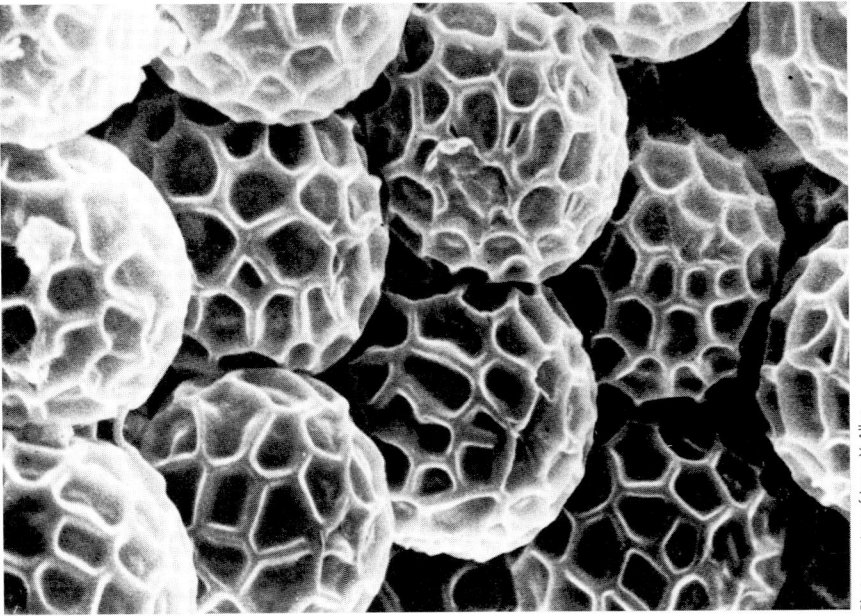

Figure 5.38 Scanning electron micrograph of teliospores of a wheat smut fungus.

LICHENS (not a division, but rather a symbiotic association of an alga and a fungus)

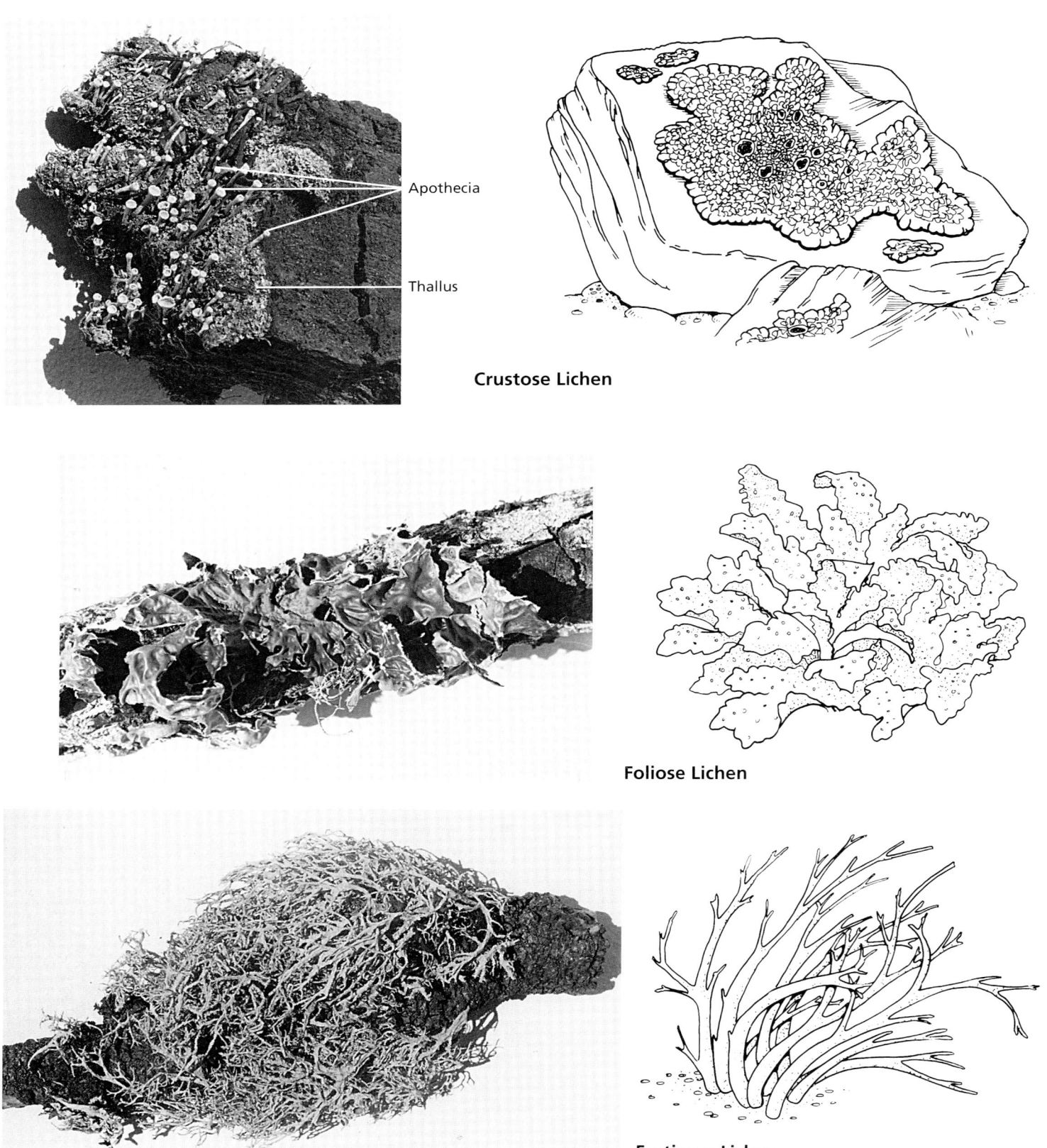

Crustose Lichen

Foliose Lichen

Fruticose Lichen

Figure 5.39 Although highly diverse in structure and appearance, the many kinds of lichens are grouped into three broad categories.

CHAPTER 5 — Kingdom Fungi

LICHENS (not a division, but rather a symbiotic association of an alga and a fungus)

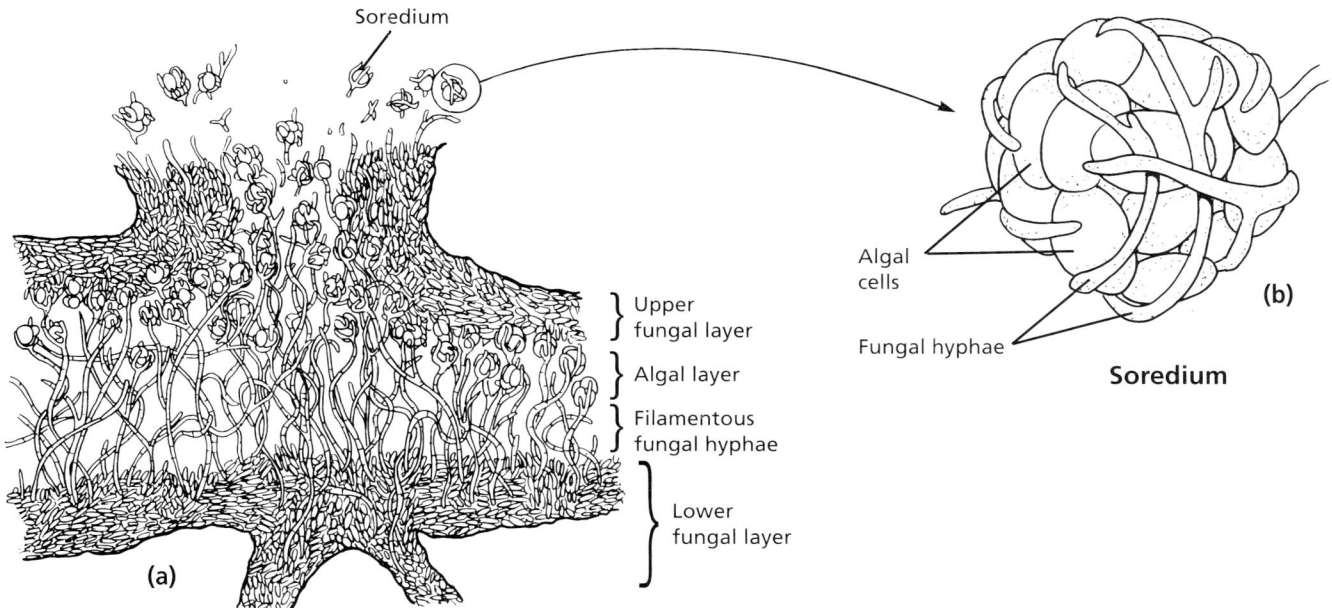

Figure 5.40 Many lichens reproduce by producing soredia, which are small bodies containing both algal and fungal cells. (a) lichen thallus; (b) soredium.

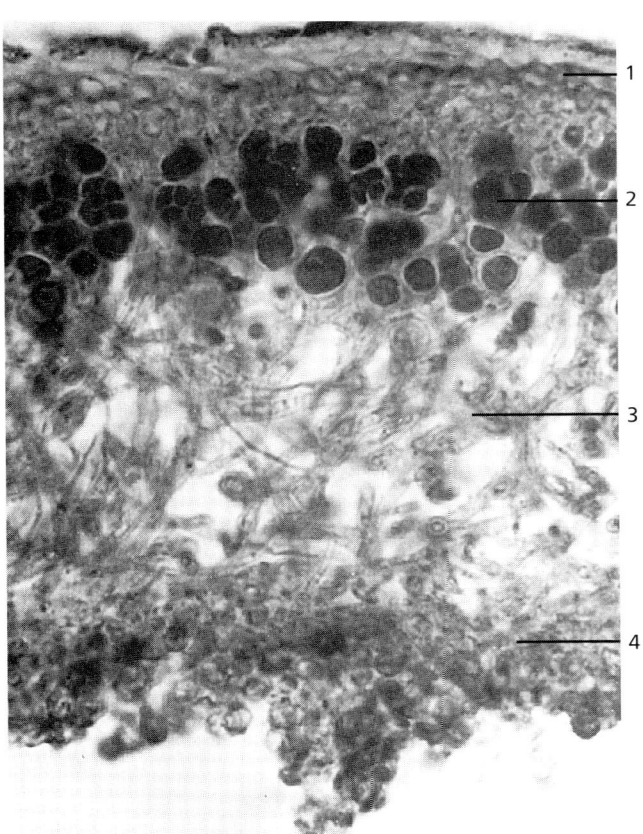

Figure 5.41 The lichen thallus is often constructed of distinct algal and fungal layers. This cross section of a lichen thallus clearly shows these layers. (X100)

1. Upper fungal layer
2. Algal layer
3. Filamentous fungal hyphae
4. Lower fungal layer

Figure 5.42 Closeup of a crustose lichen on the bark of a tree.

1. Lichen

Kingdom Plantae: Division Bryophyta (Bryophytes)

The division Bryophyta includes the liverworts, hornworts, and mosses. Some bryologists divide this division into three divisions: Hepaticophyta, Anthocerotophyta, and Bryophyta (mosses only). These plants inhabit damp, shady environments and are found worldwide. Though simple plants, bryophytes are old and successful. Devonian fossil bryophytes have been found, some 375-400 million years old. Currently, 16,000 species are known. Because many bryophytes are sensitive to sulfur dioxide, they cannot live in air polluted environments.

Although most bryophytes are only 1–2 cm in height, a few species may grow to 15–20 cm. The adult, free-living plant is haploid and produces gametes. It is referred to as a *gametophyte*. The gametophytes of many bryophytes have a thick, waxy *cuticle* that protects the plant from desiccation. Lacking roots, gametophytes are attached to the substrate by elongated single cells or multicellular filaments known as *rhizoids*. Rhizoids of one gametophyte often extend horizontally intertwining with rhizoids of other gametophytes, forming loose colonies of individual plants that are effective in holding moisture.

In some liverworts, the gametophyte is dorsoventrally flattened, while in others it is upright and "leafy." With the possible exception of certain mosses, bryophytes lack vascular tissue. Transport within the stem is through diffusion, capillary action, and cytoplasmic streaming. Lacking true leaves, some bryophytes have leaflike extensions that collect moisture and assist in reproduction. *Stomata* (for gas exchange) are present on the sporophytes of hornworts and mosses. Unlike stomata of typical flowering plants, however, the stomata of some bryophytes are openings surrounded by a single, doughnut-shaped guard cell.

Like all true plants, bryophytes have *alternation of generations*. This means that their reproductive cycle has a haploid (n) phase in which the gametophyte produces gametes. After fusion in pairs, the gametes form a zygote. The zygote germinates, producing a diploid ($2n$) sporophyte. Through meiotic division, the *spores* produced from the sporophytes complete the cycle by giving rise to new gametophytes.

Most bryophytes have separate male and female gametophytes. The male gametophytes have *antheridia* which produce flagellated *sperm* cells. The female gametophytes have *archegonia* in which *eggs* are produced.

Water is essential for fertilization because sperm produced by the antheridia swim to the archegonia. In some species of bryophytes, raindrops may disperse the sperm and insects may play a limited role in dispersal. The diploid zygote develops into an embryonic sporophyte within the protective jacket of the archegonia.

During the next stage of diploid development, a *stalk* or *seta* forms to free the sporophyte from the archegonium. A spore-producing capsule, or sporangium, forms at the tip of the stalk. The spores, produced through meiosis, are haploid cells that disperse when the sporangium burst. As the spores germinate, a threadlike *protonema* is produced that gives rise to a haploid gametophyte, completing the life cycle.

TABLE 6.1
Representatives of the Division Bryophyta

Classes and Representative Kinds	Characteristics
Class Hepaticae (liverworts)	Flat or leafy gametophytes, single-celled rhizoids; simple sporophytes that lack stomata
Class Anthocerotae (hornworts)	Flat, lobed gametophytes; more complex sporophytes with stomata
Class Musci (mosses)	Leafy gametophytes, multicellular rhizoids; sporophytes with stomata

CLASS HEPATICAE (liverworts)

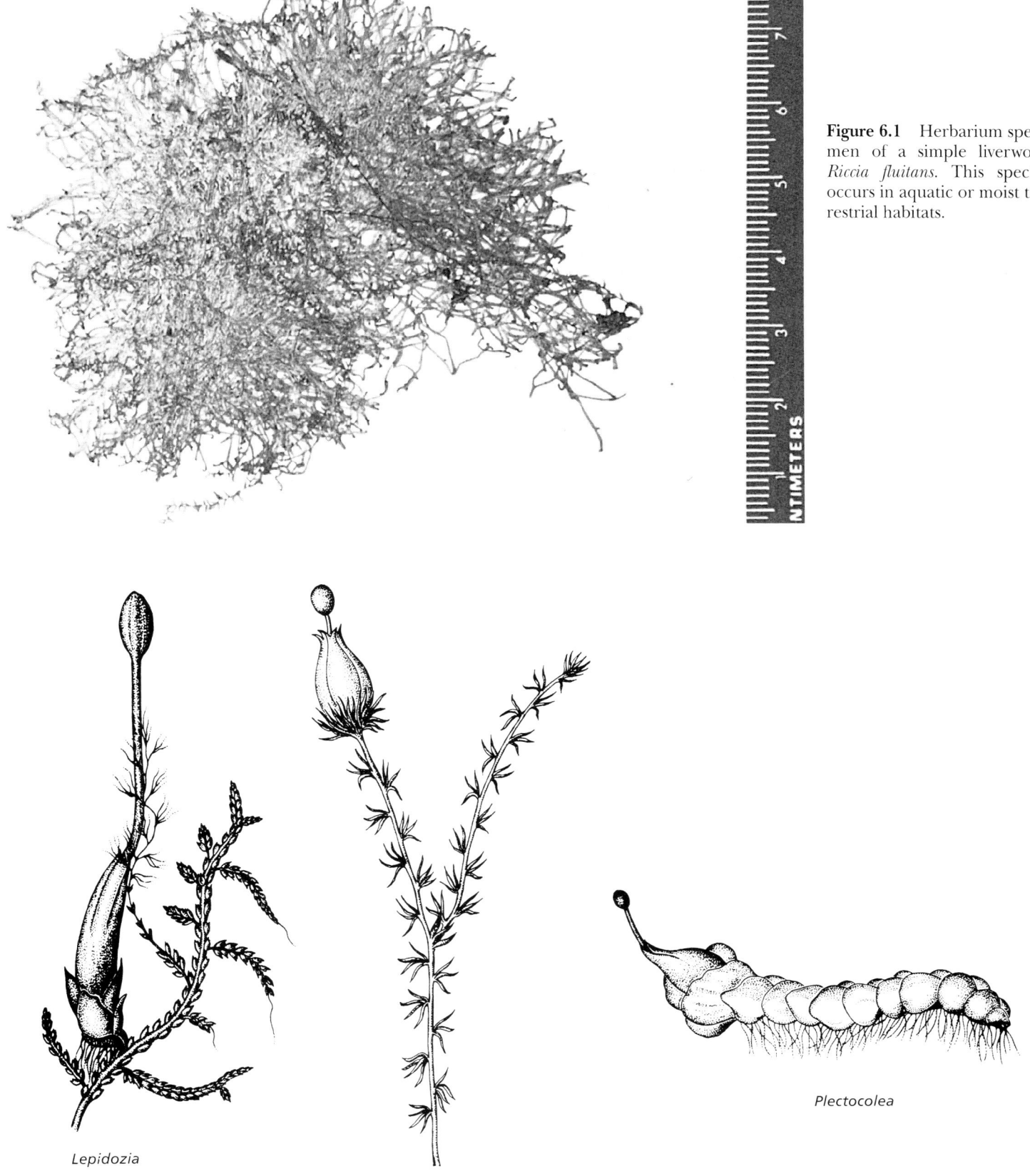

Figure 6.1 Herbarium specimen of a simple liverwort, *Riccia fluitans*. This species occurs in aquatic or moist terrestrial habitats.

Figure 6.2 Examples of three genera of leafy liverworts, showing the gametophyte with attached sporophyte plants.

CLASS HEPATICAE (liverworts)

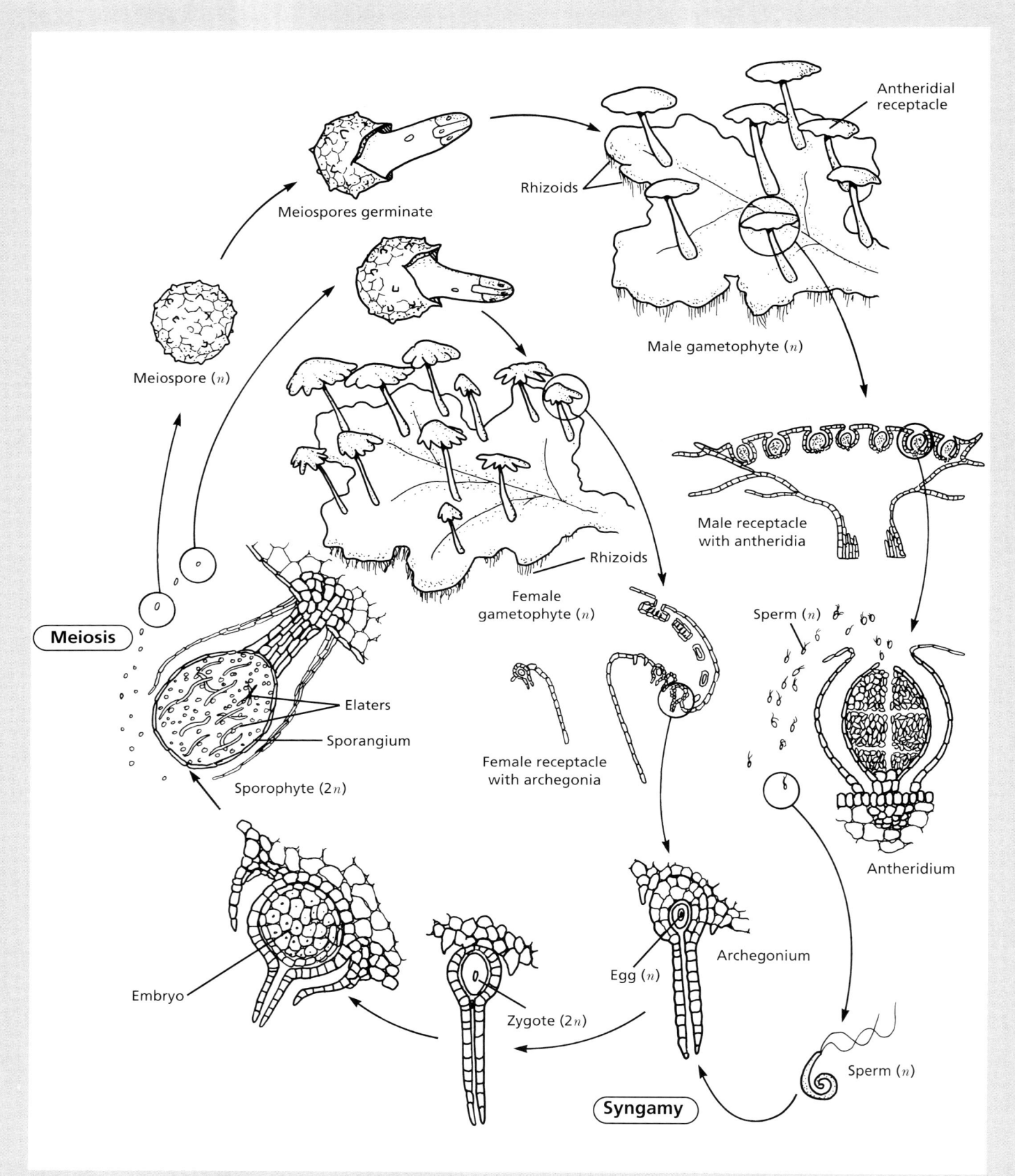

Figure 6.3 Life cycle of the thalloid liverwort, *Marchantia*.

CLASS HEPATICAE (liverworts)

Figure 6.4 Photograph of several gametophyte plants of the common liverwort, *Marchantia*.
1. Antheridial receptacle (*n*)
2. Gametophyte thallus (*n*)
3. Gemmae cup (*n*)

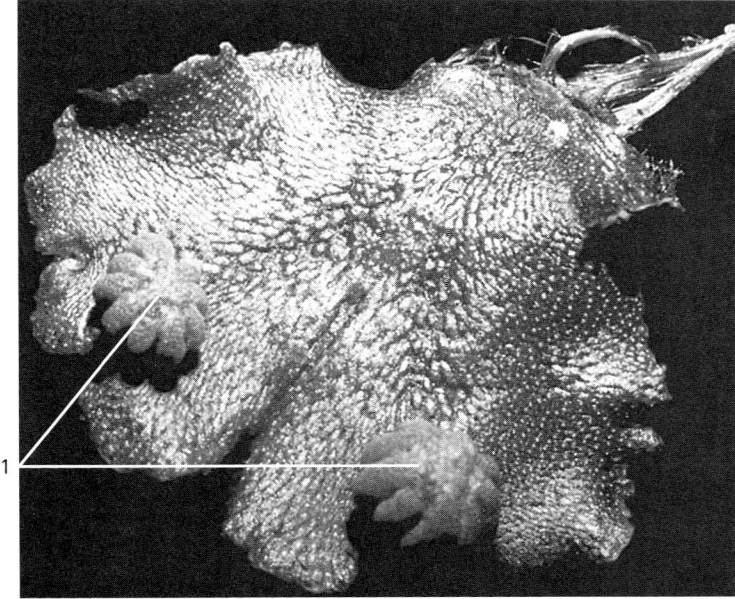

Figure 6.6 Dorsal view of a female gametophyte of the liverwort, *Marchantia*.
1. Archegonial receptacles

Figure 6.5 Ventral view of the common liverwort, *Marchantia*, showing numerous rhizoids.
1. Rhizoids

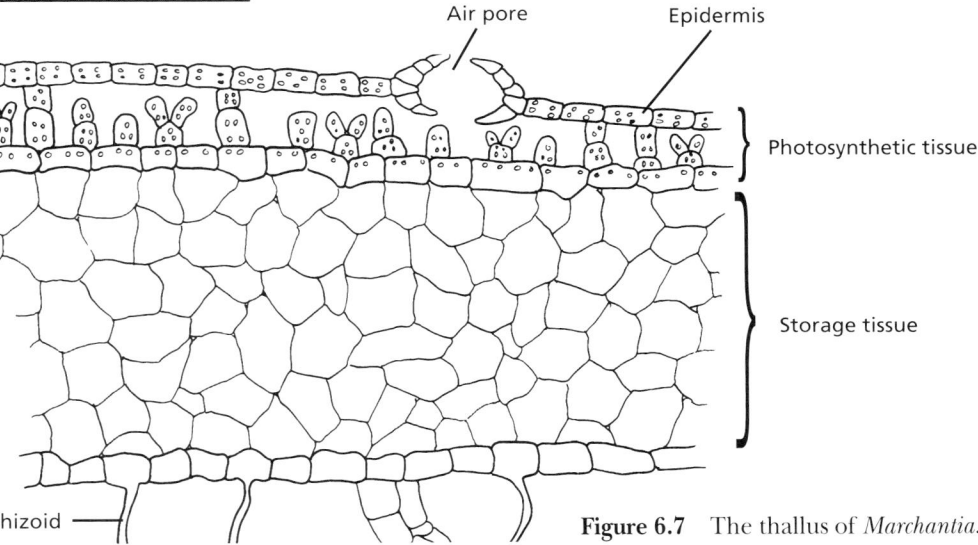

Figure 6.7 The thallus of *Marchantia*.

CLASS HEPATICAE (liverworts)

Photo courtesy of James V. Allen

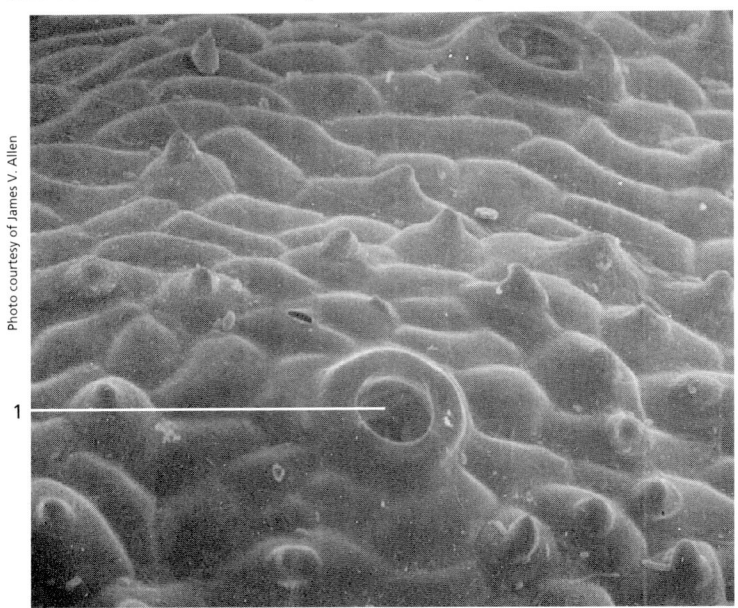

Figure 6.8 Scanning electron micrograph of the thallus of a liverwort, *Marchantia*. (X1000)

1. Air pore

hoto courtesy of JW. M. Hess

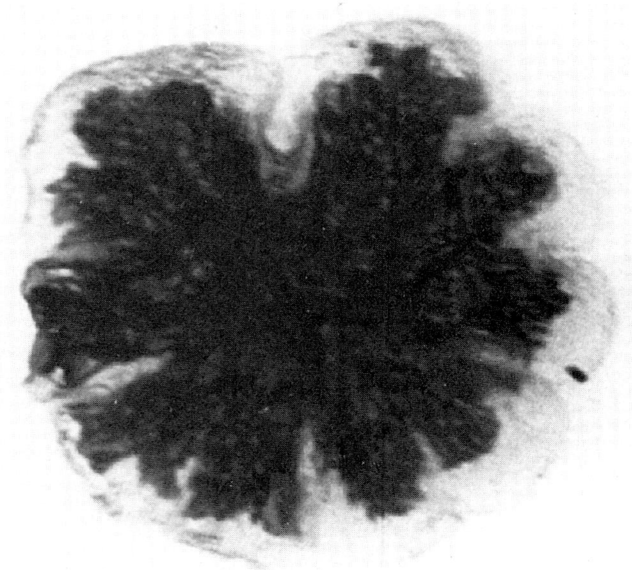

Figure 6.9 Dorsal surface of a male receptacle from a liverwort, *Marchantia*. (X10)

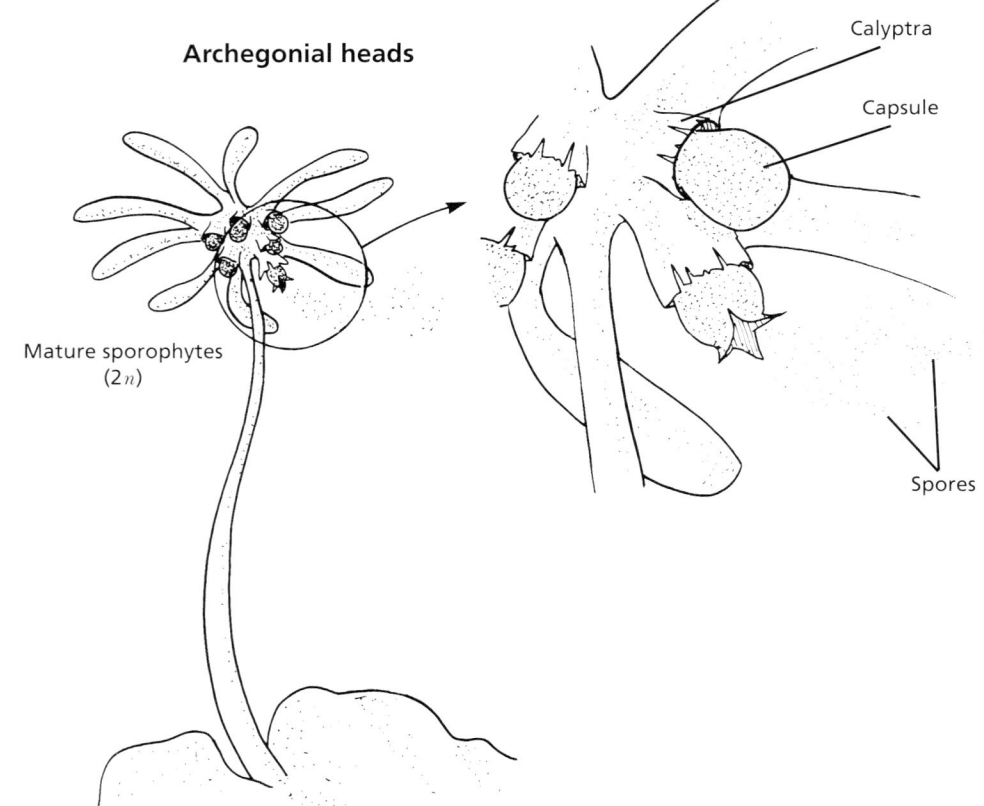

Figure 6.10 The sporophyte of *Marchantia*. in place on the archegonial receptacle.

CHAPTER 6 — Kingdom Plantae: Division Bryophytes (Bryophyta)

CLASS HEPATICAE (liverworts)

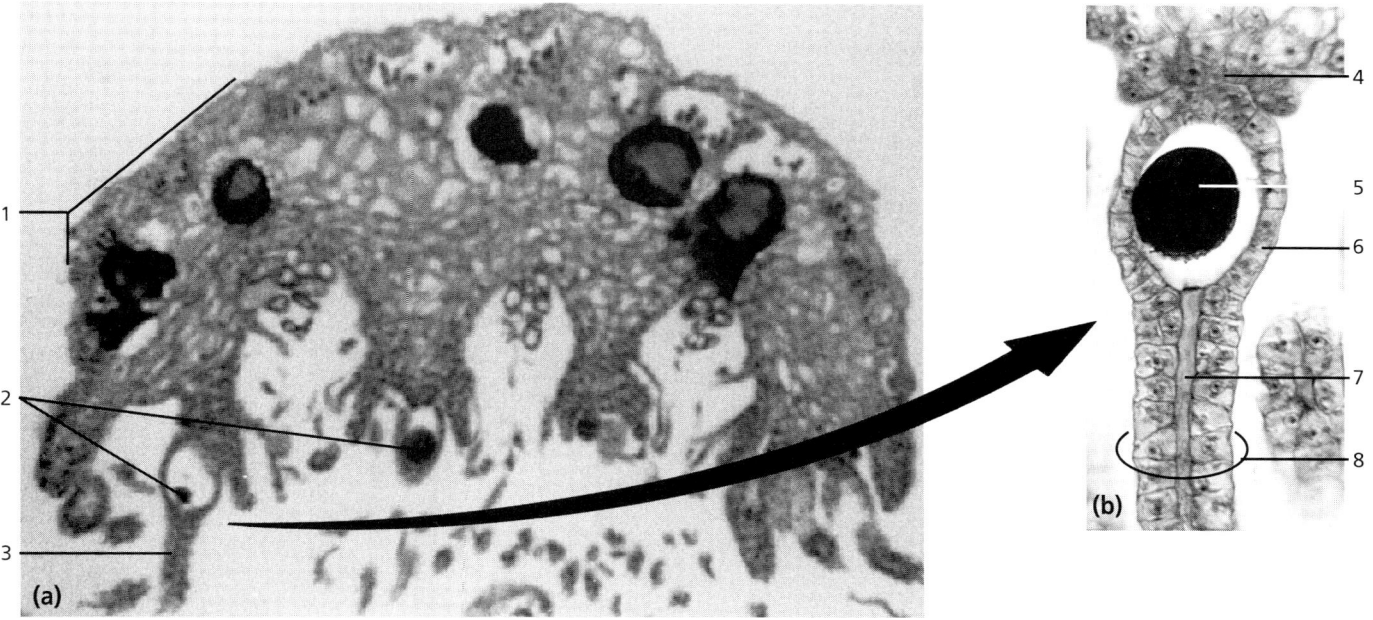

Figure 6.11 The archegonial receptacle (a) of a liverwort, *Marchantia*, in longitudinal section. (X40) The archegonium (b) showing an egg. (X240)

1. Archegonial receptacle
2. Eggs
3. Neck of archegonium
4. Stalk of archegonium
5. Egg
6. Venter of archegonium
7. Neck canal cells
8. Neck of archegonium

Figure 6.12 The female receptacle of a liverwort, *Marchantia*, in dorsal view. (X10)

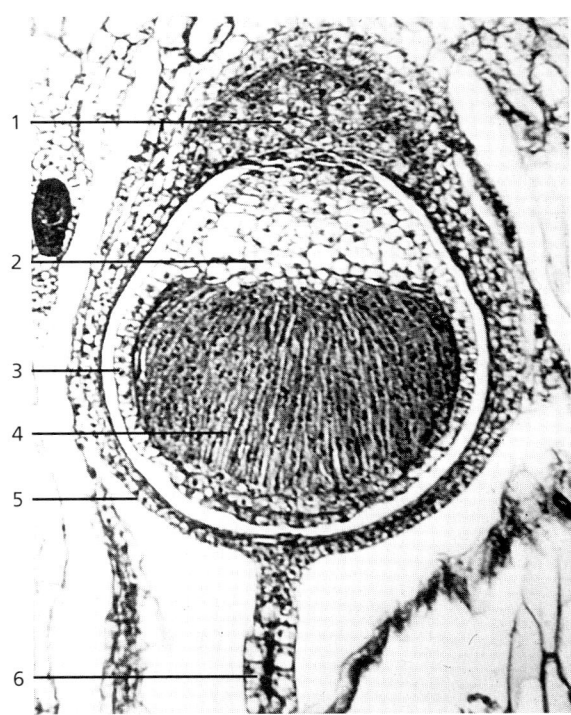

Figure 6.13 Young sporophyte of a liverwort, *Marchantia*, in longitudinal view. (X240)

1. Foot (2*n*)
2. Seta (stalk) (2*n*)
3. Young capsule (2*n*)
4. Sporogenous tissue (2*n*)
5. Enlarged archegonium (calyptra) (*n*)
6. Neck of archegonium (*n*)

CLASS HEPATICAE (liverworts)

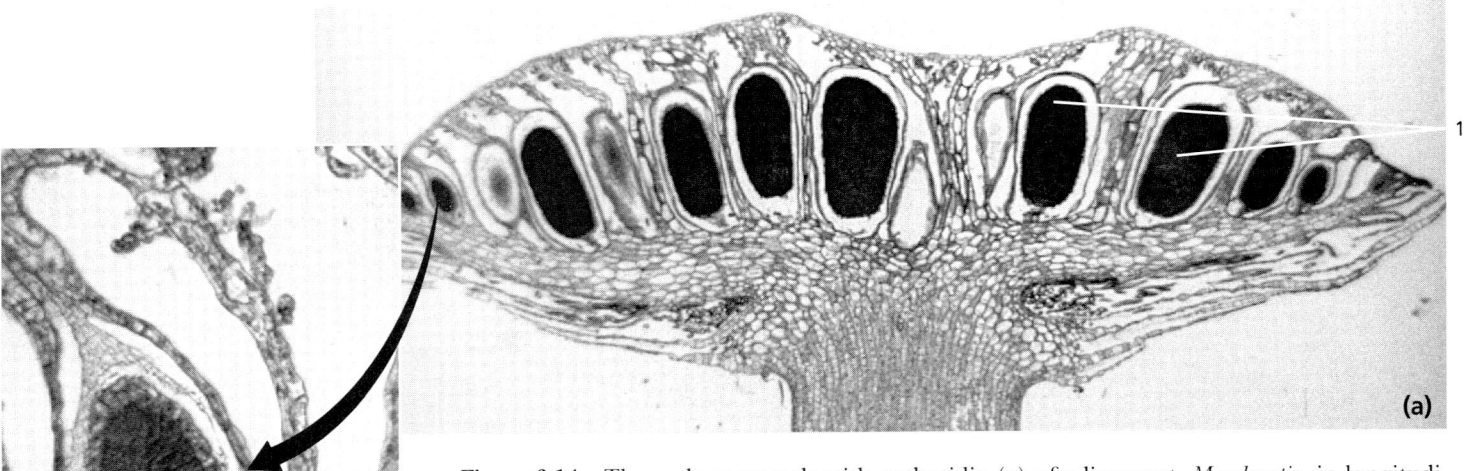

Figure 6.14 The male receptacle with antheridia (a) of a liverwort, *Marchantia*, in longitudinal section. (X40) The antheridial head (b) showing a developing antheridium. (X200)

1. Antheridia
2. Spermatogenous tissue
3. Antheridium

Figure 6.15 A sporophyte of the leafy liverwort, *Porella*. (X4)

1. Sporophyte ($2n$)
2. Capsule ($2n$)
3. Seta (stalk) ($2n$)
4. Gametophyte (n)

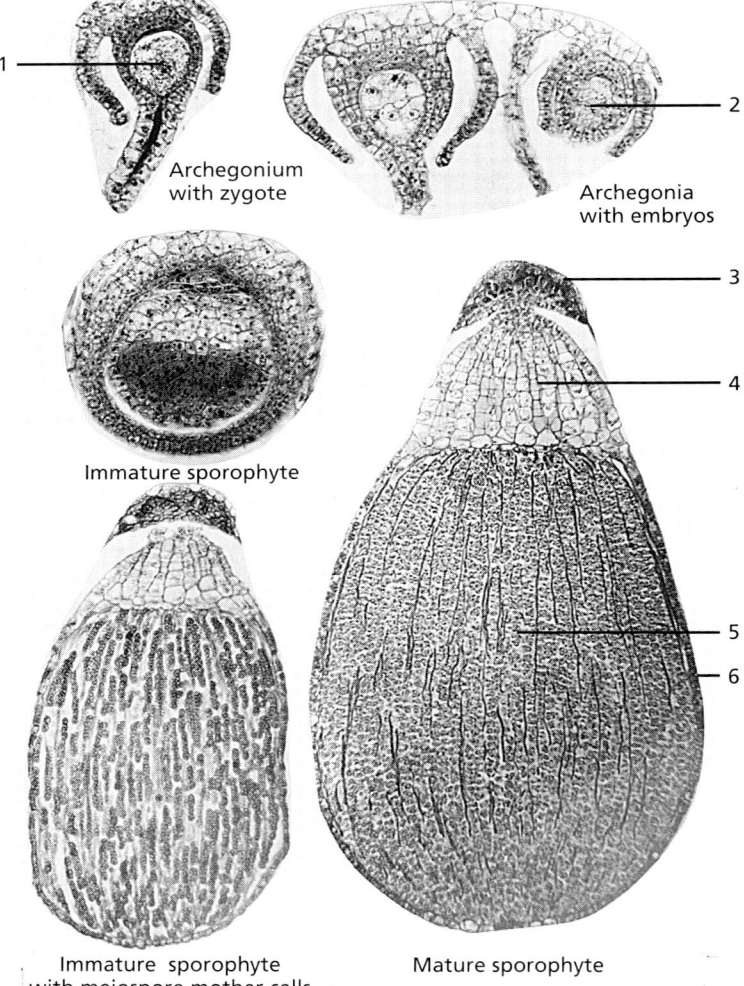

Figure 6.16 Developmental stages of the sporophyte of *Marchantia*. (X65)

1. Zygote ($2n$)
2. Embryo ($2n$)
3. Foot
4. Seta (stalk)
5. Spores (n) and elaters ($2n$)
6. Sporangium (capsule)

CLASS HEPATICAE (liverworts)

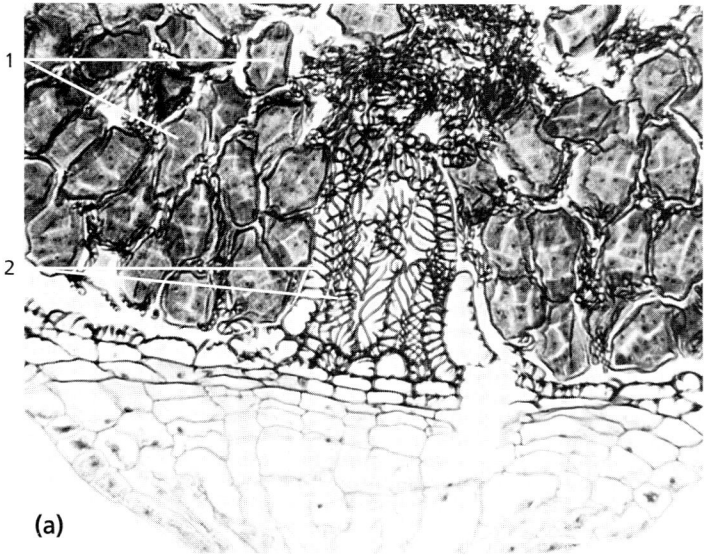
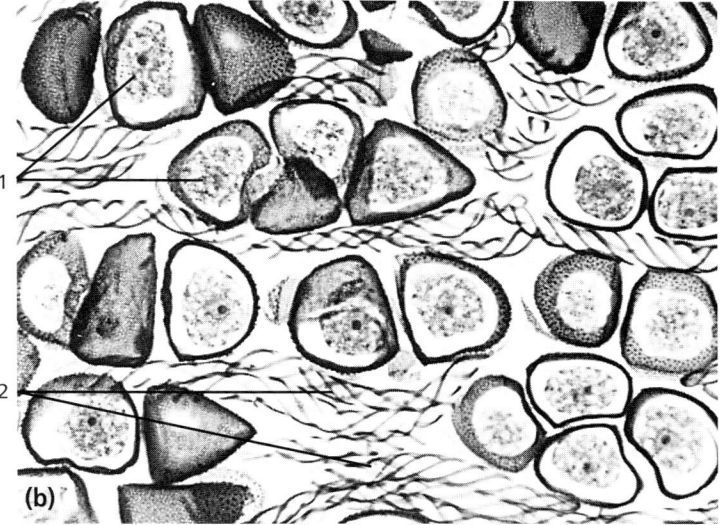

Figure 6.17 A capsule from the leafy liverwort *Pelia;* (a) in longitudinal view and (b) in transverse section. (X430)
1. Spores   2. Elaters

CLASS ANTHOCEROTAE (hornworts)

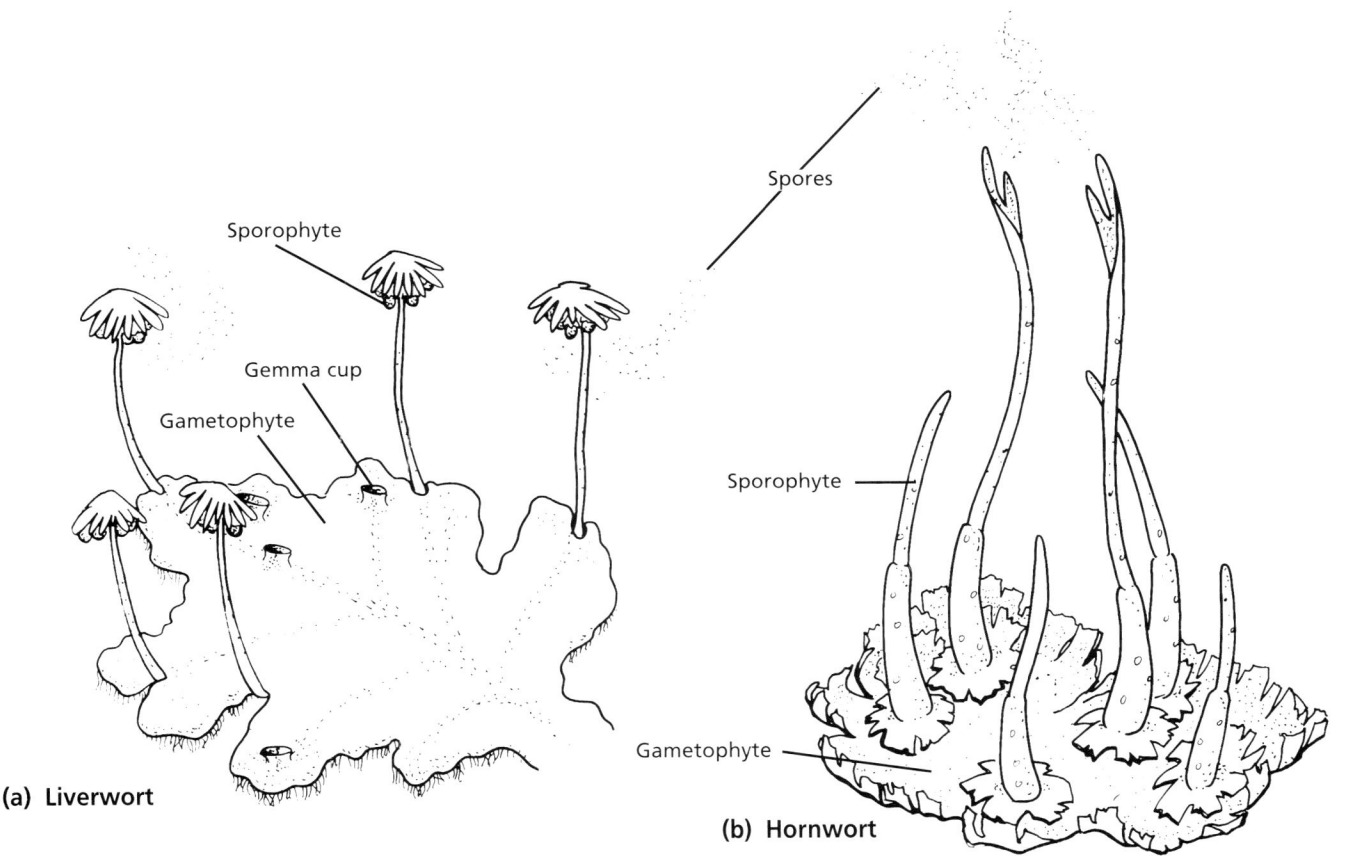

Figure 6.18 A comparison of the sporophytes and gametophytes of (a) the liverwort, *Marchantia*, and (b) the hornwort, *Anthoceros*.

CLASS ANTHOCEROTAE (hornworts)

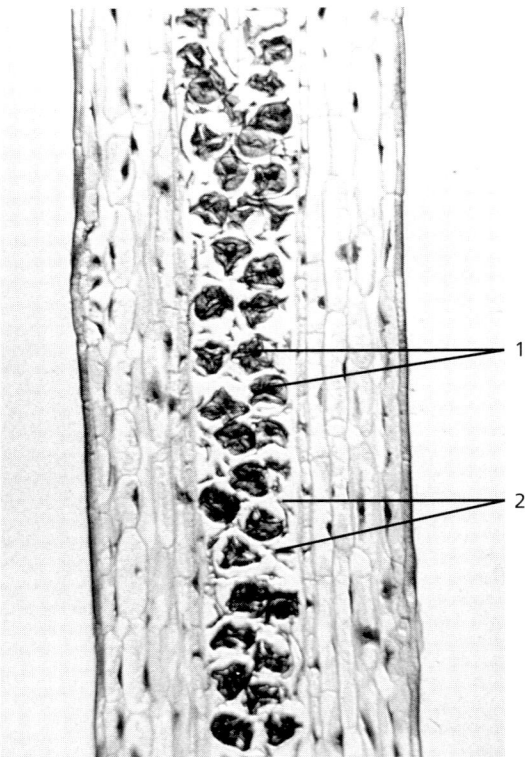

Figure 6.19 Hornwort, *Anthoceros*, a longitudinal section of the sporangium of a sporophyte. (X100)

1. Spores
2. Elater-like structures

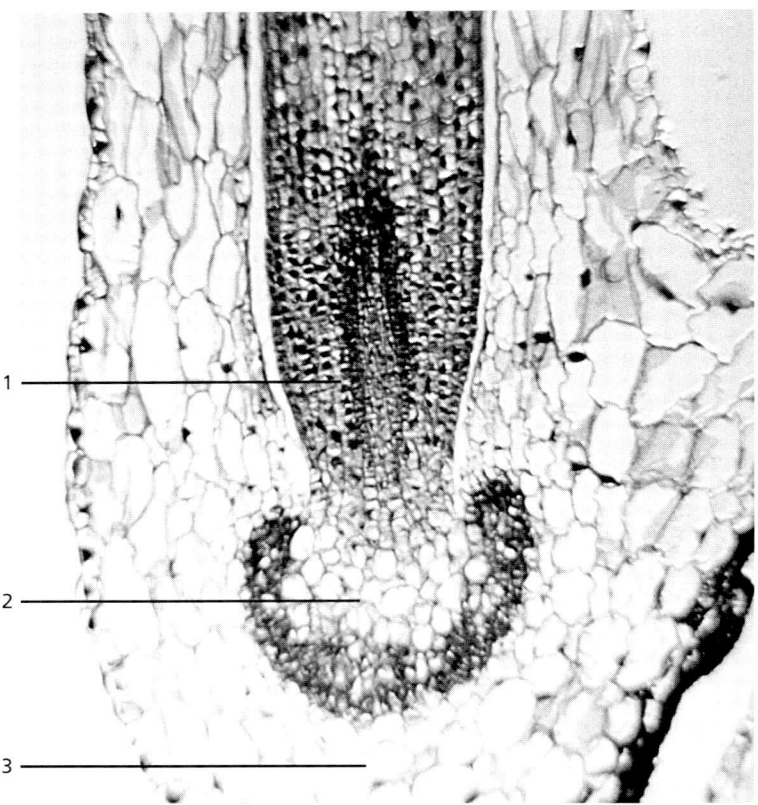

Figure 6.20 Longitudinal section of a portion of the sporophyte plant of the hornwort *Anthoceros*. (X100)

1. Meristematic region of sporophyte ($2n$)
2. Foot ($2n$)
3. Gametophyte (n)

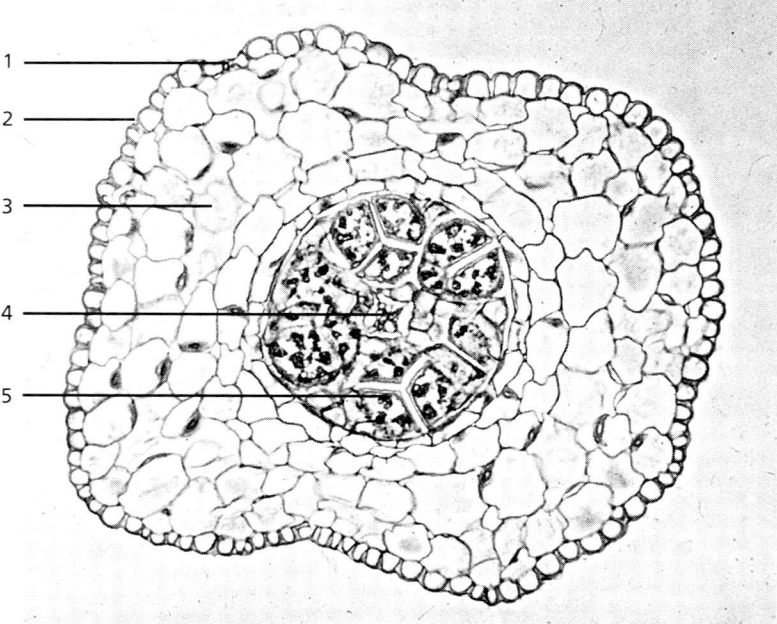

Figure 6.21 Hornwort *Anthoceros*; transverse section through the capsule of a sporophyte. (X100)

1. Stoma
2. Epidermis
3. Photosynthetic tissue
4. Columella
5. Spore

CHAPTER 6 — Kingdom Plantae: Division Bryophytes (Bryophyta)

CLASS MUSCI (mosses)

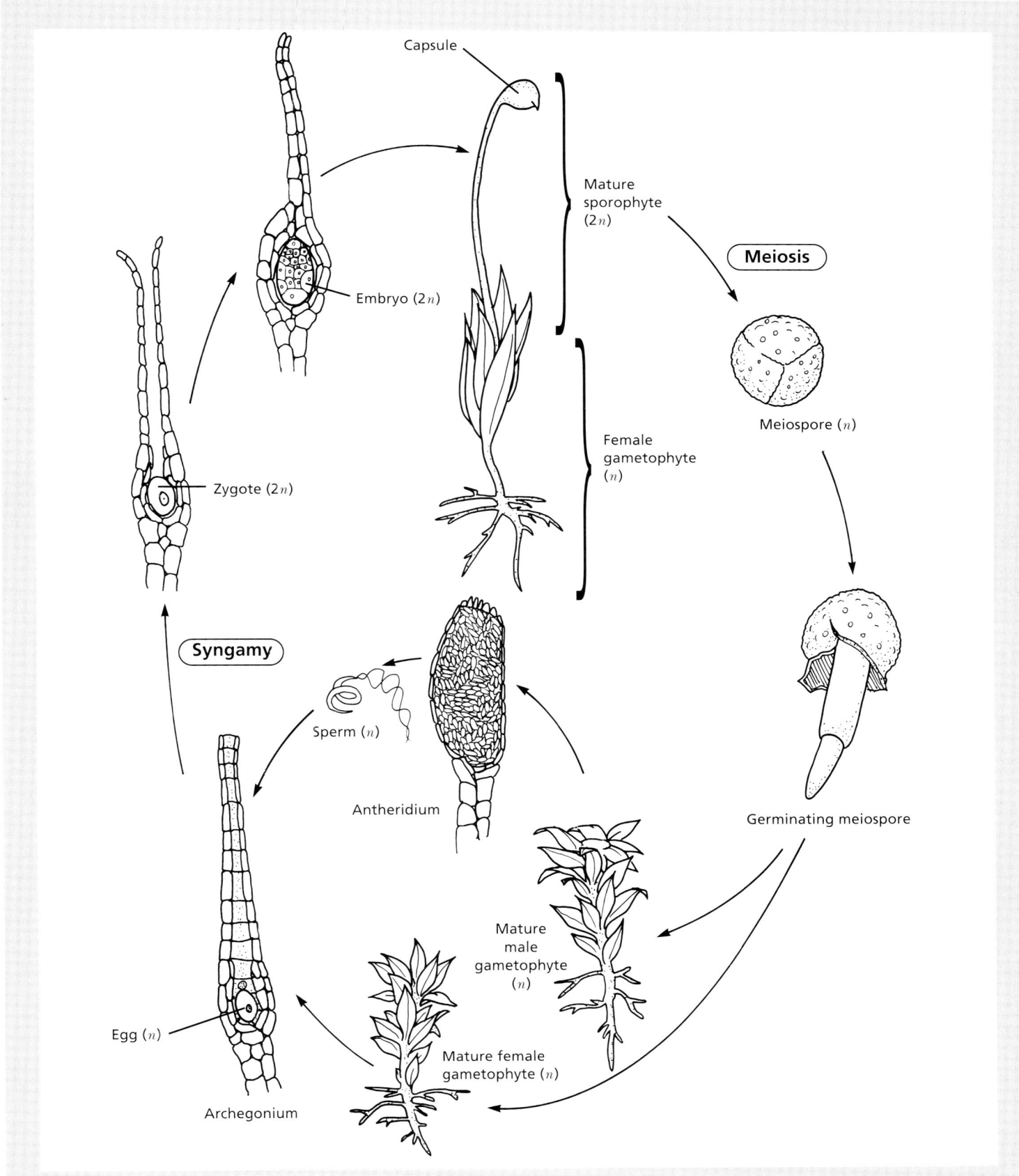

Figure 6.22 Life cycle of a moss.

CLASS MUSCI (mosses)

Figure 6.23 Herbarium specimen of peat moss, *Sphagnum*. *Sphagnum* grows in boggy habitats and is ecologically important in forming extensive peat bogs, lowering the pH of their own environment.

Figure 6.24 Gametophyte plants of peat moss, *Sphagnum*, showing attached sporophytes.
1. Sporophyte
2. Pseudopodium
3. Gametophytes

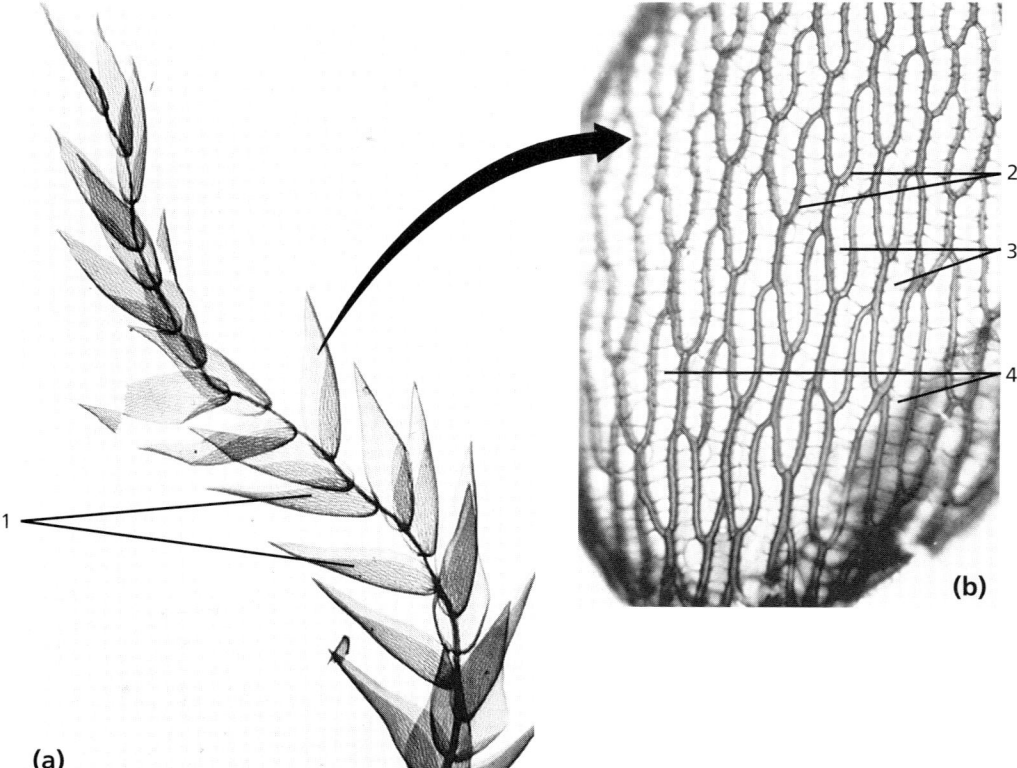

Figure 6.25 A gametophyte (a) of peat moss, *Sphagnum*. (X4) A magnified view of a leaf (b), showing the dead cell chambers that aid in water storage. (X40)

1. Leaves
2. Photosynthetic cells
3. Dead cells
4. Pores

CHAPTER 6 — Kingdom Plantae: Division Bryophytes (Bryophyta)

CLASS MUSCI (mosses)

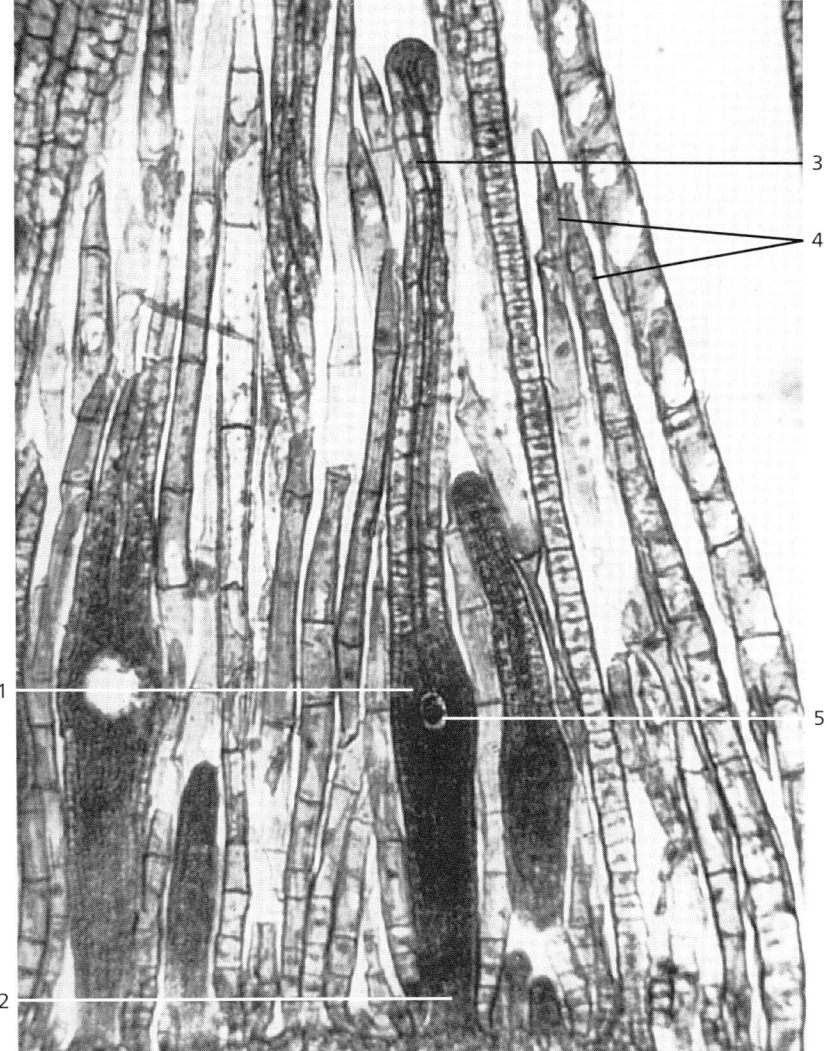

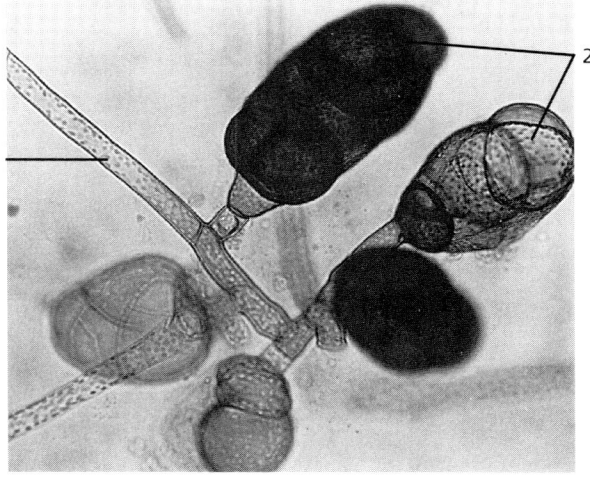

Figure 6.26 The archegonial head of the moss, *Mnium*, in longitudinal section. The paraphyses are non-reproductive filaments that support the archegonia. (X180)

1. Venter 2. Stalk 3. Neck 4. Paraphyses 5. Egg

Figure 6.27 The gametophyte of a moss develops from buds along the protonema. Several buds developing from a protonema are illustrated here. (X430)

1. Protonema 2. Buds

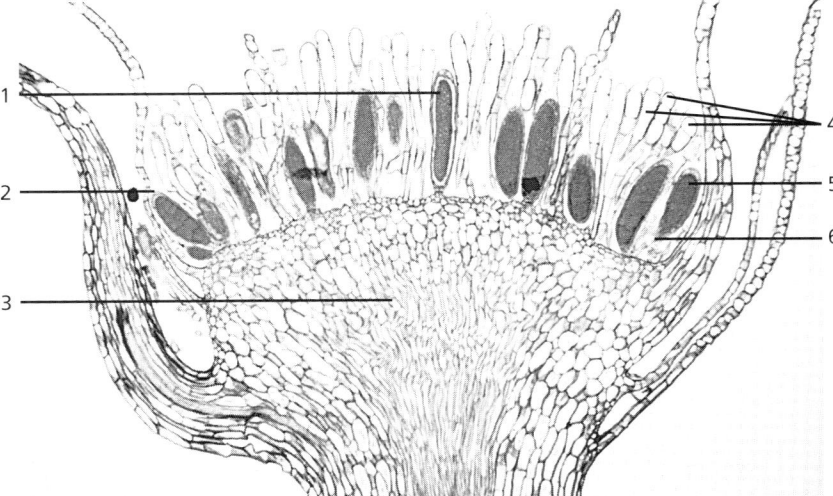

Figure 6.28 The antheridial head of the moss, *Mnium*, in longitudinal section. (X65)

1. Spermatogenous tissue
2. Sterile jacket layer
3. Male gametophyte (*n*)
4. Paraphyses (sterile filaments)
5. Antheridium (*n*)
6. Stalk

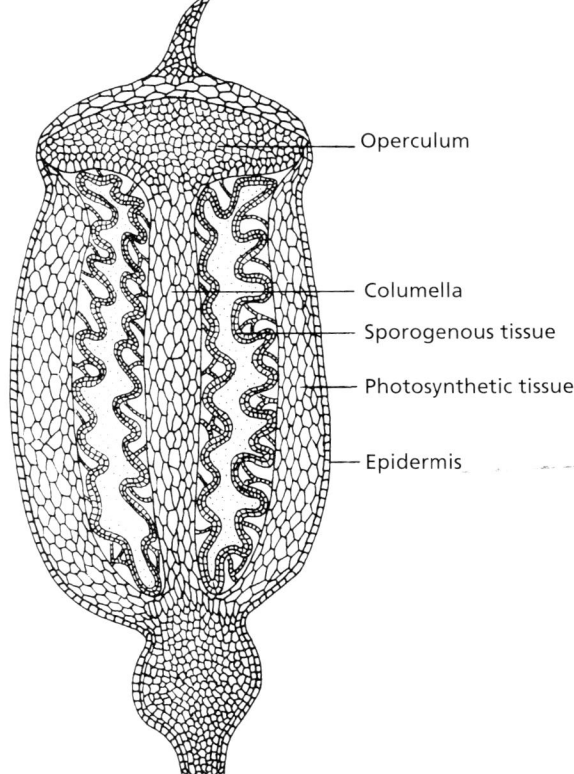

Figure 6.29 A longitudinal section through the capsule of a moss. The sporogenous tissue surrounding the columella produces meiospores.

CLASS MUSCI (mosses)

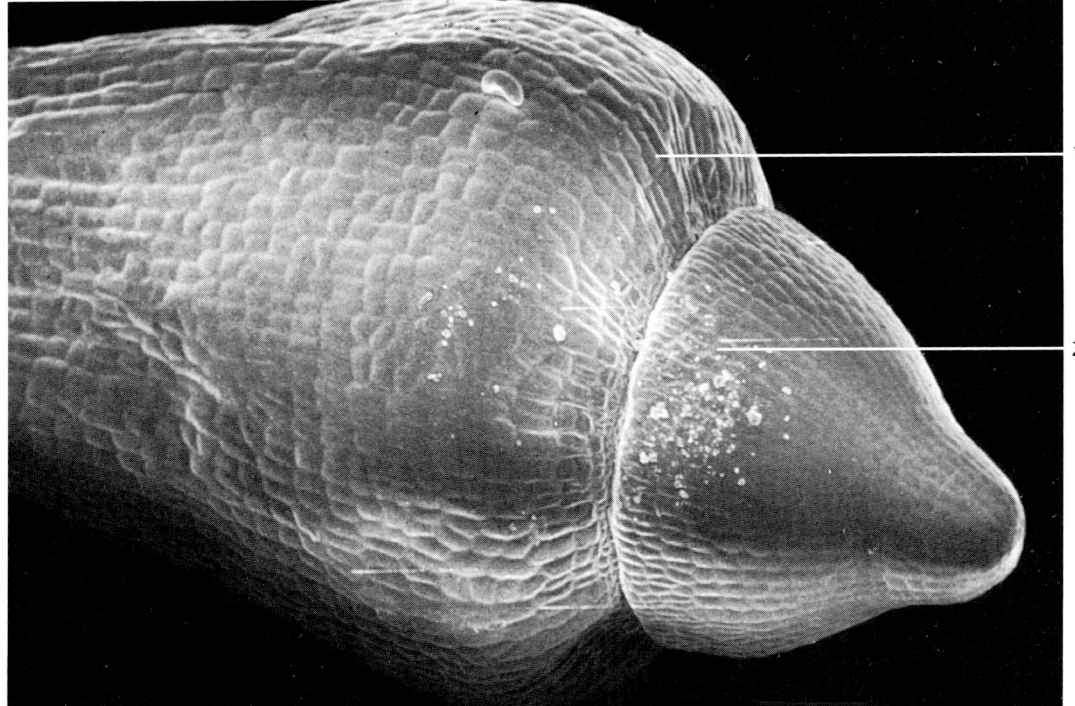

Figure 6.30 Scanning electron micrograph of the sporophyte capsule of the moss *Mnium*. (X50)
1. Capsule
2. Operculum

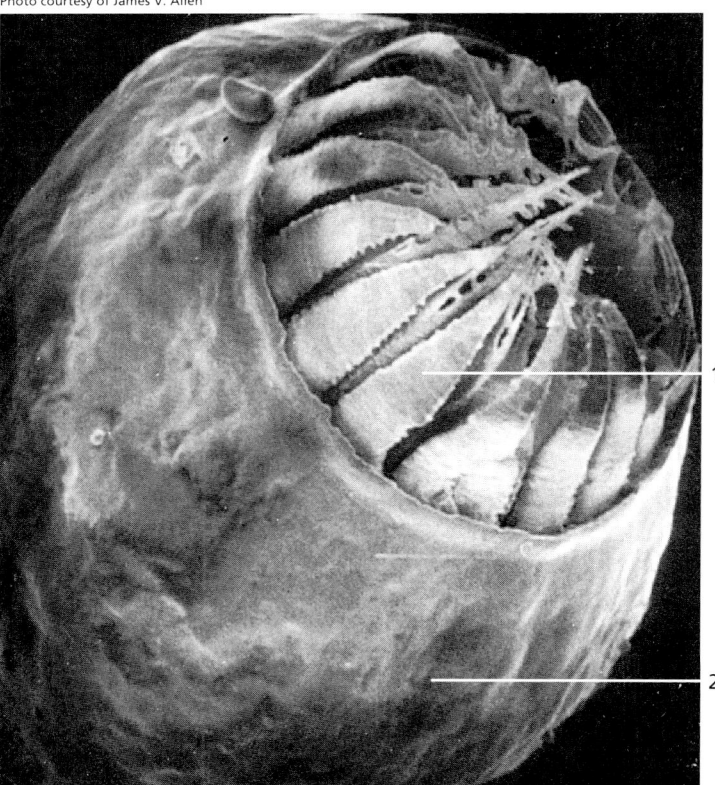

Figure 6.31 Scanning electron micrograph of the peristome of the moss *Mnium*. The operculum is absent in the specimen. (X75)
1. Peristome   2. Capsule

Figure 6.32 Scanning electron micrograph of the peristome of the moss *Mnium*. (X150)
1. Outer teeth of peristome   2. Inner teeth of peristome   3. Capsule

Kingdom Plantae: Seedless Vascular Plants

Four divisions of vascular plants do not produce seeds. These are: Psilotophyta (sometimes spelled Psilophyta; the whisk ferns); Lycophyta (club mosses and quillworts); Sphenophyta (horsetails); and Pterophyta (ferns) (Table 7.1). These plants inhabit damp, shady environments and are found worldwide. Such plants were particularly abundant during the Carboniferous Period, 350-375 million years ago. At that time, many species were large and treelike. Such remains, along with other plants were compacted, forming coal beds. Several thousand species of seedless vascular plants are now known.

Psilopsids are homosporous vascular plants represented by only two living genera, *Psilotum* and *Tmesipteris*, and several species. *Psilotum* is a tropical or subtropical plant. In the United States it occurs in Arizona, Texas, Louisiana, Florida, and Hawaii. It grows readily in greenhouses and may be considered a greenhouse weed. *Tmesipteris* occurs in Australia and islands of the South Pacific, including New Zealand and New Caledonia. Both genera are homosporous, have simple sporophytes, and have no histological distinction between the rhizomes and stems. *Tmesipteris* is an epiphyte on tree ferns and other plants.

Lycophytes are represented by five living genera and about 1,000 species. The most familiar lycophyte is the club moss, *Lycopodium*. The nearly 200 species of this genus are found worldwide in tropical to Arctic regions. They are homosporous. Species of the closely-related *Selaginella* are heterosporous.

The 15 living species of horsetails are included in the single genus, *Equisetum*. Three species of *Equisetum* are tropical, and eleven occur in the United States and Canada. Horsetails have jointed stems and scalelike leaves. The stems are further characterized by prominent nodes and elevated siliceous ribs. Horsetails are homosporous, and sporangia develop in a cone at the apex of the stem.

About 12,000 living species of ferns are known. Many have large, feathery leaves, called *fronds*. Both roots and fronds grow out of an underground stem called the *rhizome*. As the fronds develop, they appear to be rolled-up and hence are called *fiddleheads*. Spores produced by sporangia on the underside of the frond of a fern are dispersed by the wind to suitable, moist habitats for germination. Spores germinate, to become a small gametophyte or *prothallus*. A gametophyte has antheridia that produce sperm and archegonia that produce eggs. Spiral-shaped sperm swim from the antheridia to an archegonium, where fertilization occurs. A zygote forms within the archegonium; but it soon grows through the gametophyte, takes root, and becomes a mature sporophyte fern plant.

TABLE 7.1
The Seedless Vascular Plants

Divisions and Representative Kinds	Characteristics
Division Psilotophyta (whisk ferns)	Plants small; true roots and leaves absent; rhizome and rhizoids present; homosporous
Division Lycophyta (club mosses, quillworts, spike mosses)	Homosporous and heterosporous; many are epiphytes in the tropics, but not in temperate climates where cold winters occur
Division Sphenophyta (horsetails)	Epidermis embedded with silica; tips of stems bear cone-like structures containing sporangia; homosporous
Division Pterophyta (ferns)	Plants often large and conspicuous; leaves complex, rhizome common; mostly homosporous, a few heterosporous

DIVISION PSILOTOPHYTA (whisk ferns)

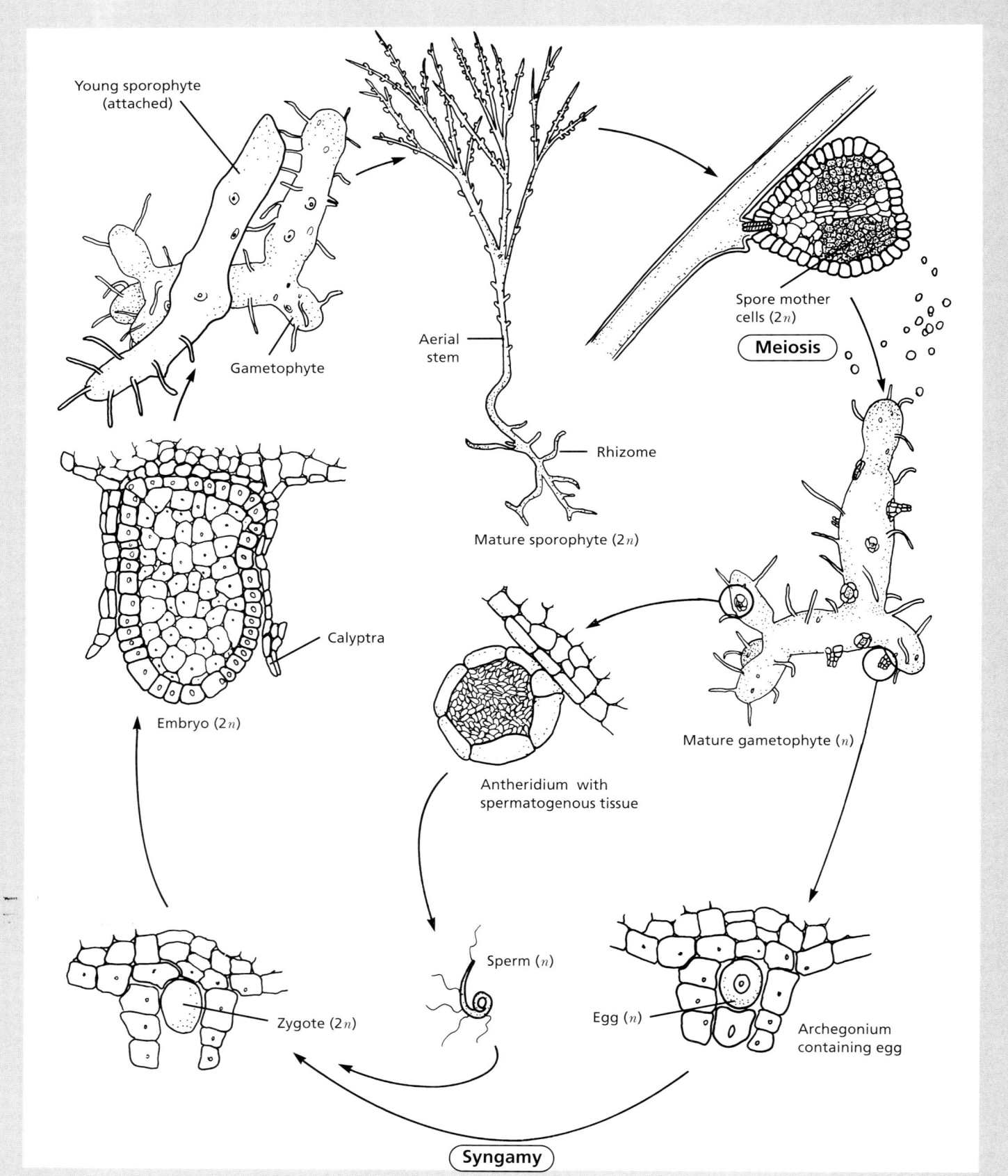

Figure 7.1 Life cycle of the whisk fern, *Psilotum*.

DIVISION PSILOTOPHYTA (whisk ferns)

Figure 7.2 The whisk fern, *Psilotum nudum*, is a simple vascular plant lacking true leaves and roots.

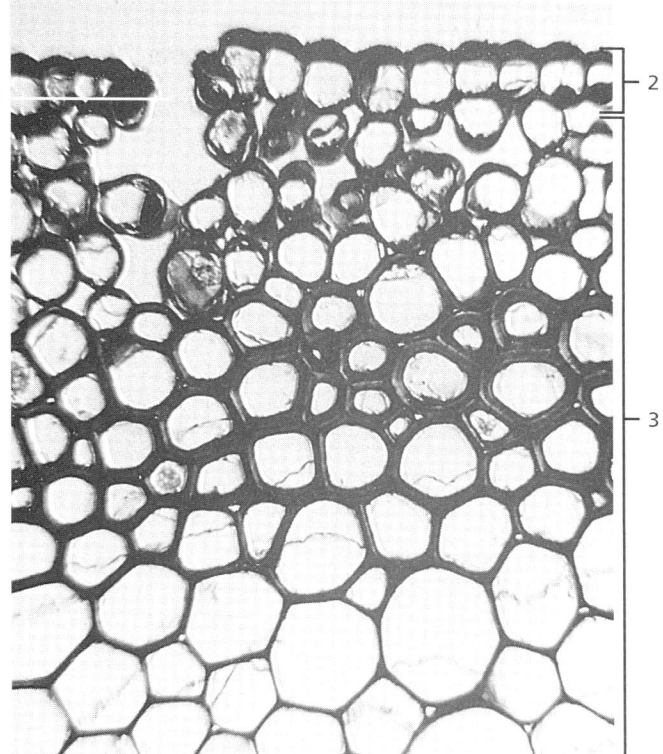

Figure 7.3 A photomicrograph of a scalelike outgrowth from the branch of the whisk fern, *Psilotum nudum*. (X430)

1. Stoma   2. Epidermis   3. Ground tissue

Figure 7.4 The sporophyte of the whisk fern, *Psilotum nudum*. The branches of the sporophyte support sporangia, which produce the spores.

1. Sporangia   2. Branch

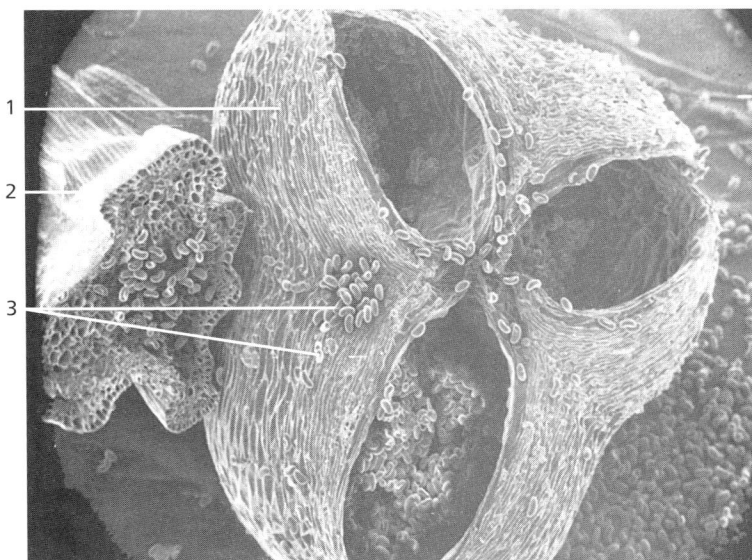

Photo courtesy of James V. Allen

Figure 7.5 Scanning electron micrograph of a ruptured synangium (3 fused sporangia) of *Psilotum*, which is spilling spores. (X75)

1. Sporangium (often called a synangium)
2. Branch
3. Spores (*n*)

DIVISION PSILOTOPHYTA (whisk ferns)

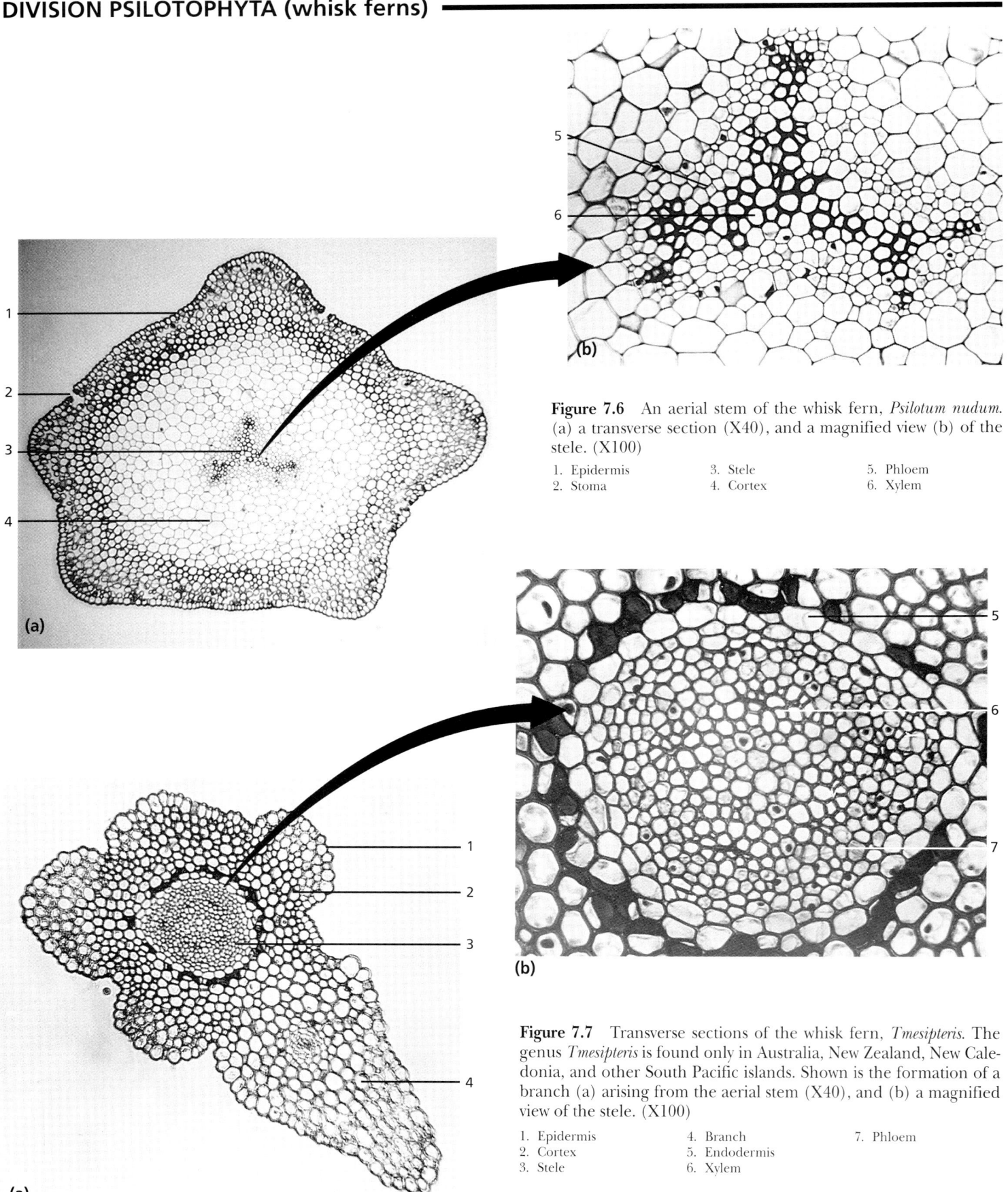

Figure 7.6 An aerial stem of the whisk fern, *Psilotum nudum*. (a) a transverse section (X40), and a magnified view (b) of the stele. (X100)

1. Epidermis
2. Stoma
3. Stele
4. Cortex
5. Phloem
6. Xylem

Figure 7.7 Transverse sections of the whisk fern, *Tmesipteris*. The genus *Tmesipteris* is found only in Australia, New Zealand, New Caledonia, and other South Pacific islands. Shown is the formation of a branch (a) arising from the aerial stem (X40), and (b) a magnified view of the stele. (X100)

1. Epidermis
2. Cortex
3. Stele
4. Branch
5. Endodermis
6. Xylem
7. Phloem

DIVISION LYCOPHYTA (club mosses, quillworts, spike mosses)

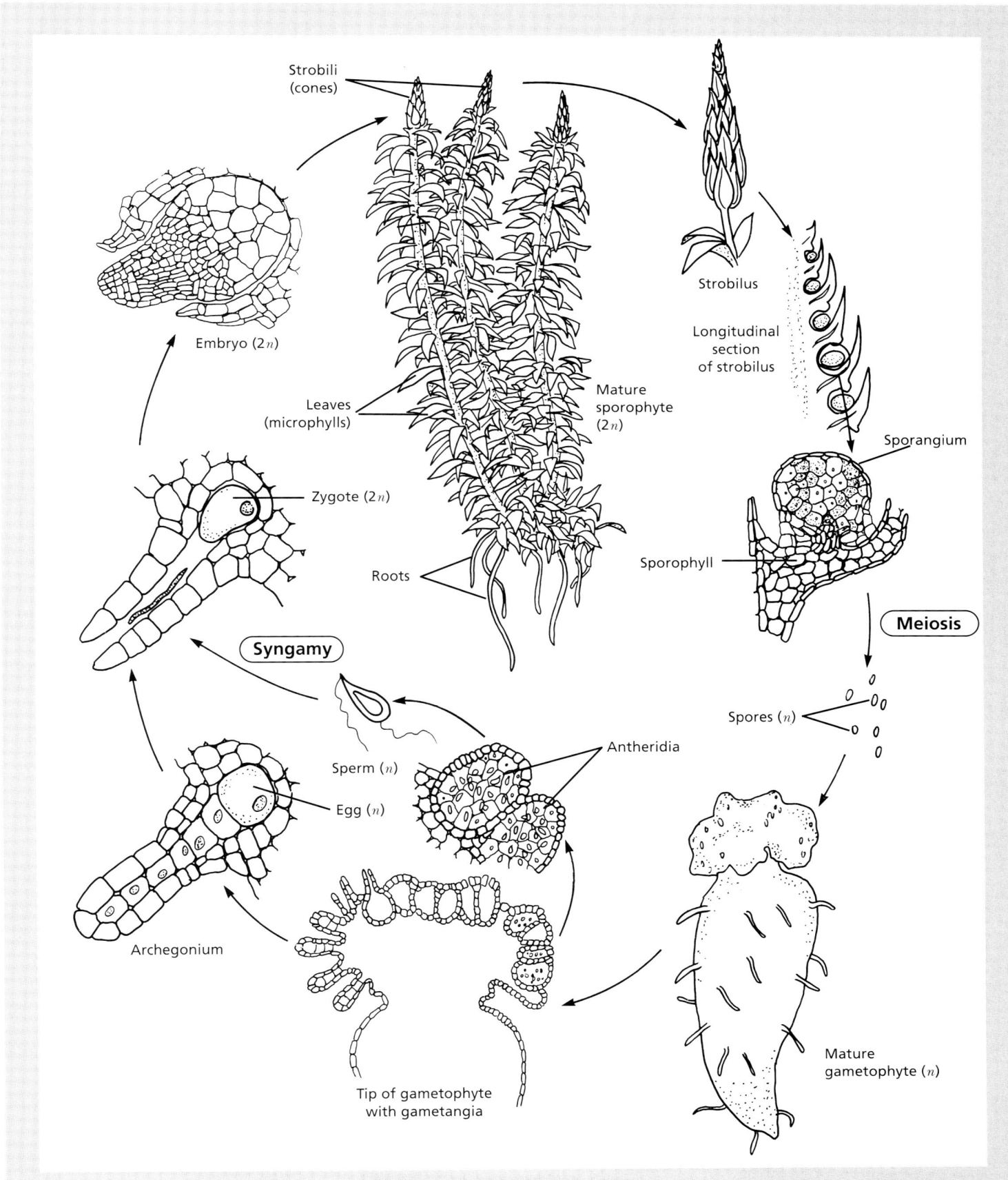

Figure 7.8 Life cycle of the club moss, *Lycopodium*.

DIVISION LYCOPHYTA (club mosses, quillworts, spike mosses)

Figure 7.9 A club moss, *Lycopodium*.
1. Strobilus 2. Leaves (microphylls) 3. Aerial stem

Figure 7.10 An enlargement of a branch tip of *Lycopodium*, showing sporangia. (X10)
1. Sporangia 2. Sporophylls (leaves with attached sporangia)

Figure 7.11 Herbarium specimen of a lycopod, *Lycopodium clavatum*. *Lycopodium* occurs from the arctic to the tropics. Being evergreen, they may be most obvious during winter.

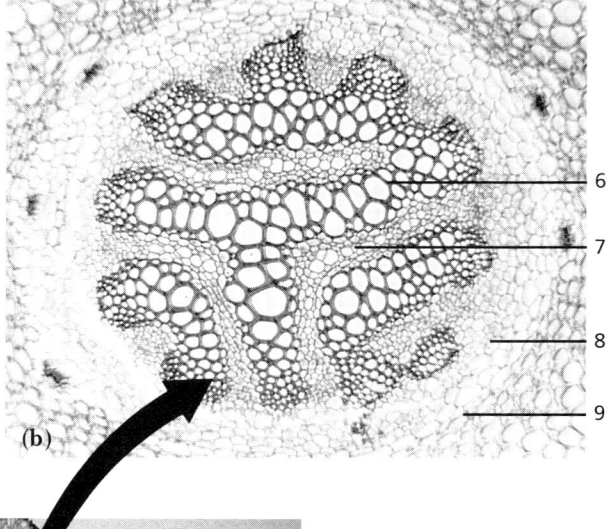

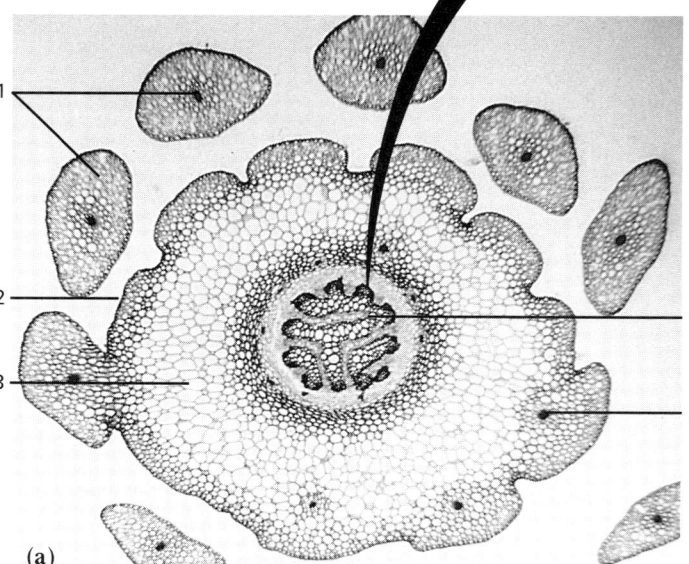

Figure 7.12 An aerial stem of the club moss, *Lycopodium* shown in transverse section (a) (X100), and a magnified view (b) of the stele. (X200)
1. Leaves (microphylls)
2. Epidermis
3. Cortex
4. Stele
5. Leaf trace
6. Xylem
7. Phloem
8. Pericycle
9. Endodermis

DIVISION LYCOPHYTA (club mosses, quillworts, spike mosses)

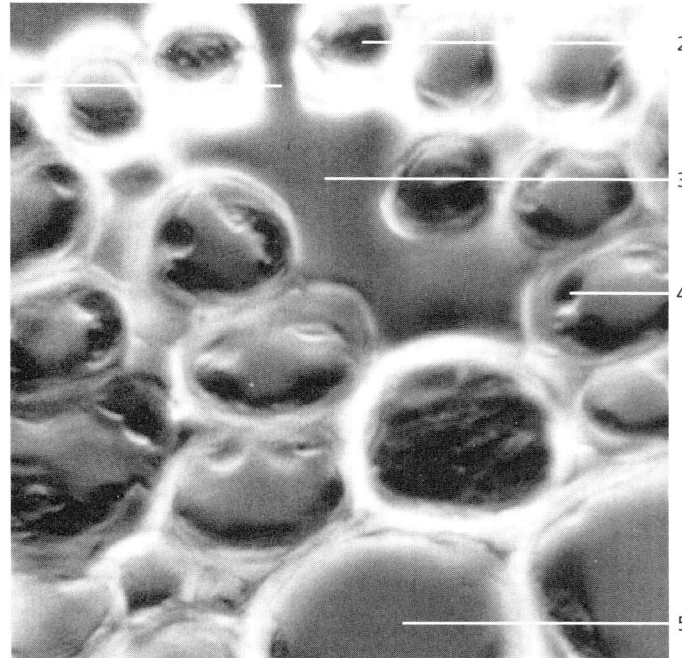

Figure 7.13 Phase contrast micrograph of *Lycopodium* stem. (X600)

1. Stoma
2. Guard cell
3. Stomatal chamber
4. Chloroplast
5. Cortex parenchyma cell

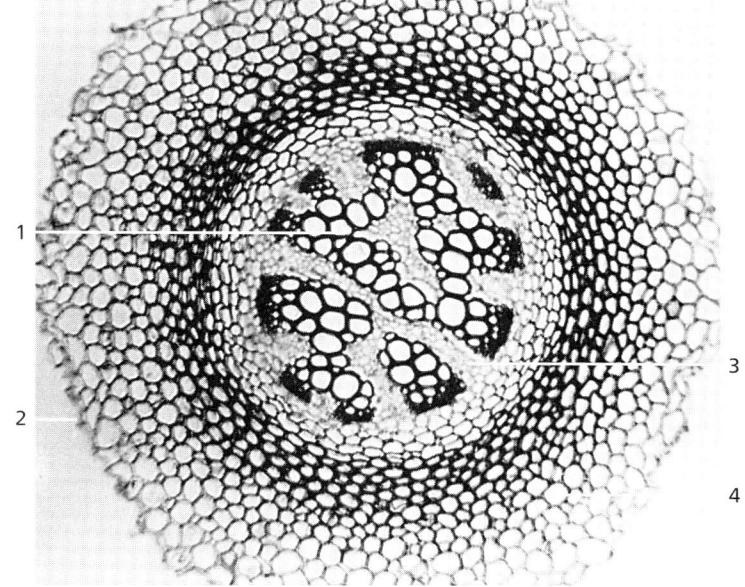

Figure 7.14 *Lycopodium*, rhizome. A transverse section of a rhizome of *Lycopodium* is similar to an aerial stem, but lacks the microphylls. (X40)

1. Xylem
2. Epidermis
3. Phloem
4. Cortex

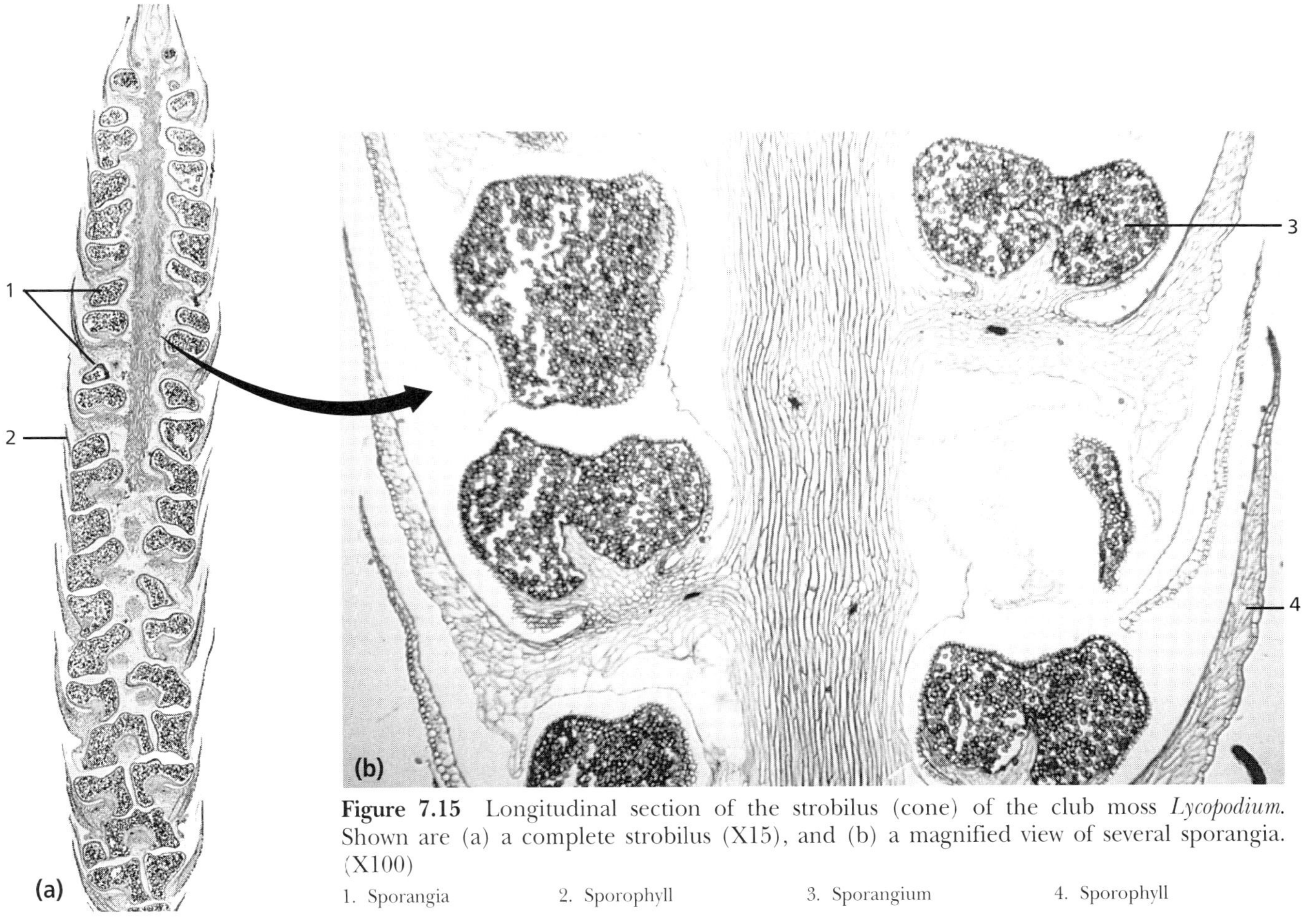

Figure 7.15 Longitudinal section of the strobilus (cone) of the club moss *Lycopodium*. Shown are (a) a complete strobilus (X15), and (b) a magnified view of several sporangia. (X100)

1. Sporangia
2. Sporophyll
3. Sporangium
4. Sporophyll

DIVISION LYCOPHYTA (club mosses, quillworts, spike mosses)

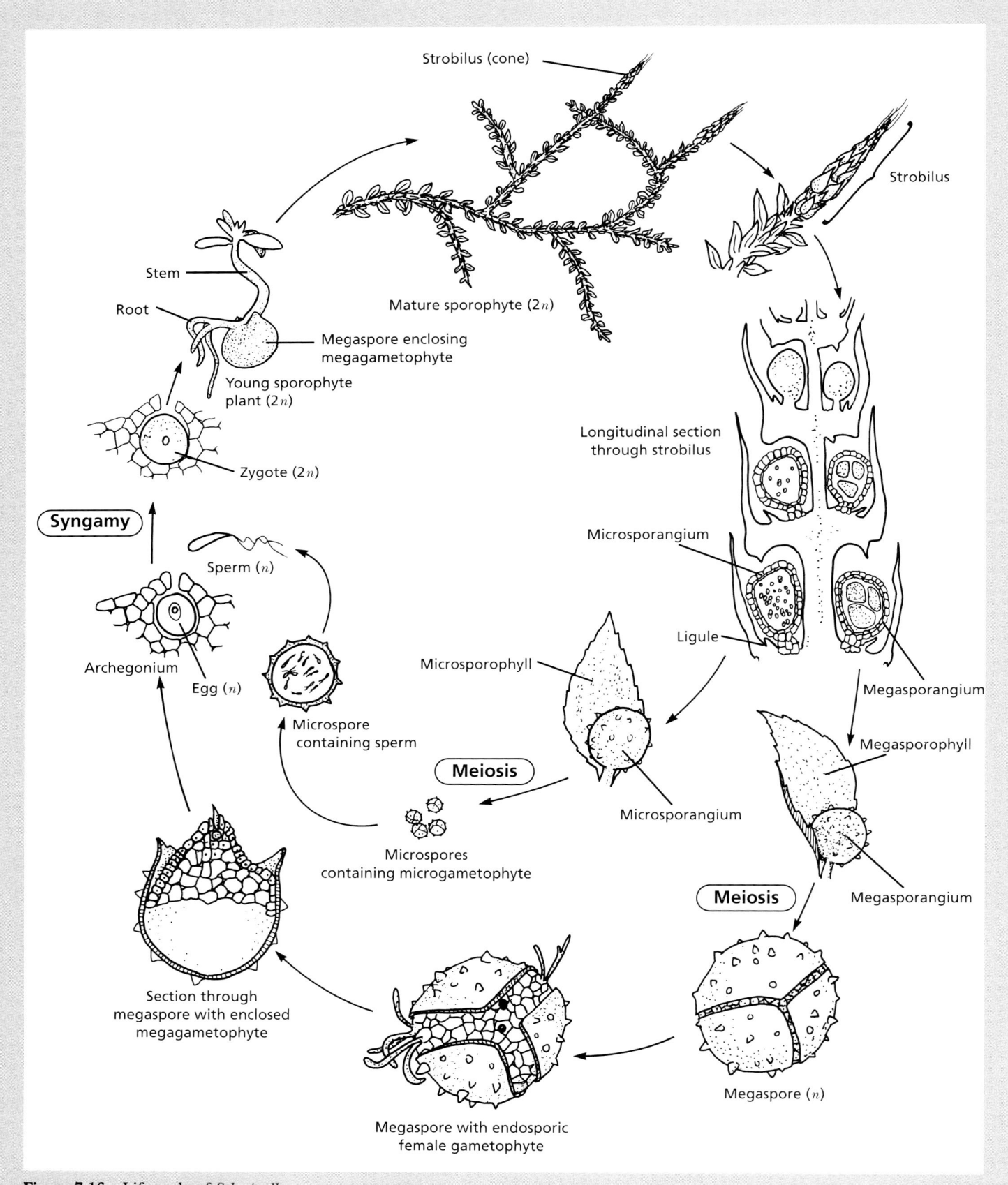

Figure 7.16 Life cycle of *Selaginella*.

DIVISION LYCOPHYTA (club mosses, quillworts, spike mosses)

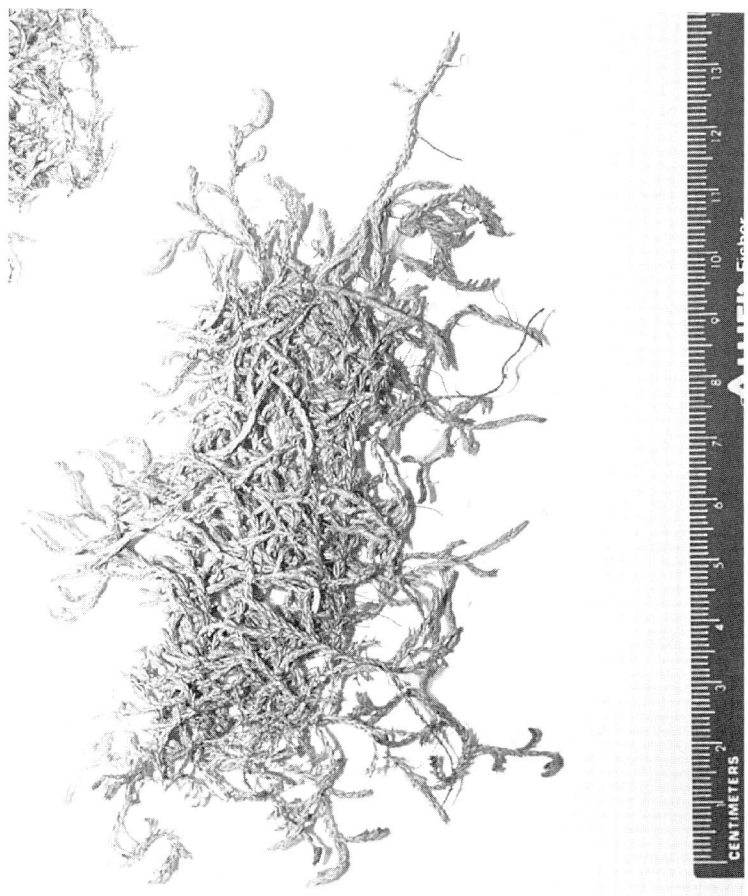

Figure 7.17 Herbarium specimen of a lycopod, *Selaginella*. These lycopods are mainly tropical in distribution. Some are found in arid regions, however, where they are dormant during dry seasons.

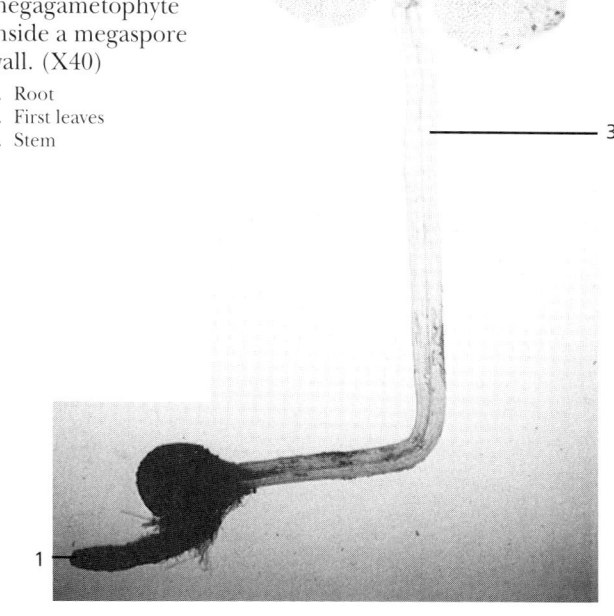

Figure 7.18 Young sporophyte of *Selaginella*, growing from a megagametophyte inside a megaspore wall. (X40)

1. Root
2. First leaves
3. Stem

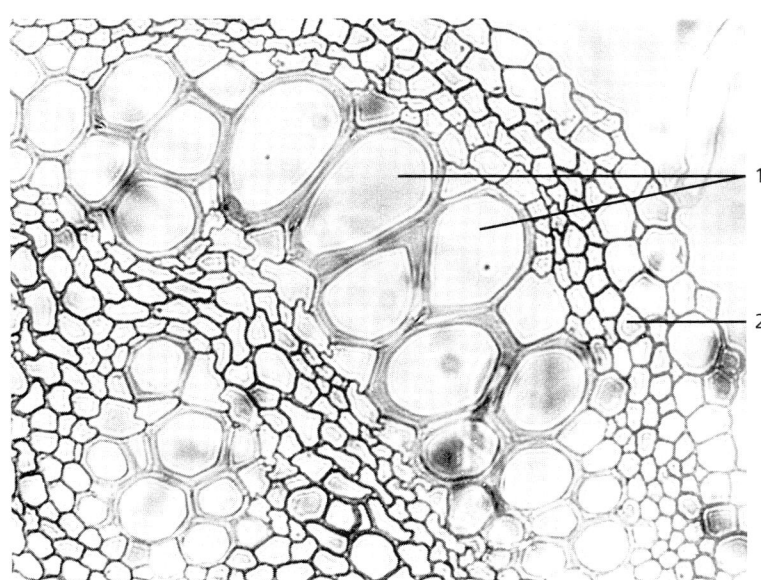

Figure 7.19 Vascular tissue in the stele of *Selaginella*, in transverse view. (X150)

1. Xylem   2. Phloem

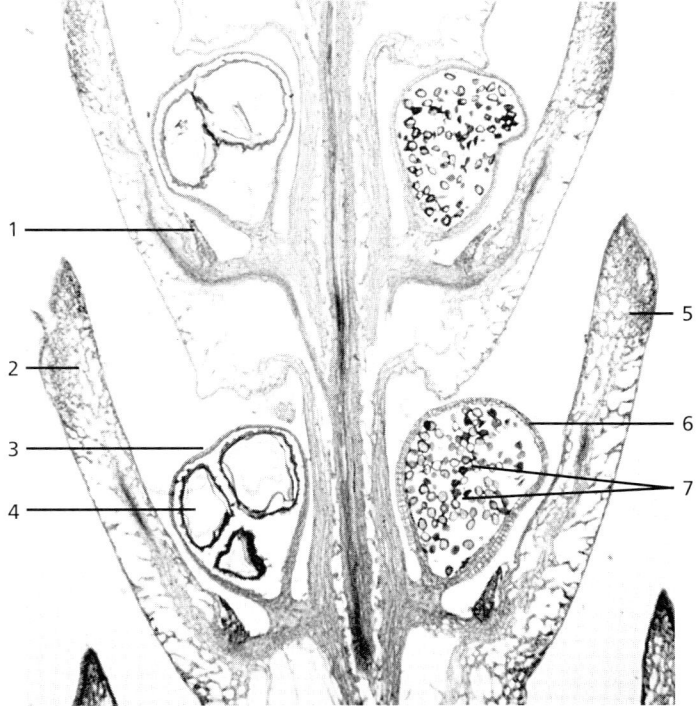

Figure 7.20 Longitudinal view of a strobilus of *Selaginella*. (X40)

1. Ligule
2. Megasporophyll
3. Megasporangium
4. Megaspore (*n*)
5. Microsporophyll
6. Microsporangium
7. Microspores (*n*)

DIVISION LYCOPHYTA (club mosses, quillworts, spike mosses)

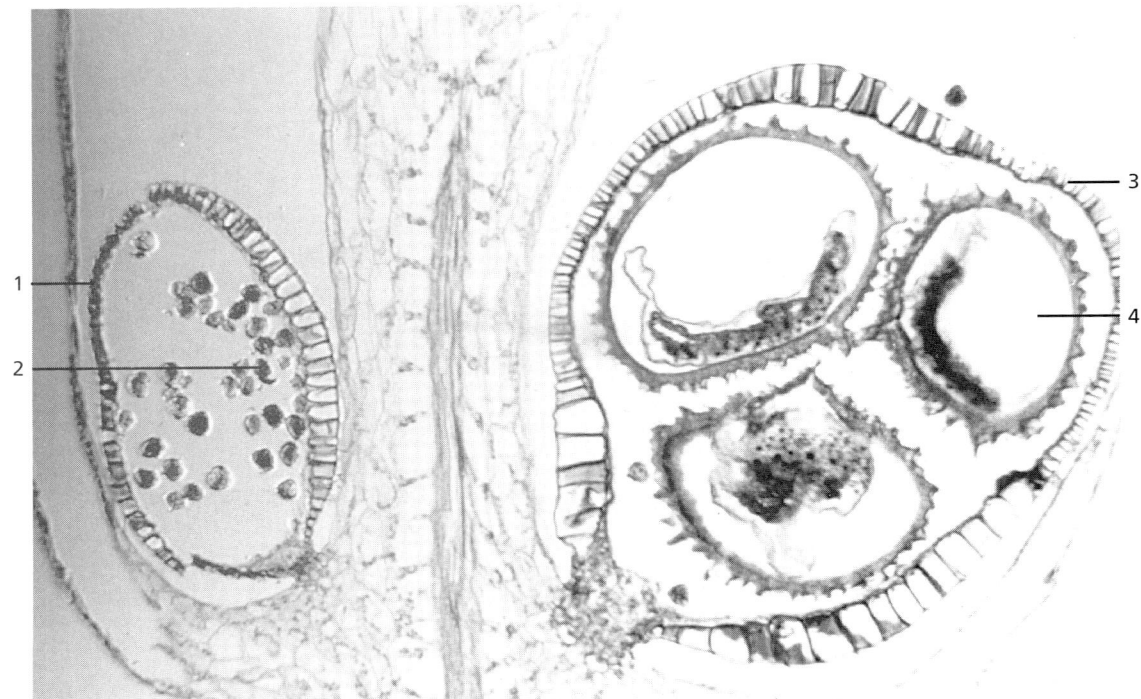

Figure 7.21 Longitudinal view through a strobilus of *Selaginella* showing microspores and megaspores. (X100)
1. Microsporangium
2. Microspores
3. Megasporangium
4. Megaspore

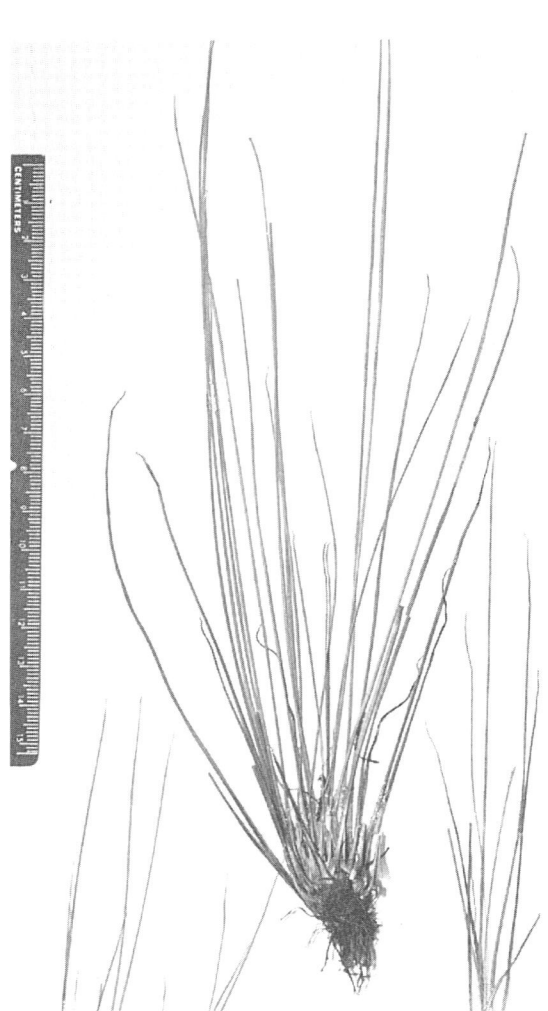

Figure 7.22 Herbarium specimen of a quillwort, *Isoetes melanopoda*. The quillwort, an aquatic plant, has a small, underground stem and quill-like leaves. It is heterosporous. The megasporangia and microsporangia are located at the bases of different leaves.

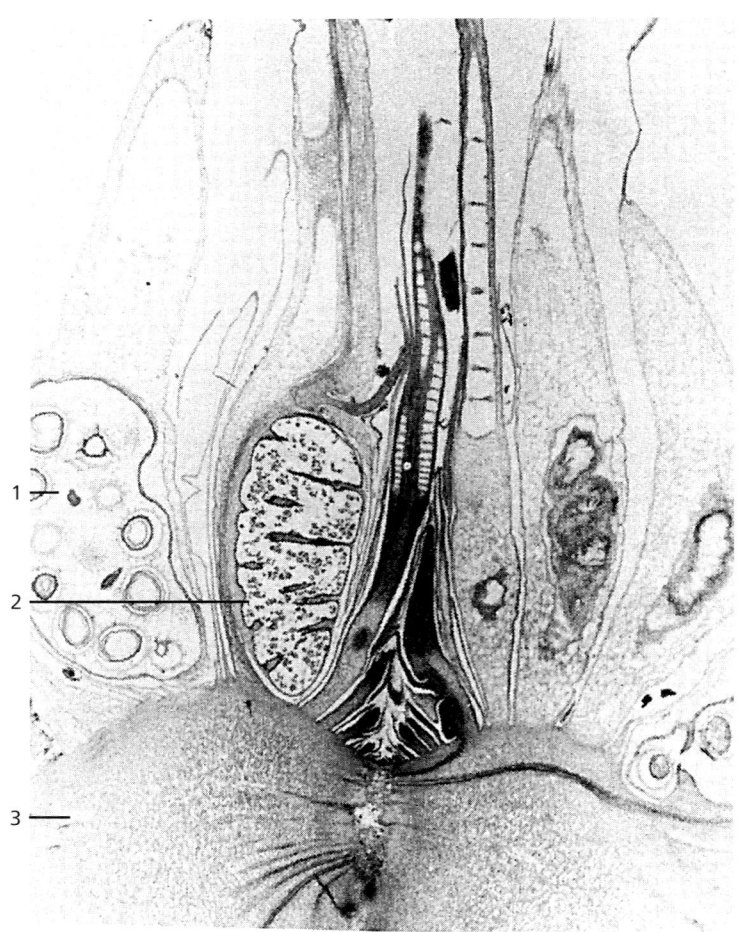

Figure 7.23 A longitudinal section of *Isoetes*, an aquatic, grass-like lycopod.
1. Megasporangium   2. Microsporangium   3. Corm

DIVISION SPHENOPHYTA (horsetails)

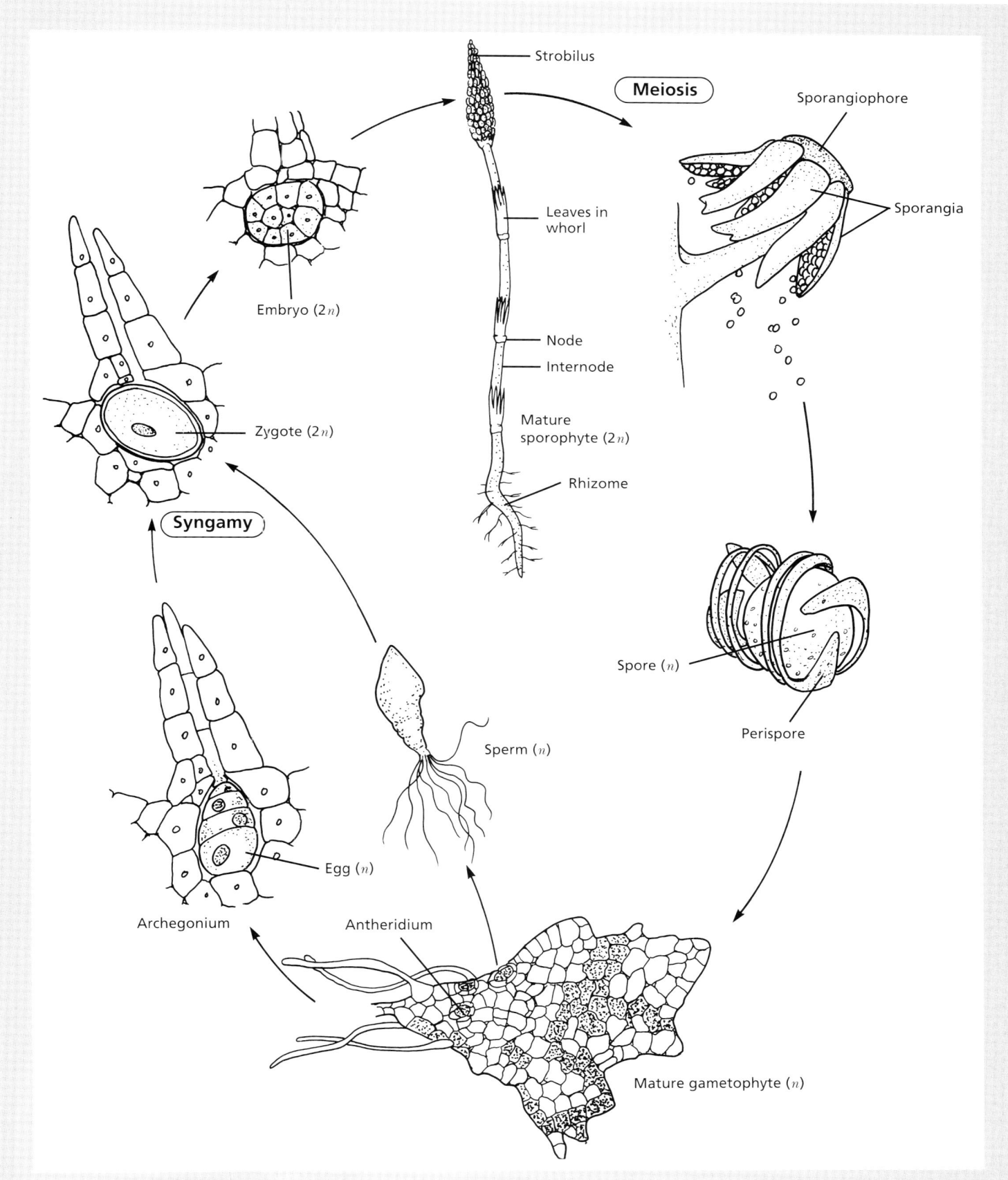

Figure 7.24 Life cycle of the horsetail, *Equisetum*.

DIVISION SPHENOPHYTA (horsetails)

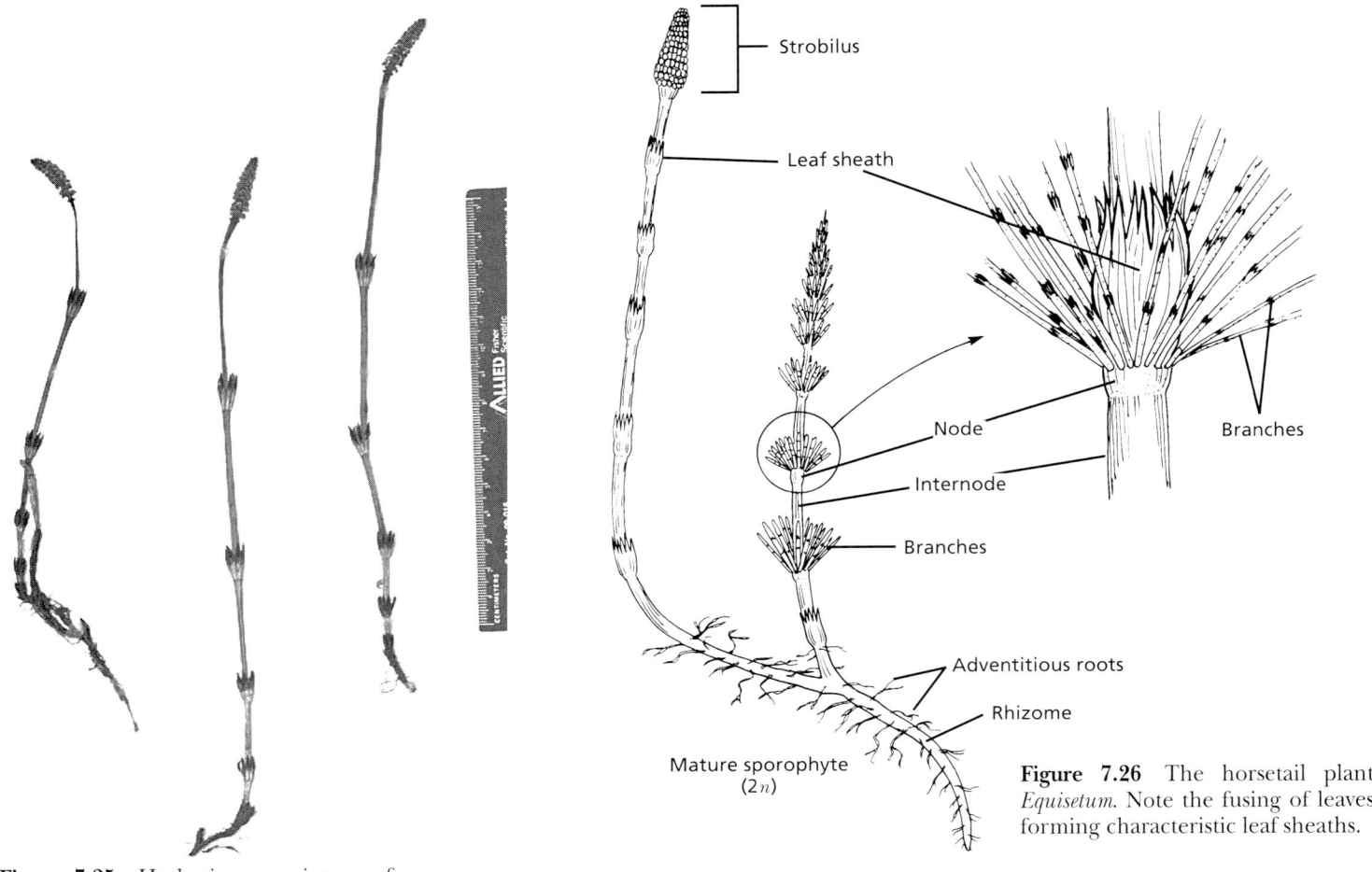

Figure 7.25 Herbarium specimens of the horsetail plant, *Equisetum arvense*. Species of Sphenophyta were abundant throughout tropical regions during the Paleozoic Era, some 300 million years ago. Currently, Sphenophyta are represented by this single living genus.

Figure 7.26 The horsetail plant *Equisetum*. Note the fusing of leaves forming characteristic leaf sheaths.

Figure 7.27 Meadow horsetail, *Equisetum*, showing (a) a segment of the stem, (b) a mature strobilus, and (c) a strobilus shedding its spores. (X2)

1. Stem
2. Whorl of leaves
3. Separated sporangiophores revealing sporangia
4. Sporangia shedding spores

DIVISION SPHENOPHYTA (horsetails)

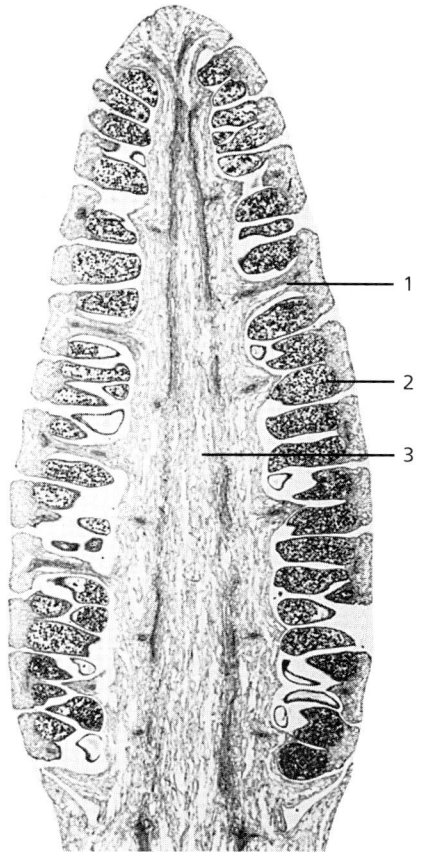

Figure 7.28 *Equisetum*, longitudinal section of the strobilus. (X2)
1. Sporangiophore
2. Sporangium
3. Strobilus axis

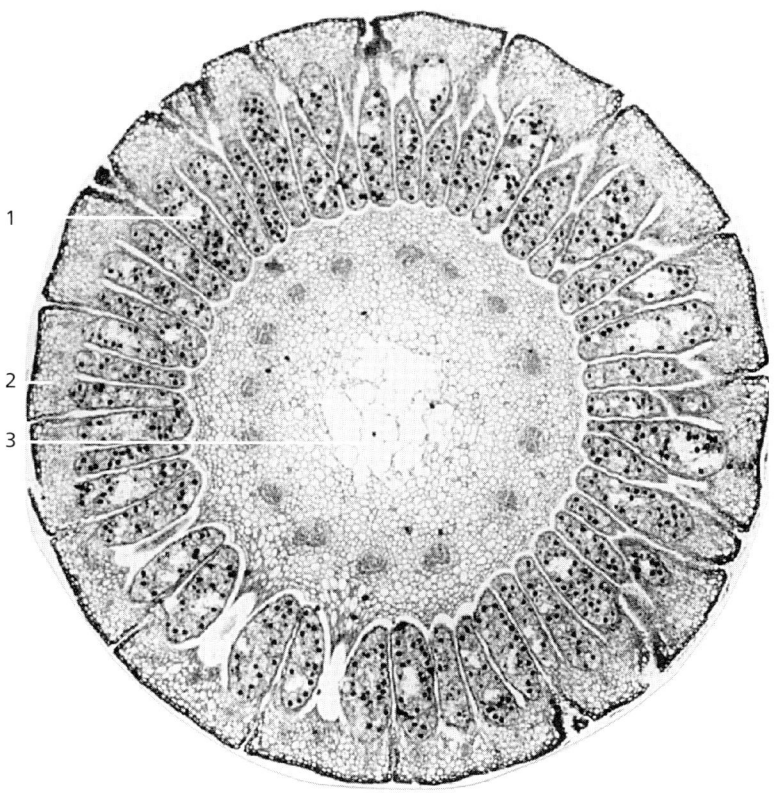

Figure 7.29 *Equisetum*, a transverse section of the strobilus. (X20)
1. Sporangium
2. Sporangiophore
3. Strobilus axis

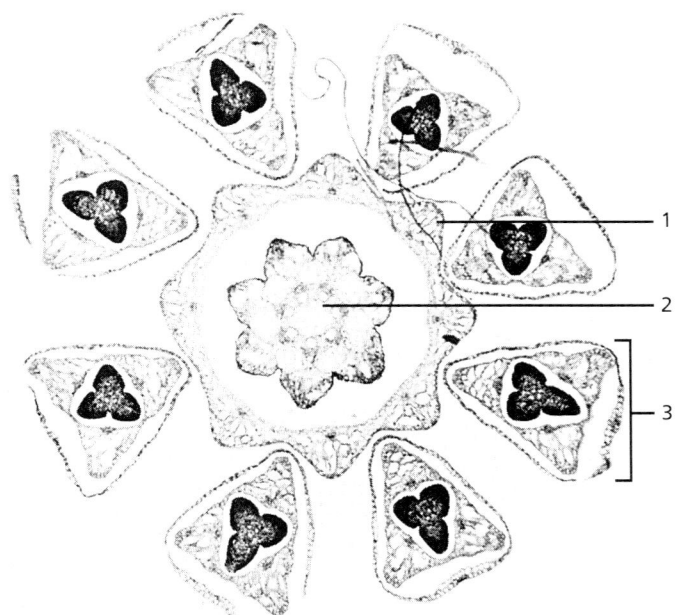

Figure 7.30 A transverse section of the stem of *Equisetum*, just above a node. (X10)
1. Leaf sheath
2. Main stem
3. Branch

Figure 7.31 Scanning electron micrograph of the spores of *Equisetum* and the elaters coiled about them. (X900)
1. Spore
2. Elater

DIVISION PTEROPHYTA (ferns)

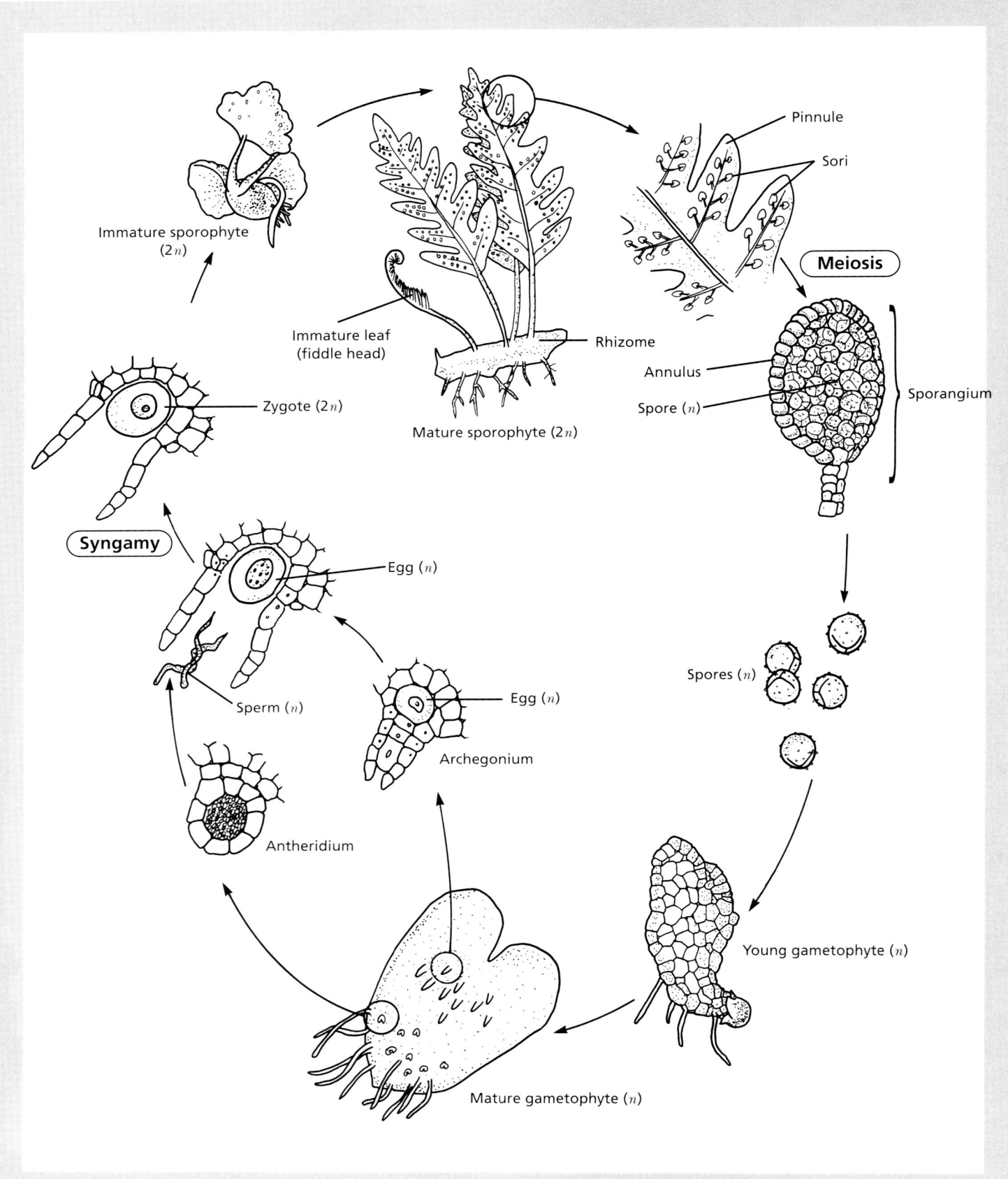

Figure 7.32 Life cycle of a fern.

DIVISION PTEROPHYTA (ferns)

Figure 7.33 Herbarium specimen of a primitive fern *Botrychium multifidum*.

Figure 7.34 The water fern, *Azolla*, is a floating fresh-water plant found throughout Europe and the United States. *Azolla* may become bright orange or red during fall.

Figure 7.35 *Asplenium*, a fern which occurs throughout the Northern Hemisphere.

1. Immature leaf (fiddle head)

DIVISION PTEROPHYTA (ferns)

Figure 7.36 Fronds of the staghorn fern, *Platycerium alcicorne*.

Figure 7.37 The fern *Cyrtomium*, showing sori on the underside of the pinnae.
1. Pinna   2. Sori

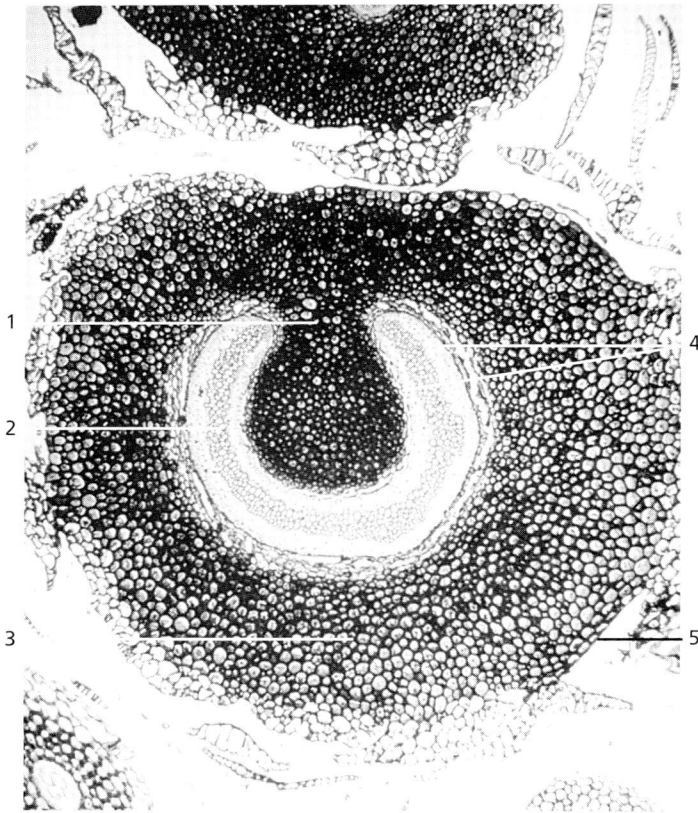

Figure 7.38 Stele of the maidenhair fern, *Adiantum*. (X40)
1. Leaf gap
2. Xylem
3. Cortex
4. Phloem
5. Epidermis

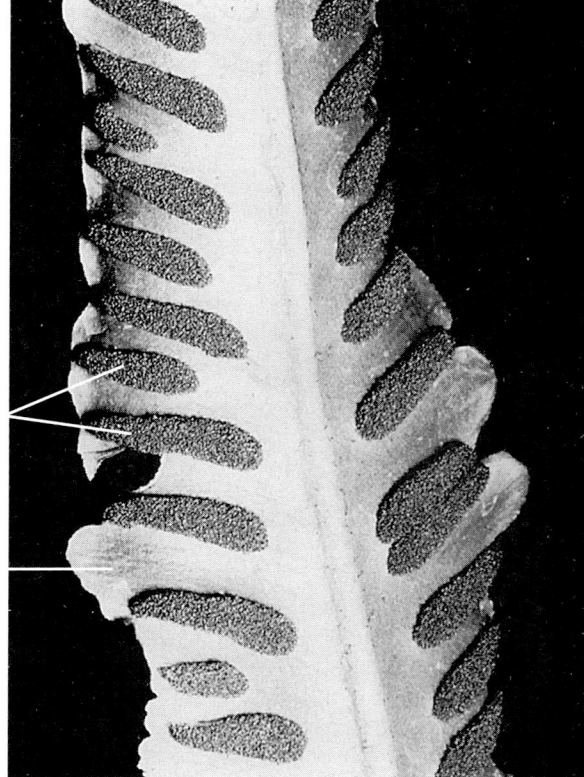

Figure 7.39 The fern *Asplenium*, showing the sori on the undersurface of the pinna.
1. Sori   2. Pinna

CHAPTER 7 — Kingdom Plantae: Seedless Vascular Plants

DIVISION PTEROPHYTA (ferns)

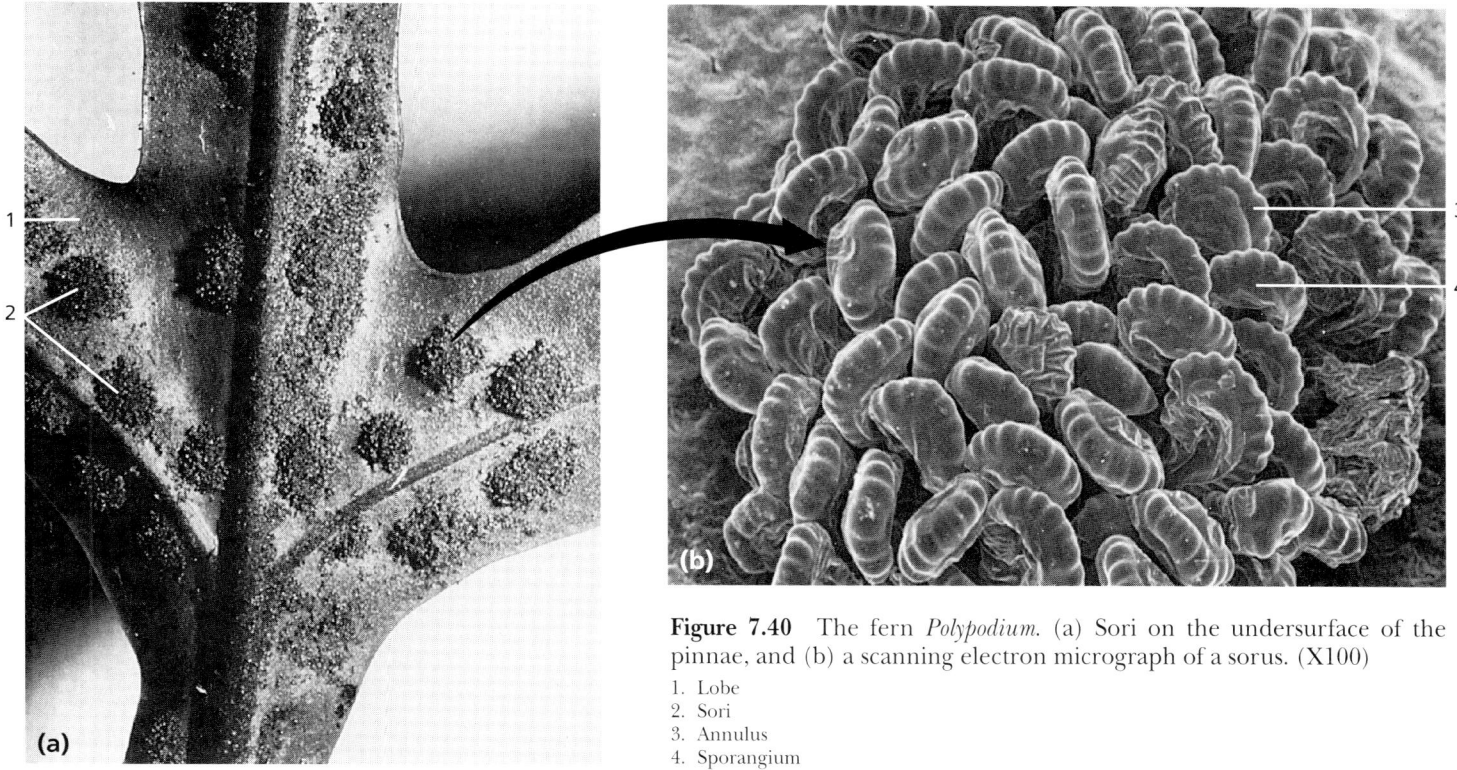

Figure 7.40 The fern *Polypodium*. (a) Sori on the undersurface of the pinnae, and (b) a scanning electron micrograph of a sorus. (X100)
1. Lobe
2. Sori
3. Annulus
4. Sporangium

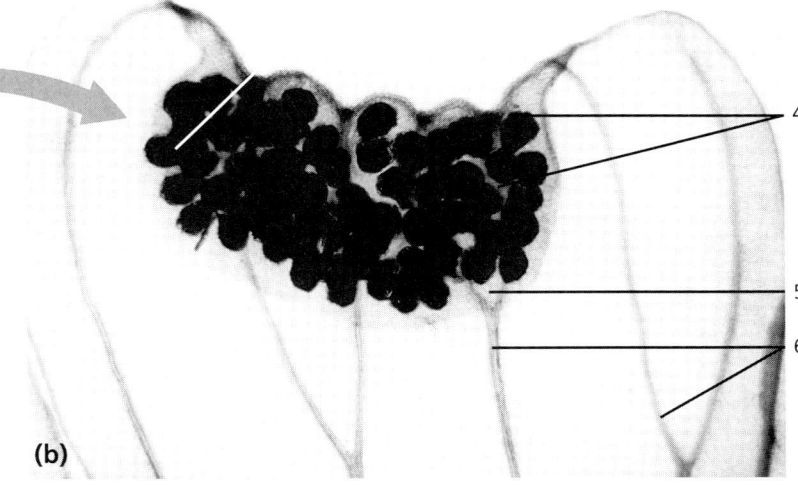

Figure 7.41 The maidenhair fern *Adiantum*. (a) Pinnae and sori, and (b) a magnified view (X100) of the tip of a pinna folded under to form a false indusium that encloses the sorus.
1. Sori
2. False indusium
3. Pinnule
4. Sporangia with spores
5. False indusium enclosing a sorus
6. Vascular tissue of the pinnule

DIVISION PTEROPHYTA (ferns)

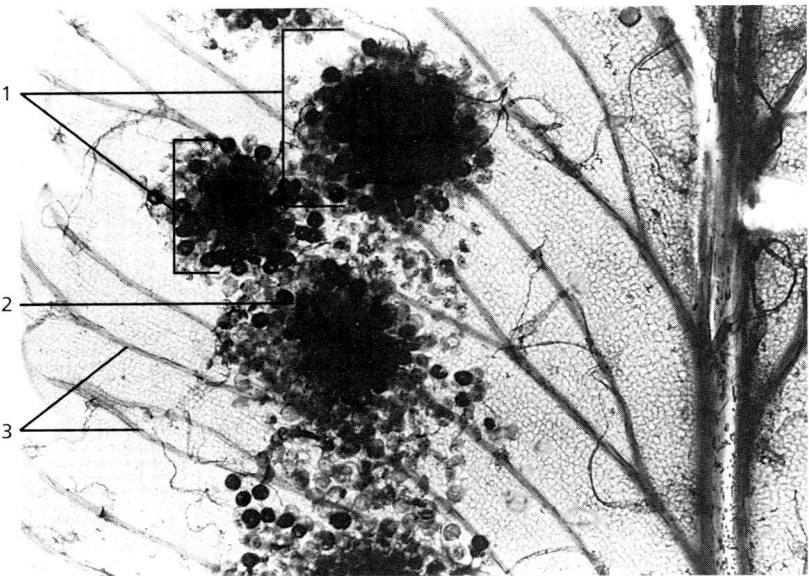

Figure 7.42 Sori of the fern *Polystichum*.
1. Sori
2. Sporangia
3. Veins of pinna

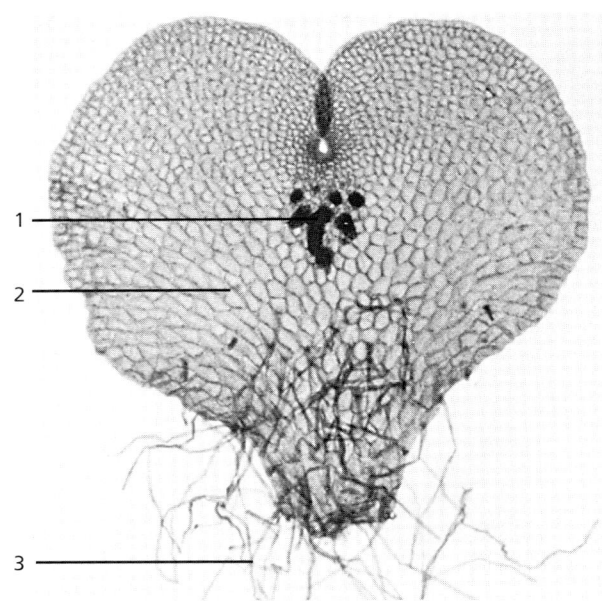

Figure 7.43 Fern gametophyte showing archegonia. (X40)
1. Archegonia
2. Gametophyte (prothallus)
3. Rhizoids

Figure 7.44 Sorus of a homosporous fern, *Cyrtomium falcatum*. (X100)
1. Spores (*n*)
2. Indusium (2*n*)
3. Epidermis of pinnule
4. Mesophyll tissue of pinnule
5. Sporangia (2*n*)

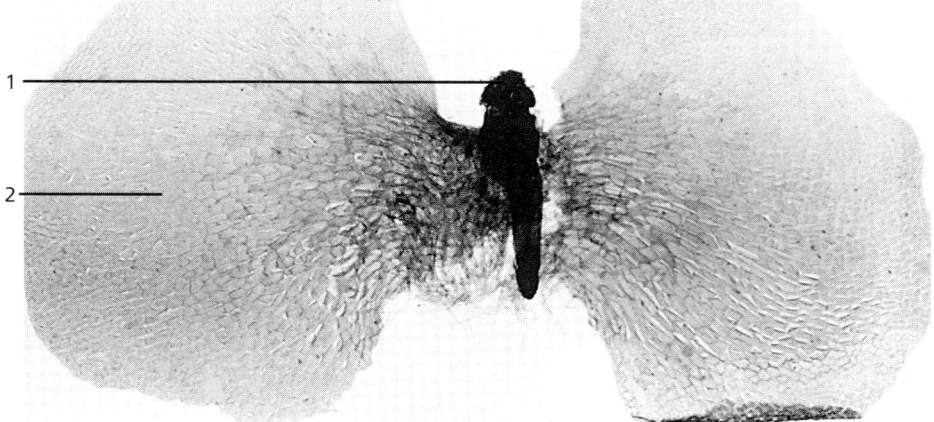

Figure 7.45 Fern gametophyte with a young sporophyte attached. (X40)
1. Young sporophyte
2. Gametophyte

DIVISION PTEROPHYTA (ferns)

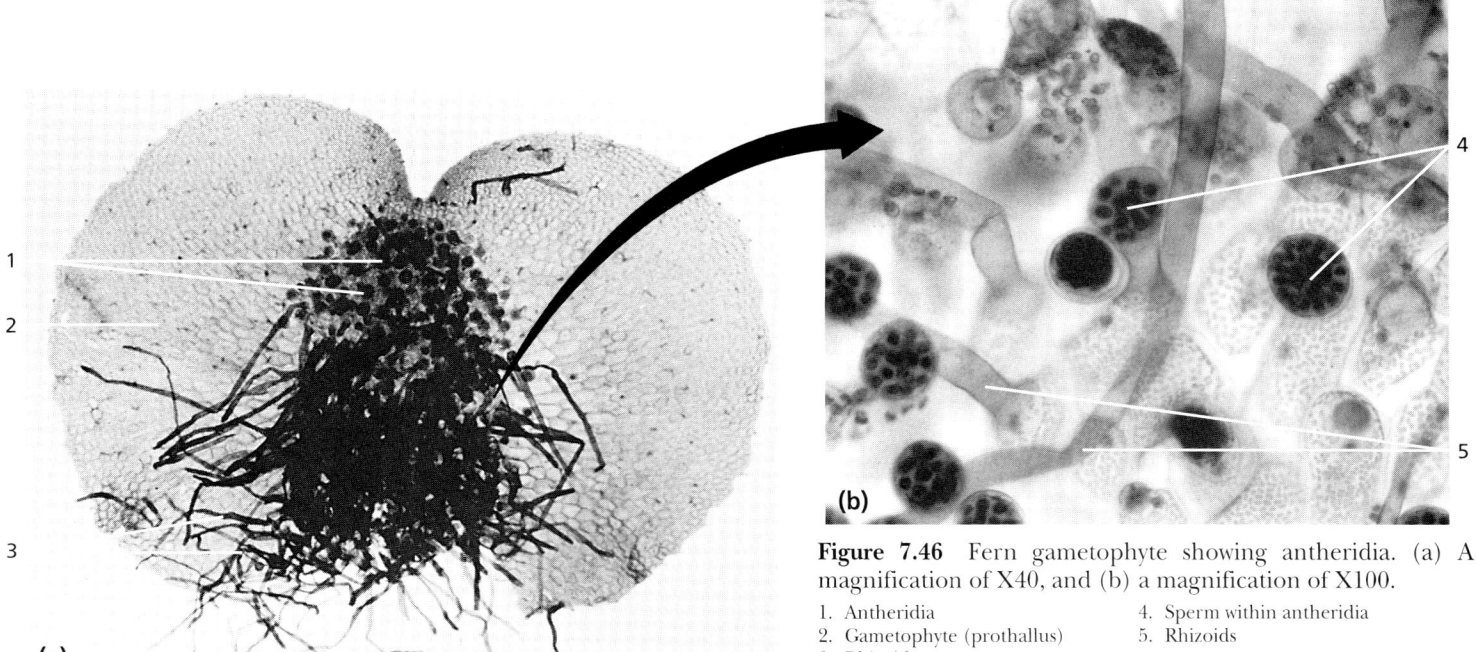

Figure 7.46 Fern gametophyte showing antheridia. (a) A magnification of X40, and (b) a magnification of X100.
1. Antheridia
2. Gametophyte (prothallus)
3. Rhizoids
4. Sperm within antheridia
5. Rhizoids

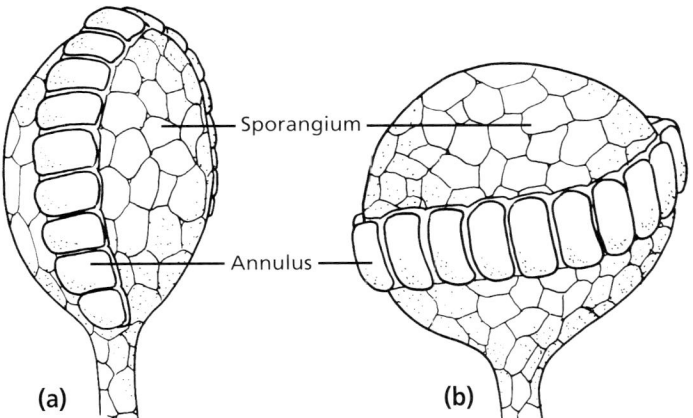

Figure 7.47 The position of the annulus is a feature of importance in fern classification. The sporangium depicted in (a) has a vertical annulus while the sporangium depicted in (b) has a horizontal annulus.

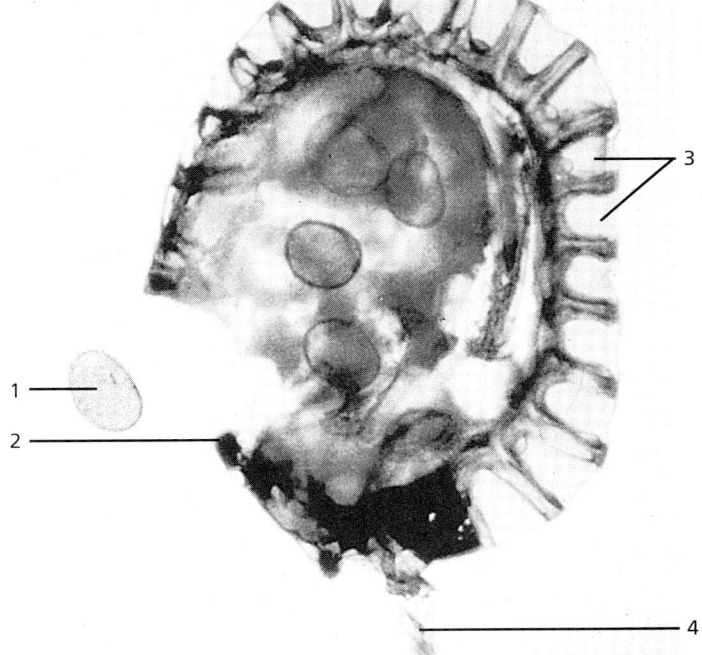

Figure 7.48 A sporangium of the fern *Cyrtomium*, discharging a spore. (X430)
1. Spore (*n*)   2. Lip cell   3. Annulus   4. Stalk

DIVISION PTEROPHYTA (ferns)

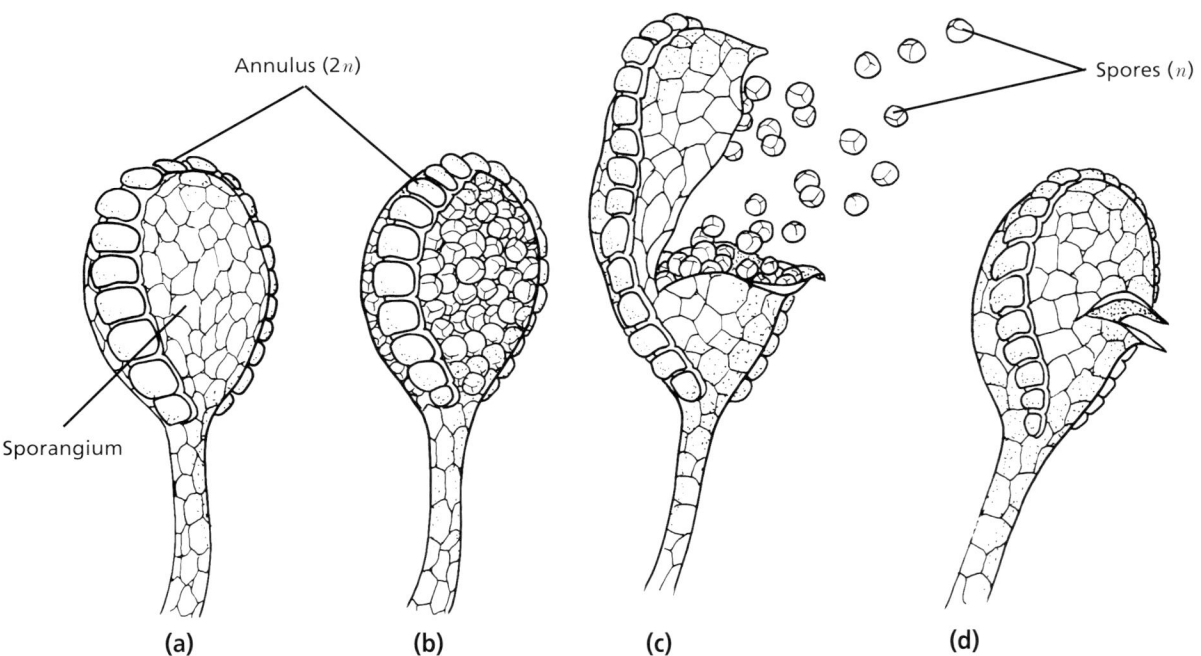

Figure 7.49 Spore dispersal from a mature sporangium. As the cells of the annulus lose water, they dry and shrink, causing the sporangium to split releasing its spores. (a) A mature sporangium in side view; (b) enclosed spores within sporangium; (c) contracted annulus, ruptured sporangium, and released spores; (d) a sporangium following release of its spores.

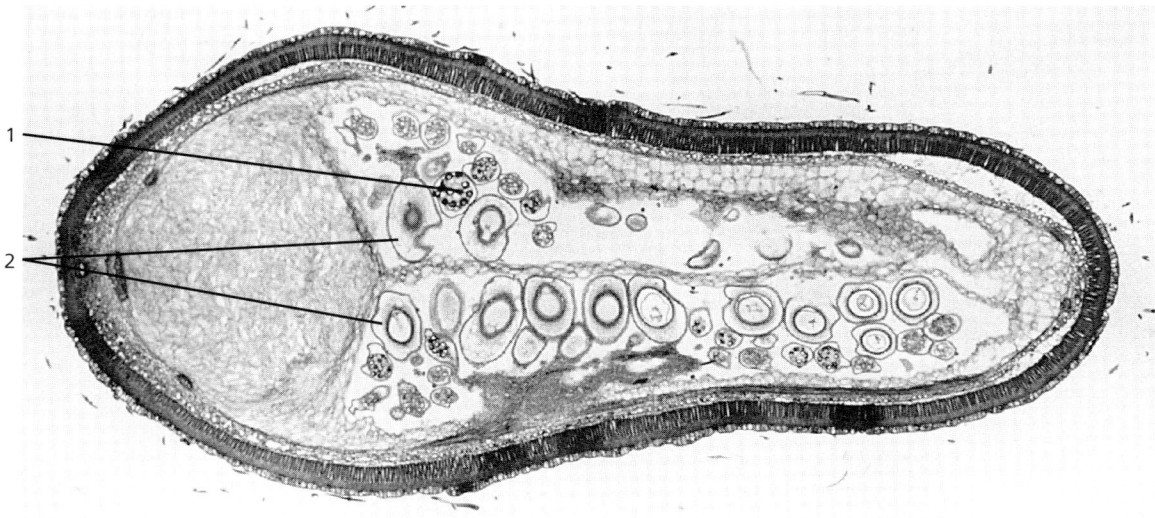

Figure 7.50 The water fern, *Marsilea*, belongs to one of the two living orders of heterosporous ferns. This is a cross section of a fertile pinnule.
1. Microsporangia with microspores
2. Megasporangia with megaspores

Kingdom Plantae: Gymnosperms (Exposed Seed Plants)

Gymnosperms (exposed seed plants) include plants in four divisions: Cycadophyta (cycads); Ginkophyta (ginkgos); Coniferophyta (conifers); and Gnetophyta (gnetophytes) (Table 8.1). Gymnosperms arose in the late Devonian period, about 375 million years ago. They dominated land floras throughout most of the Mesozoic Era until the late Cretaceous Period, some 100 million years ago. Currently, about 65 genera and 720 species are known.

Reproduction in seed plants is well adapted to a land existence. Seeds develop from protective structures known as *ovules*, which mature to produce protective and nutritive layers around the embryo. Gymnosperms produce their seeds in protective cones, (in contrast to angiosperms (see Chapter 9) which produce their seeds in a protective *fruits*). In the life cycle of a gymnosperm, such as a pine, the mature sporophyte (tree) has cones that produce megaspores (that develop into female gametophyte generations), and cones that produce microspores (that develop into male gametophyte generations — mature pollen grains). Following fertilization, immature sporophyte generations are present in seeds located in the gametophyte cones. The cone opens and the seeds disperse to the ground and germinate if conditions are right. As stated earlier, reproduction in angiosperms is similar to gymnosperms except that angiosperm pollen and ovules are produced in flowers, rather than in cones and seeds develop within protective fruits.

Cycads are gymnosperms superficially resembling palm trees. They are represented by 10 extant genera and about 100 species. They are all tropical or subtropical in distribution. The ovules and seeds of cycads are exposed, and the sperm are flagellated and motile. Most cycads have short, thick trunks, or stems, with large pinnate leaves attached at the crown. The leaves are thick with sunken stomata. Most cycads are less than 3 meters tall, although, one species reaches 20 meters in height.

A single living species, *Ginkgo biloba*, comprises the division of Ginkgophyta. *Ginkgo biloba* is native to China, but has been cultivated extensively throughout the world in temperate climates. Ginkgophyta is an ancient division in existence since late Carboniferous times, 290 million years ago. The ginkgo, or maidenhair tree, grows to 35 meters tall. The leaves of ginkgos are fan-shaped, with long petioles. Because it is slow-growing and because the wood is brittle, *Ginkgo* is not a good source of lumber.

Conifers are mostly evergreen woody shrubs and trees, although not all representatives are evergreen. There are about 50 genera, with about 550 living species. Conifers lack flowers and their seeds are exposed on the surface of cone scales. Conifers often have needle-like leaves and the stomata are sunken. Conifers supply much of the timber for building and wood for the manufacture of paper, turpentine, and many other products. Conifers are also a major source of firewood.

Three genera of gnetophytes exist — *Gnetum*, *Ephedra*, and *Welwitschia* — with about 70 living species. An interesting feature of gnetophytes is the presence of vessel elements in their xylem tissue. Most flowering plants contain vessels elements as well, providing evidence to some botanists the Gnetophyta might have been ancestral to angiosperms. Each of the three genera of gnetophytes is very different. *Gnetum* is a vine or small tree resembling a flowering plant in physical appearance. It occurs in tropical rain forests. *Ephedra* is a highly branched shrub and occurs in dry or arid habitats (including the southwest deserts of the United States). *Welwitschia* does not resemble any other plant. Found in Africa, it has a short, thick, disk-shaped stem with two long leaves attached at its edge.

TABLE 8.1
Gymnosperms within the Kingdom Plantae

Divisions and Representative Kinds	Characteristics
Division Cycadophyta (cycads)	Gymnosperms with pollen and seed cones borne on different plants; motile sperm; plants mostly shrubby; leaves large, palm-like
Division Ginkgophyta (*Ginkgo*)	Gymnosperm with deciduous, fan-shaped leaves; motile sperm; large tree
Division Coniferophyta (conifers)	Woody gymnosperms, producing seeds in cones; motile sperm absent; most with needle-like leaves that lack air spaces; stomata are sunken
Division Gnetophyta (gnetophytes)	Gymnosperms that contain vessel elements; motile sperm absent

DIVISION CYCADOPHYTA (cycads)

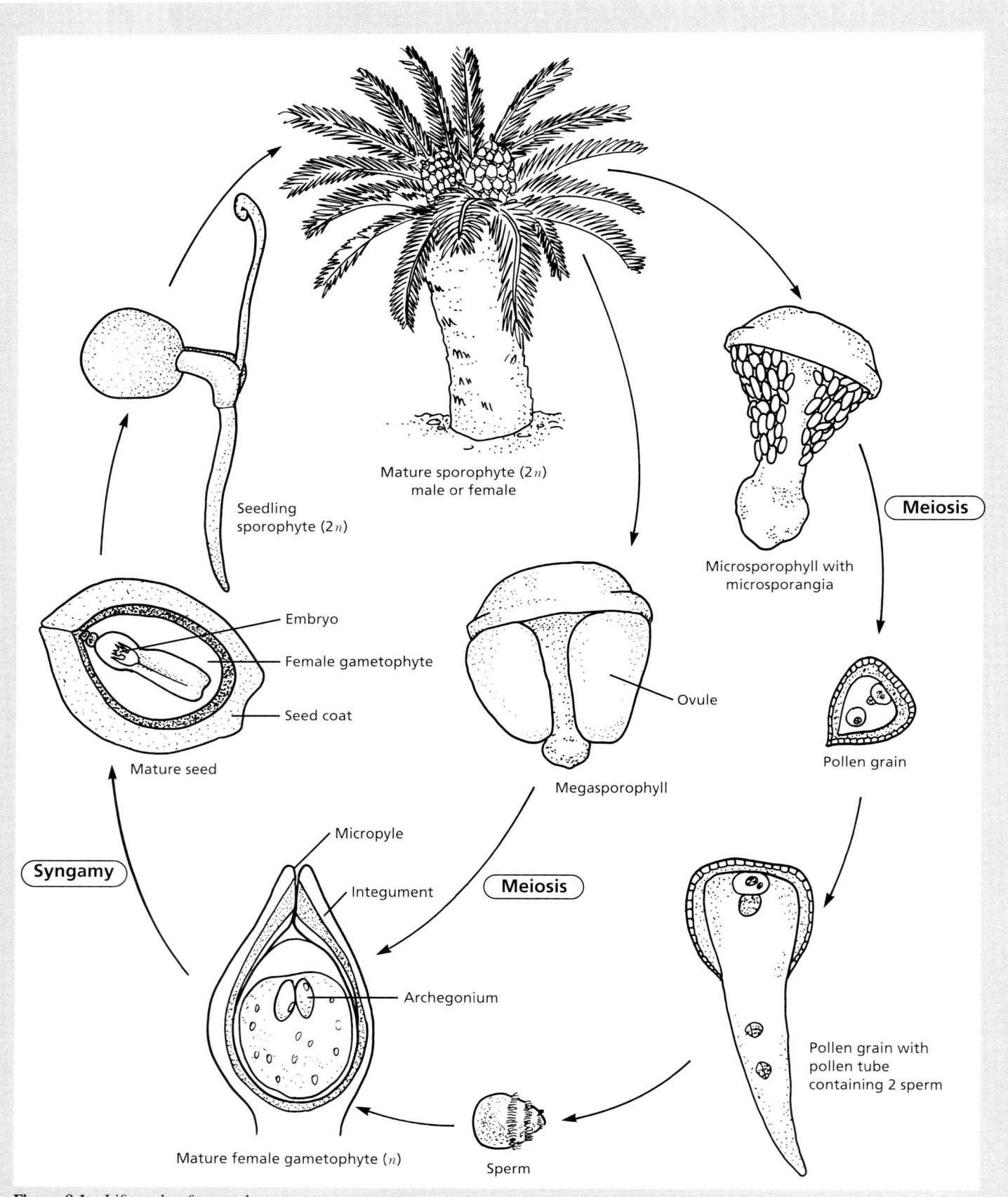

Figure 8.1 Life cycle of a cycad.

DIVISION CYCADOPHYTA (cycads)

Figure 8.2 A fossil cycad leaf impression. Fossil cycads are common in some Mesozoic rocks.

Figure 8.3 *Cycas revoluta*. Cycads were abundant during the Mesozoic Era. Currently, there are 10 living genera, with about 100 species, that are mainly found in tropical and subtropical areas. The trunk of many cycads is densely covered with petioles of shed leaves.

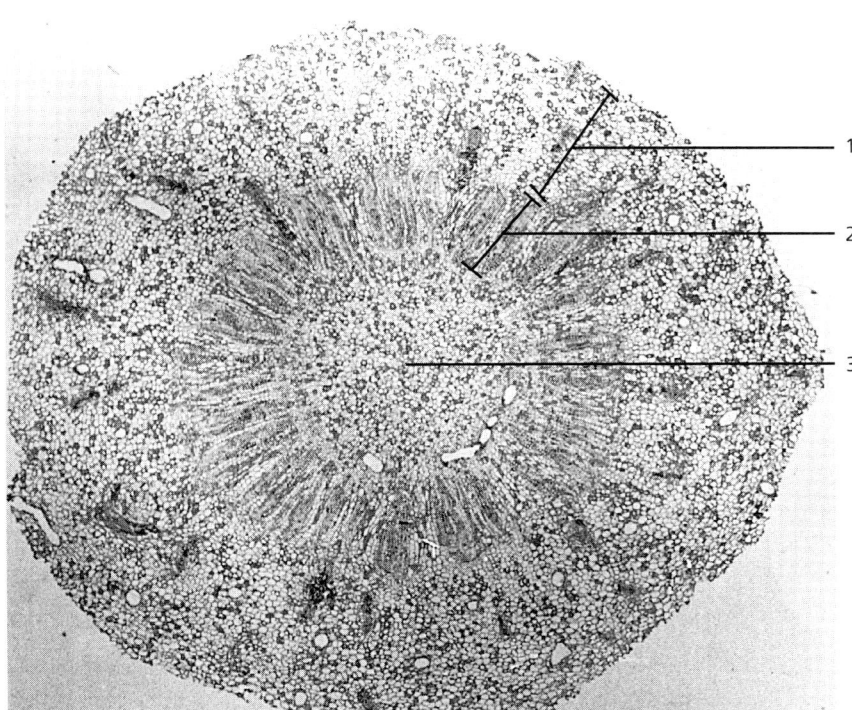

Figure 8.4 A transverse section of the stem of the cycad, *Zamia*.

1. Cortex   2. Vascular tissue   3. Pith

Figure 8.5 A young plant of the cycad *Zamia pumila*. Found in Florida, this cycad is the only species native to the United States. The rootstocks and stems of this plant were an important source of food for Seminole Native Americans.

DIVISION CYCADOPHYTA (cycads)

Figure 8.6 Microsporangiate cones of *Cycas revoluta*.
1. Cones

Figure 8.8 Microsporangiate cones of the cycad, *Zamia*. (The cone on the left (a) is longitudinally sectioned.)
1. Microsporangia
2. Microsporophyll

Figure 8.7 Herbarium specimen of a leaf of *Cycads media*. Cycads are ancient plants that predate the dinosaurs.

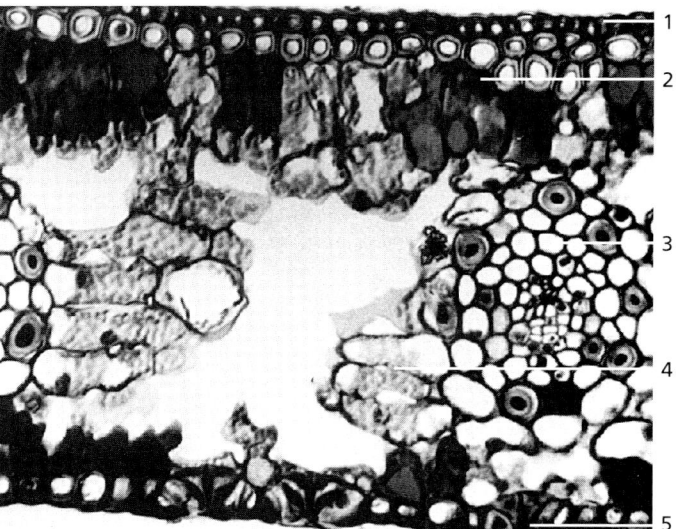

Figure 8.9 A transverse section of the leaf of the cycad, *Zamia*. (X100)
1. Upper epidermis
2. Palisade mesophyll
3. Vascular bundle (vein)
4. Spongy mesophyll
5. Lower epidermis

DIVISION CYCADOPHYTA (cycads)

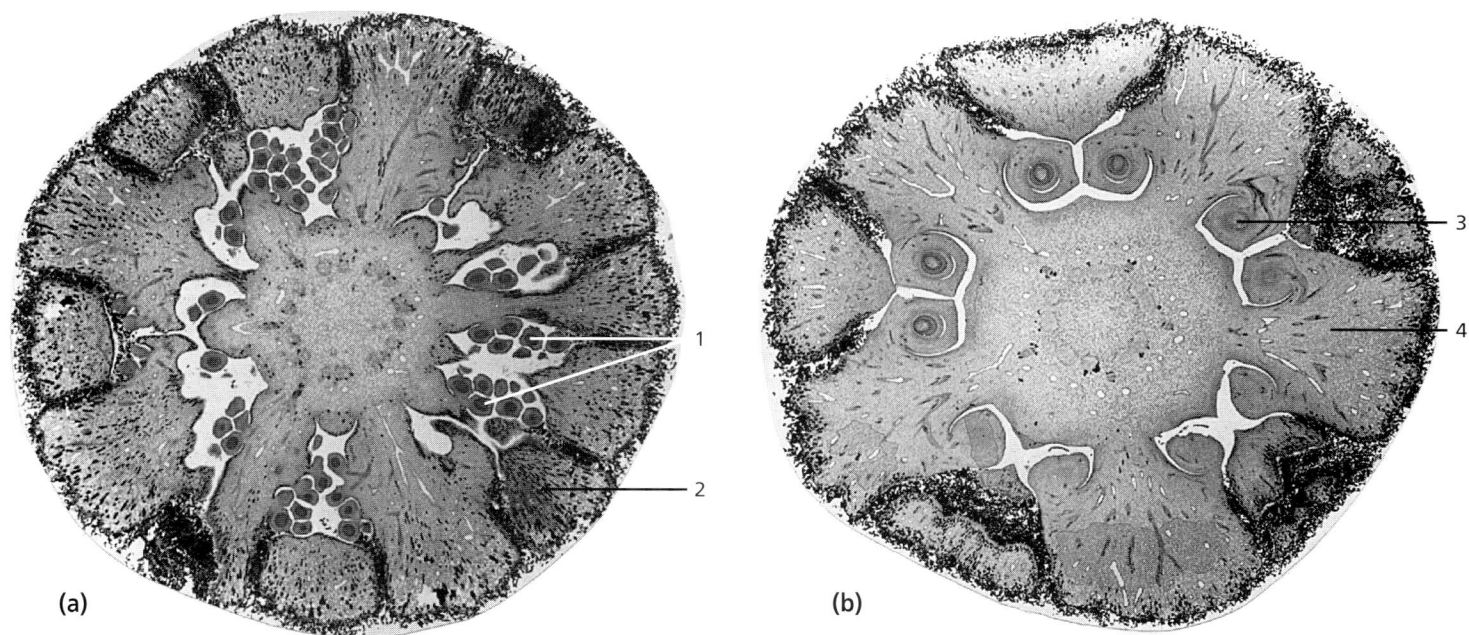

Figure 8.10 Transverse sections of a microsporangiate cone (a) and a megasporangiate cone (b) of the cycad, *Zamia*.
1. Microsporangia
2. Microsporophyll (cone scale)
3. Ovule
4. Megasporophyll (cone scale)

Figure 8.11 The ovule of the cycad *Zamia*. In (a) the ovule has two archegonia and is ready to be fertilized. In (b) the ovule has been fertilized and contains an embryo, but the seed coat has been removed. (X5)

1. Female gametophyte
2. Archegonium
3. Megasporangium (nucellus)
4. Integument
5. Embryo

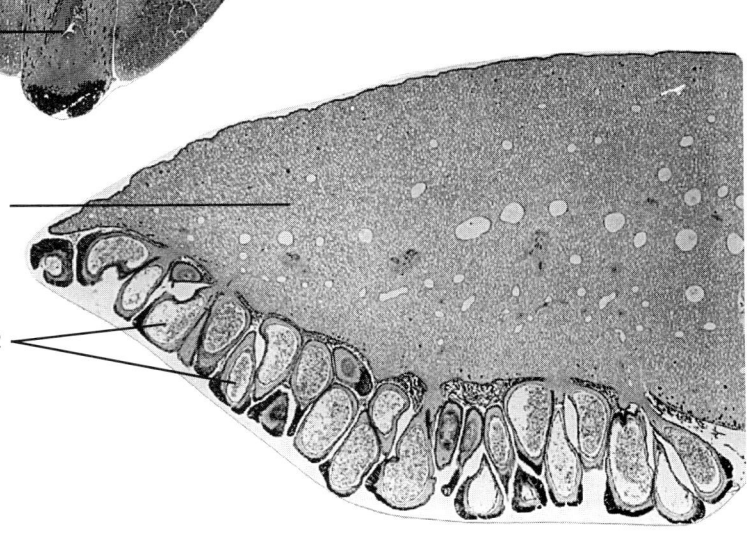

Figure 8.12 A transverse section of a microsporophyll of the cycad *Cycas*. Note that the microsporangia develop on the undersurface of the microsporophyll. (X40)

1. Microsporophyll
2. Microsporangia

DIVISION GINKGOPHYTA (*Ginkgo*)

Figure 8.13 The ginkgo, or maidenhair tree, *Ginkgo biloba*. Consisting of a central trunk with lateral branches, a mature ginkgo grows to 80 to 100 feet tall. Native to China, *Ginkgo biloba* has been introduced in temperate climates throughout the world as an interesting ornamental tree.

Figure 8.14 Herbarium specimen of *Ginkgo biloba*. As the sole living member of the division Ginkgophyta, this species was known from the fossil record before it was discovered as a living form. Able to withstand air pollution, ginkgos are often used as ornamental trees within city parks.

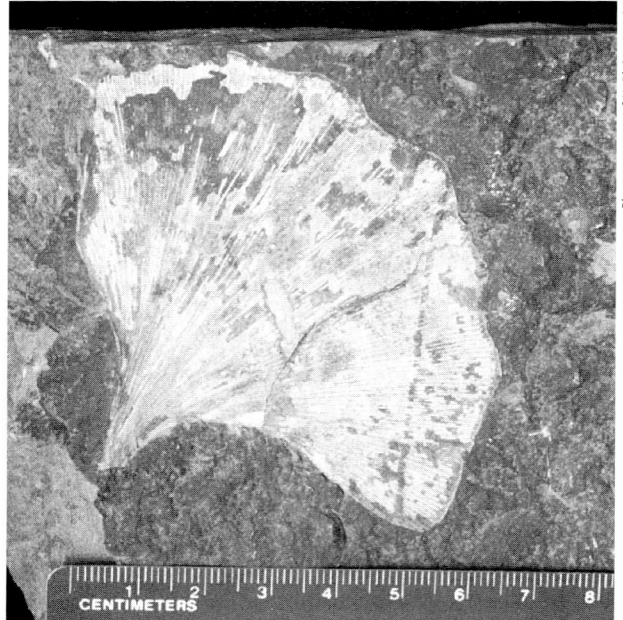

Figure 8.15 A fossil ginkgo leaf impression from paleocene rock. This specimen was found in Morton County, North Dakota.

Figure 8.16 A leaf from the ginkgo tree, *Ginkgo biloba*. Note the characteristic pattern of venation and the fan-shaped leaf.

Kingdom Plantae: Gymnosperms (Exposed Seed Plants)

DIVISION GINKGOPHYTA (*Ginkgo*)

Figure 8.17 A branch of a ginkgo tree, *Ginkgo biloba*.
1. Long shoots
2. Short shoots

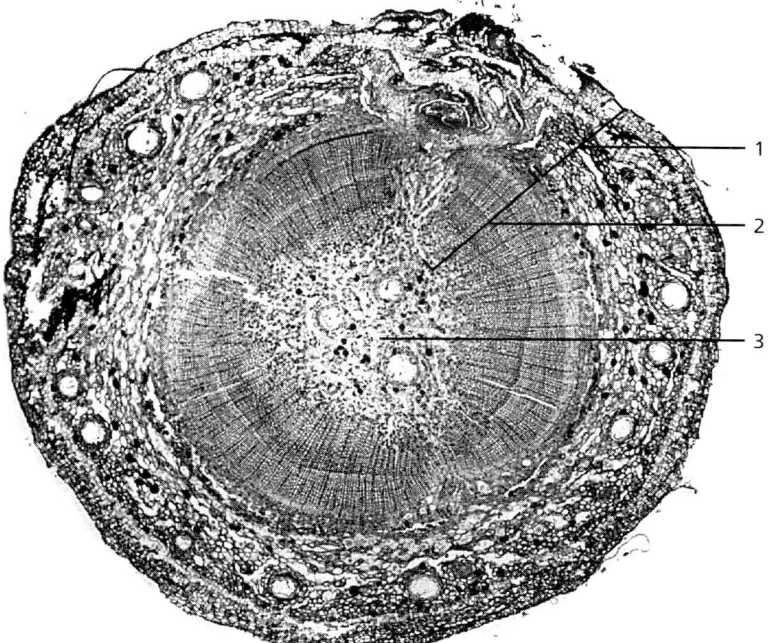

Figure 8.18 A transverse section of a short branch from the ginkgo tree, *Ginkgo biloba*. (X7)
1. Cortex
2. Vascular tissue
3. Pith

Figure 8.19 The arrangement of leaves and immature ovules on a short shoot of the ginkgo tree, *Ginkgo biloba*.
1. Leaf
2. Ovules
3. Short shoot
4. Long shoot

Figure 8.20 Microsporangiate strobili of the ginkgo tree, *Ginkgo biloba*.
1. Leaf
2. Microsporangiate strobili

DIVISION GINKGOPHYTA (*Ginkgo*)

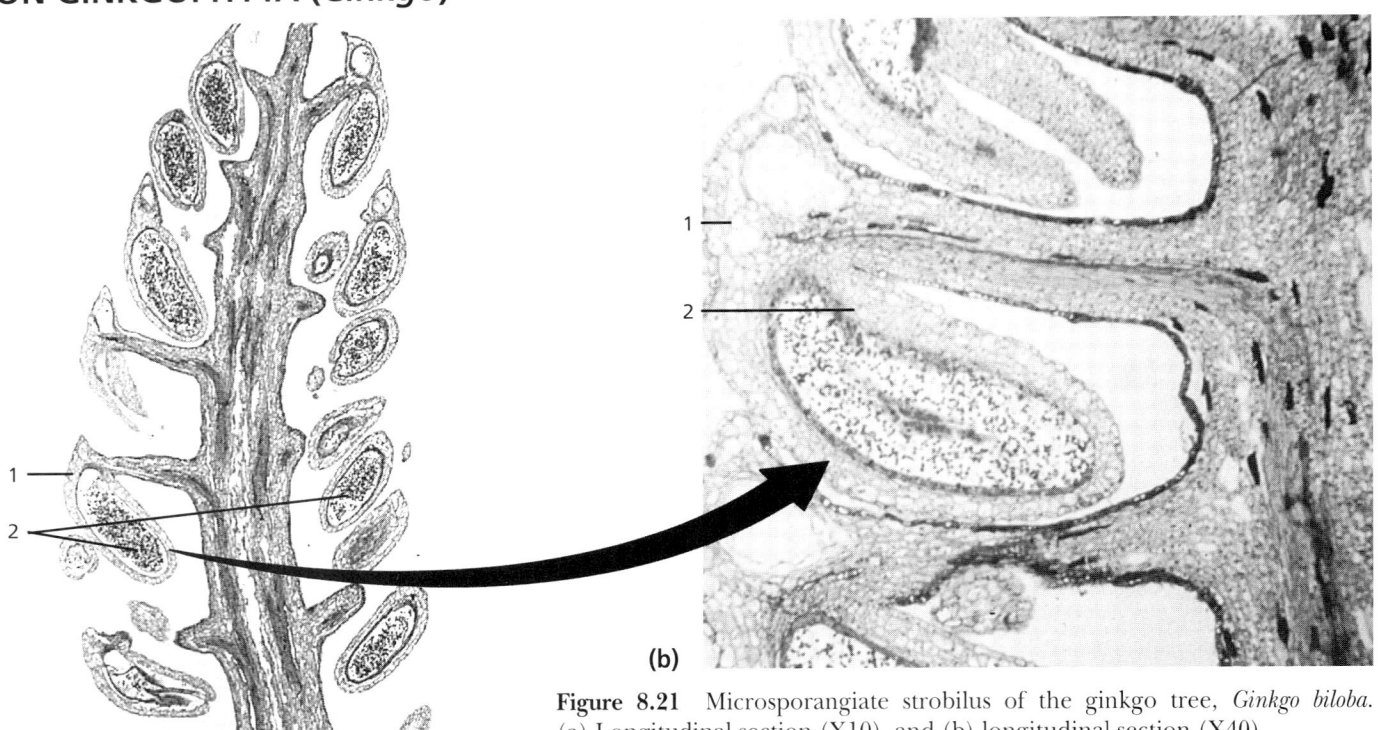

Figure 8.21 Microsporangiate strobilus of the ginkgo tree, *Ginkgo biloba*. (a) Longitudinal section (X10), and (b) longitudinal section (X40).
1. Microsporophyll 2. Microsporangia

Figure 8.22 Leaves and fleshy seeds of the ginkgo, *Ginkgo biloba*.

Figure 8.23 A mature seed of *Ginkgo* with the outer layer of the seed coat still intact.

Figure 8.24 A mature seed of *Ginkgo* with the outer layer of the seed coat removed.

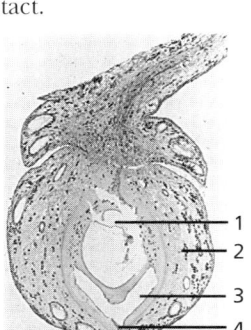

Figure 8.25 A longitudinal section of an ovule of *Ginkgo* prior to fertilization.
1. Nucellus   3. Pollen chamber
2. Integument 4. Micropyle

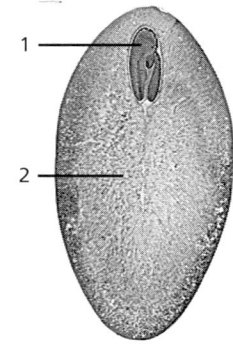

Figure 8.26 A longitudinal section of a seed of *Ginkgo*.
1. Embryo
2. Megagametophyte

DIVISION CONIFEROPHYTA (conifers)

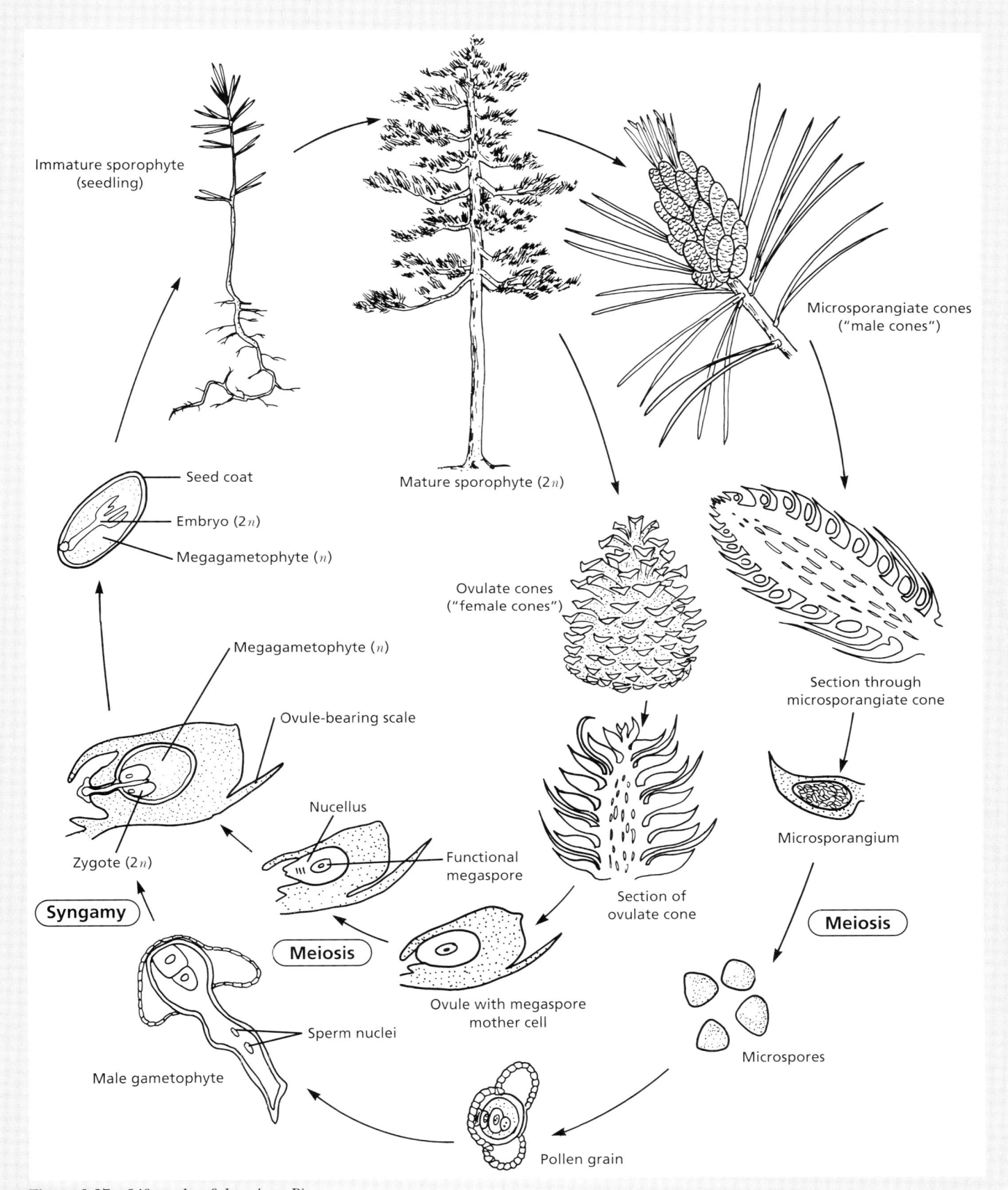

Figure 8.27 Life cycle of the pine, *Pinus*.

DIVISION CONIFEROPHYTA (conifers)

Figure 8.28 Conifer fossils from the Mesozoic Era. (a) Two small branches from the genus *Taxodium;* and (b) a petrified cone of *Pinus*.

Figure 8.29 Mature blue spruce, *Picea pungens.* Blue spruces are indigenous to the interior western United States, but they are planted widely as ornamentals.

Figure 8.30 The giant sequoia (or Sierra redwood), *Sequoiadendron giganteum,* is a member within the family Taxodiaceae. Remarkably resistant to fire, insects, and disease, some sequoias are estimated to be over two thousand years old. The diameters of some of these old trees exceed 40 feet.

CHAPTER 8 — Kingdom Plantae: Gymnosperms (Exposed Seed Plants)

DIVISION CONIFEROPHYTA (conifers)

Figure 8.31 A branch of the Norfolk Island pine, *Araucaria excelsa*, an ancient conifer frequently used as a house plant.

Figure 8.32 Piranha pine, *Araucaria auracana*, is a primitive conifer characterized by sharp, thick spine-like leaves, which persists for 10 to 15 years.

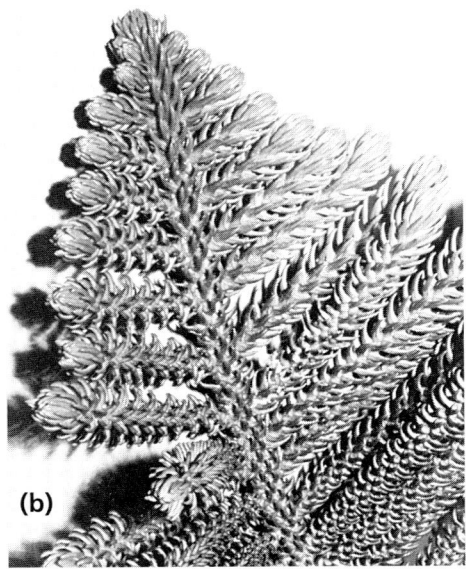

Figure 8.33 The leaves of most species of conifers are needle-shaped (a) such as these of the blue spruce, *Picea pungens*. *Araucaria* (b), however, has awl-shaped leaves, and *Podocarpus* (c) has strap-shaped leaves.

DIVISION CONIFEROPHYTA (conifers)

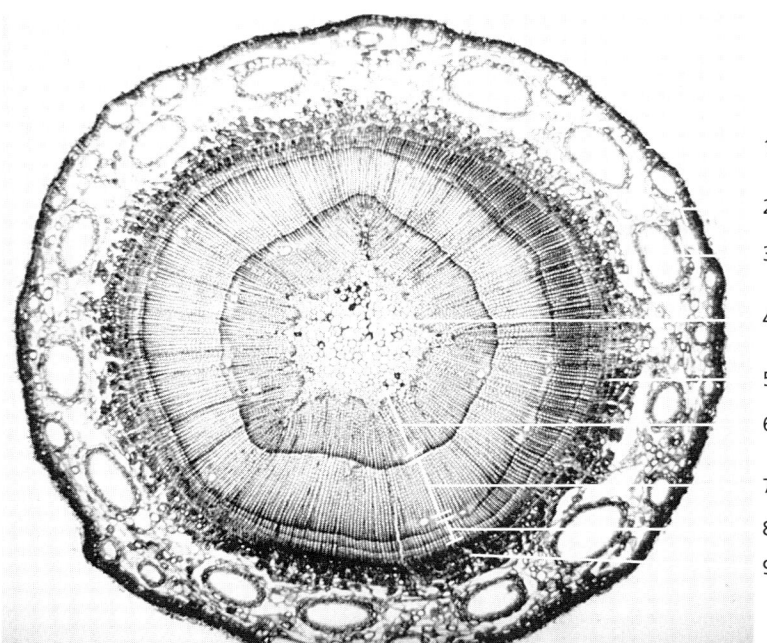

Figure 8.34 A transverse section through the stem of a young conifer, showing the arrangement of the tissue layers. (X20)

1. Periderm
2. Cortex
3. Resin duct
4. Pith
5. Cambium
6. First year's xylem
7. Second year's xylem
8. Third year's xylem
9. Phloem

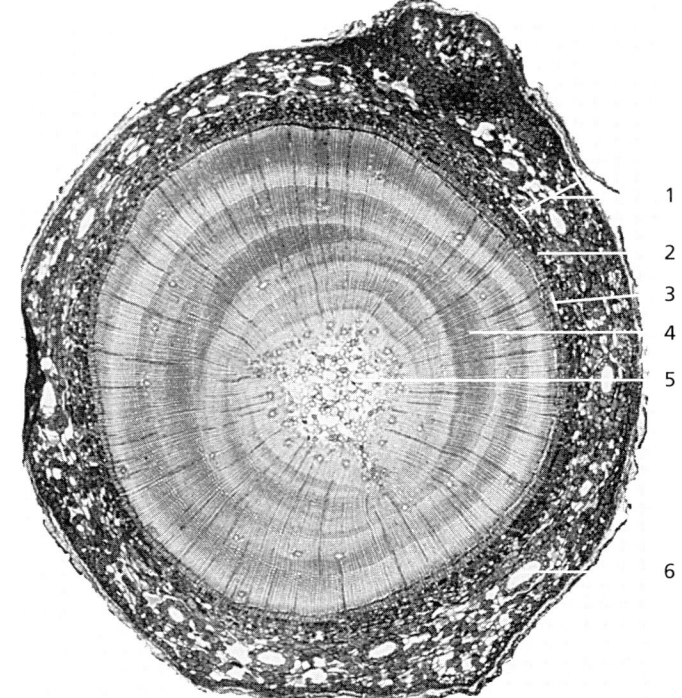

Figure 8.35 A transverse section through the stem of *Pinus*, showing secondary stem growth. (X5)

1. Bark (cortex, cork, and secondary phloem)
2. Secondary phloem
3. Vascular cambium
4. Secondary xylem
5. Pith
6. Resin duct

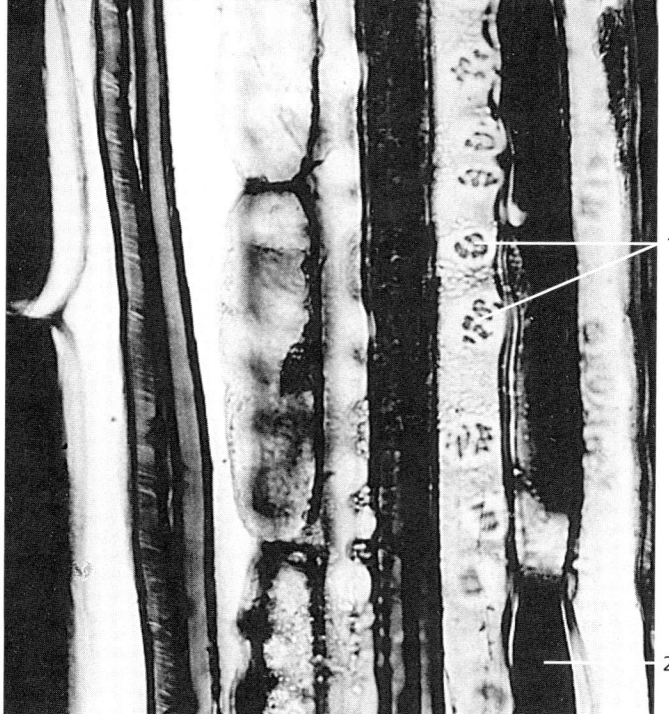

Figure 8.36 A longitudinal section through the phloem of *Pinus*. (X100)

1. Sieve areas on a sieve cell
2. Storage parenchyma

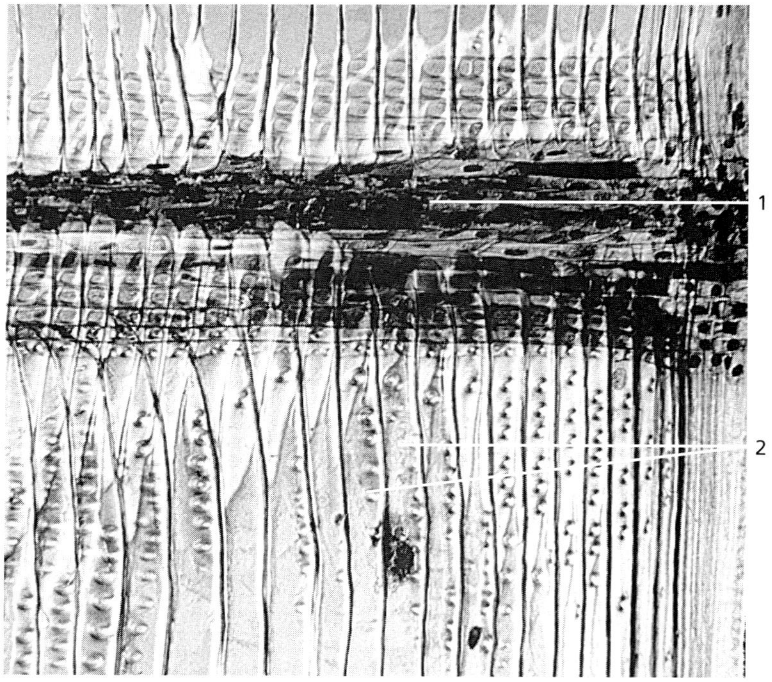

Figure 8.37 A longitudinal section through a stem of *Pinus*, cut through the xylem tissue. (X100)

1. Ray parenchyma
2. Trachieds

CHAPTER 8 — Kingdom Plantae: Gymnosperms (Exposed Seed Plants)

DIVISION CONIFEROPHYTA (conifers)

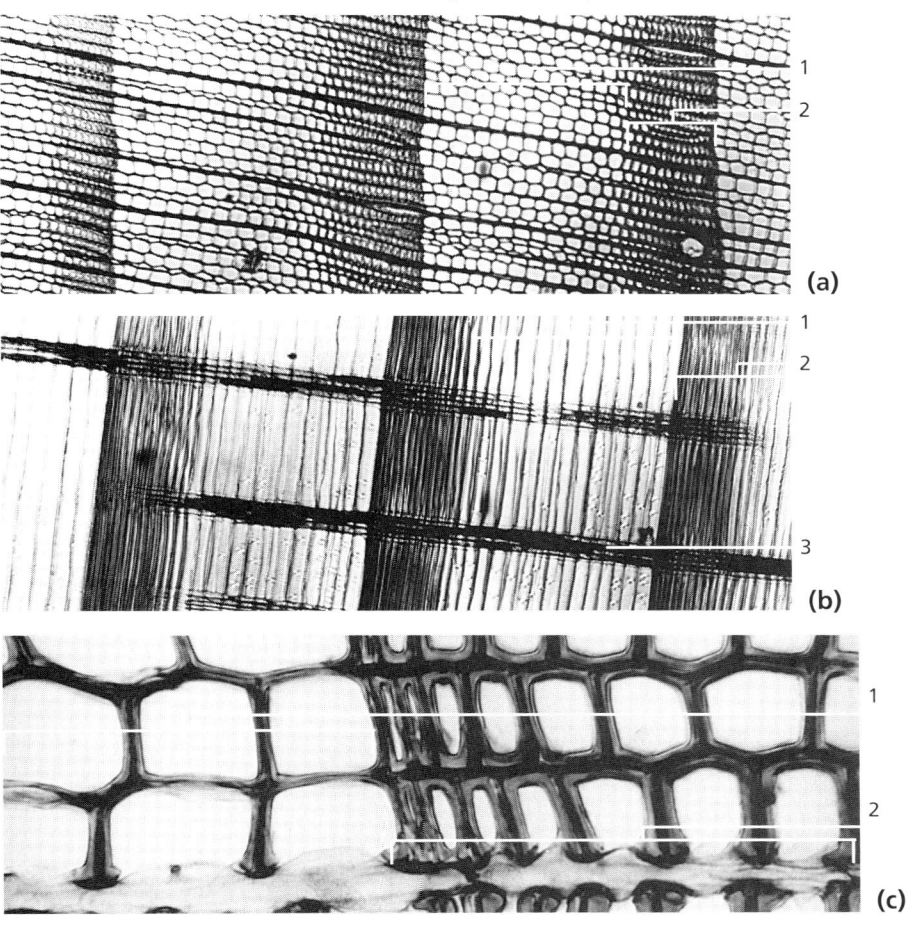

Figure 8.38 Growth rings in *Pinus;* (a) a transverse section through a *Pinus* stem (X20); (b) a radial longitudinal section through a *Pinus* stem (X20); and (c) a magnified view of the xylem cells of spring and summer wood. (X100)

1. Spring wood
2. Summer wood
3. Vascular ray

Figure 8.39 A transverse section of a leaf (needle) of *Pinus.* (X40)

1. Stoma
2. Endodermis
3. Photosynthetic mesophyll
4. Epidermis
5. Phloem
6. Xylem
7. Transfusion tissue
8. Resin duct

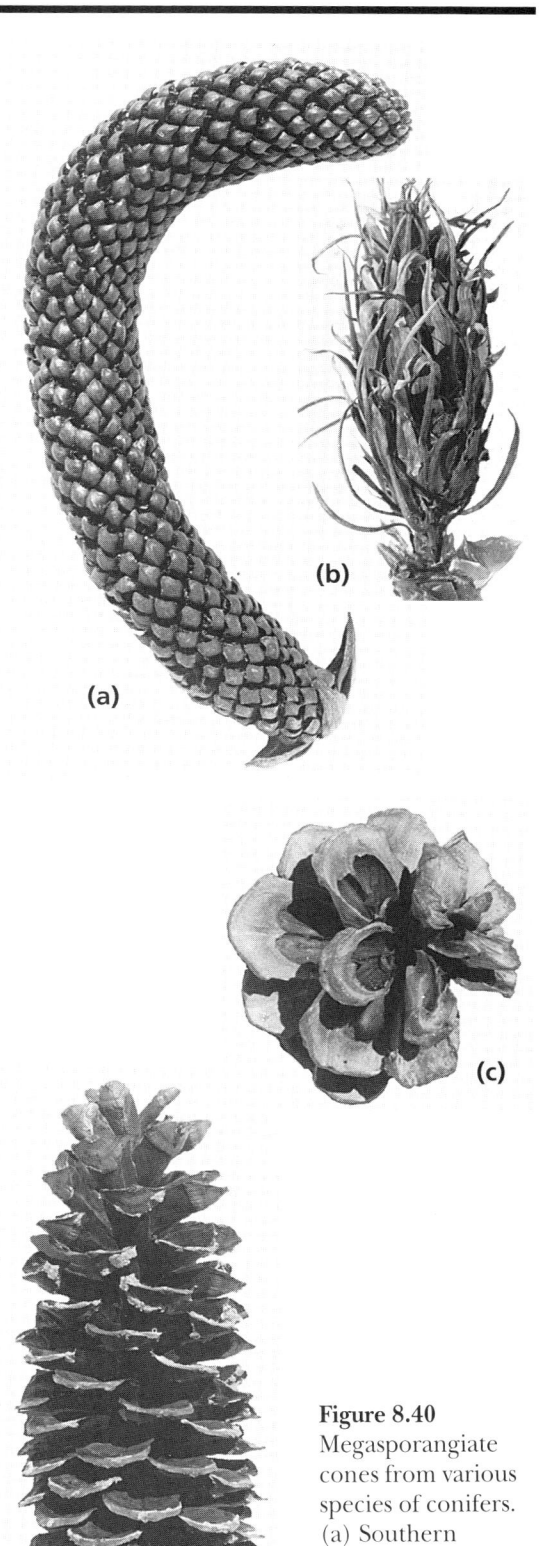

Figure 8.40 Megasporangiate cones from various species of conifers. (a) Southern hemisphere pine, *Araucaria;* (b) larch, *Larix;* (c) pinyon pine, *Pinus edulis;* and (d) sugar pine, *Pinus lambertiana.*

DIVISION CONIFEROPHYTA (conifers)

Figure 8.41 Relative positions of the cones in *Pinus*.
1. Microsporangiate cone
2. Ovulate cone

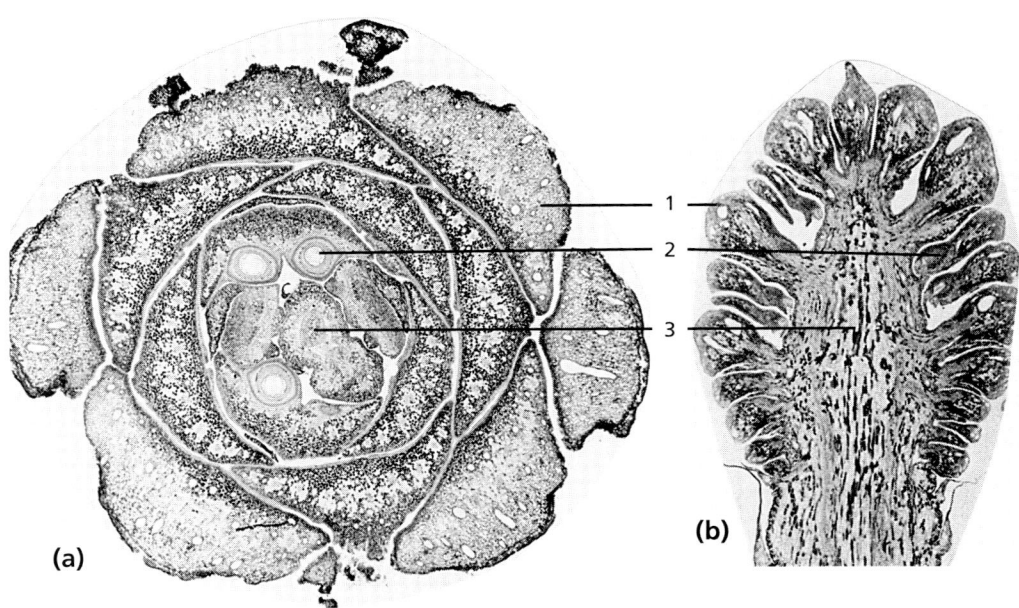

Figure 8.42 Female cones of a conifer. (a) A transverses section (X20), and (b) a longitudinal section. (X10)
1. Ovuliferous scale
2. Ovule
3. Cone axis

Figure 8.43 Microsporangiate cones of a conifer. (a) Staminate cones at end of branch; (b) a longitudinal section through a branch tip; (c) a longitudinal section through a single cone; and (d) a transverse section through a single cone.

1. Pine needles (leaves)
2. Sporophylls
3. Microsporangiate cone
4. Microsporangium
5. Sporophyll
6. Cone axis
7. Branch tip

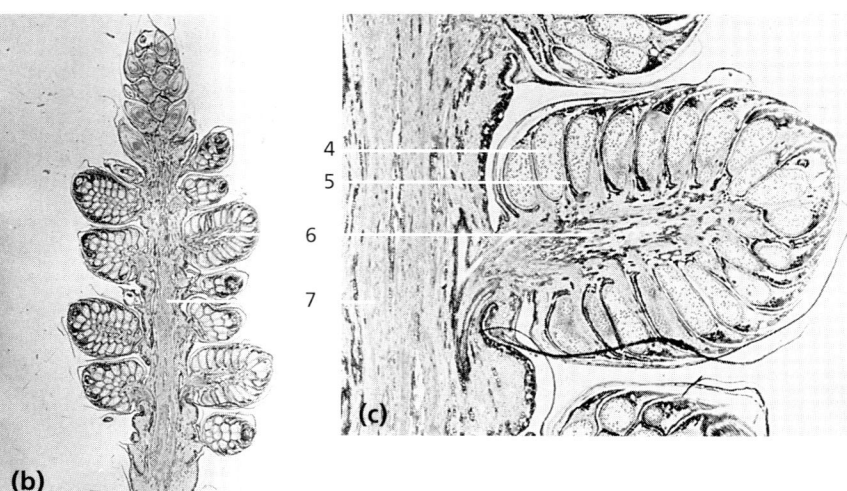

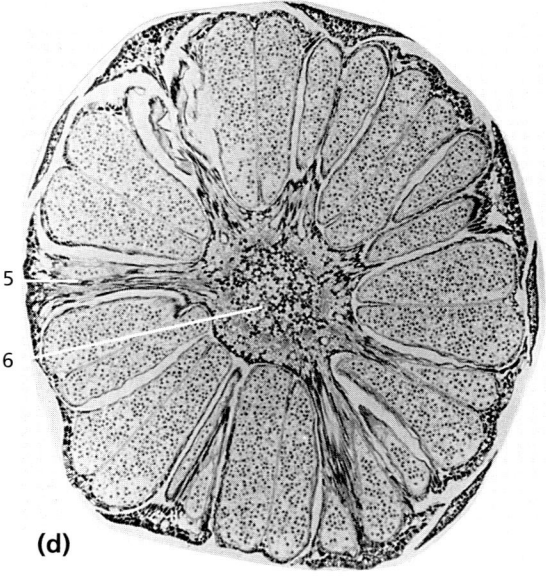

DIVISION CONIFEROPHYTA (conifers)

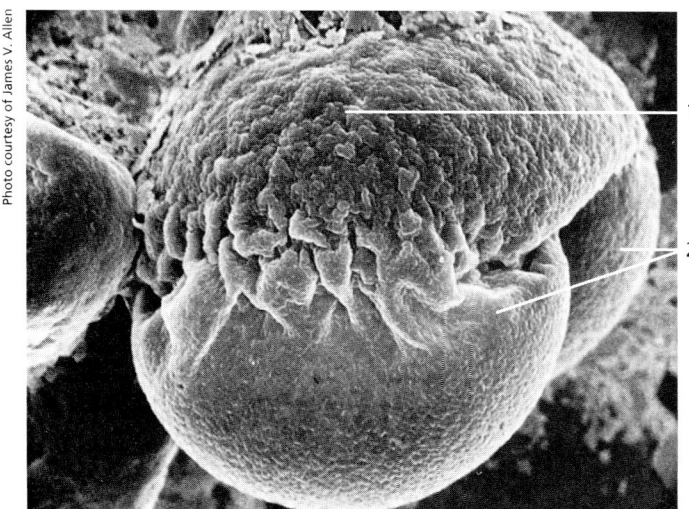

Figure 8.44 Scanning electron micrograph of a *Pinus* pollen grain with inflated bladder-like wings.
1. Pollen body
2. Wings

Figure 8.45 A young sporophyte (seedling) of a pine, *Pinus*.
1. Seedling needles
2. Seed coat

Representative Herbarium Specimens of Conifers

Figure 8.46 Herbarium specimen of *Phyllocladus alpinus*. Found in New Zealand and Australia, *Phyllocladus* is a primitive conifer. Though *Phyllocladus* superficially resembles some angiosperms, it is actually a cone-bearing species.

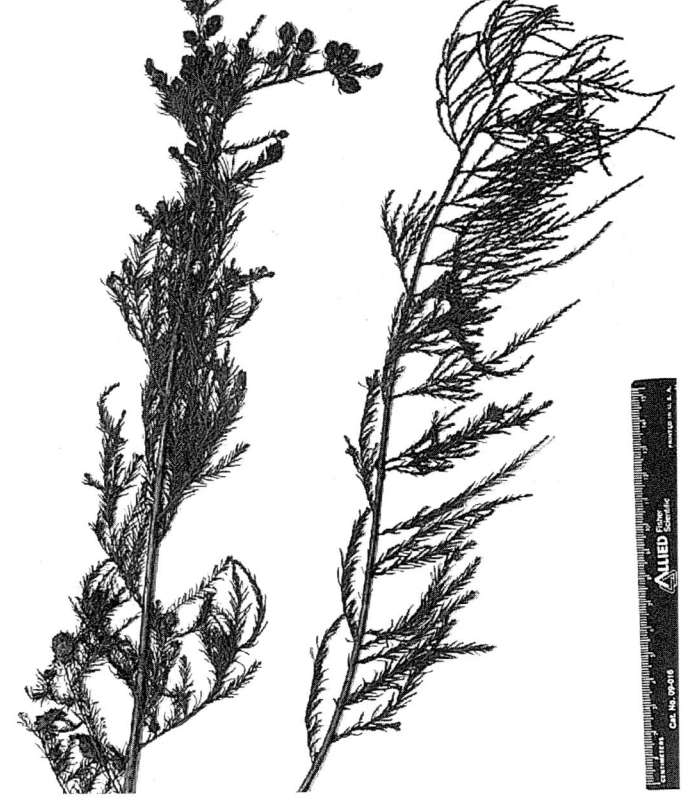

Figure 8.47 Herbarium specimen of a bald cypress, *Taxodium distichum*. The bald cypress is a deciduous member of the redwood family, Taxodiaceae. Abundant during the Tertiary Period, it is now distributed in the southern United States and Mexico.

Representative Herbarium Specimens of Conifers

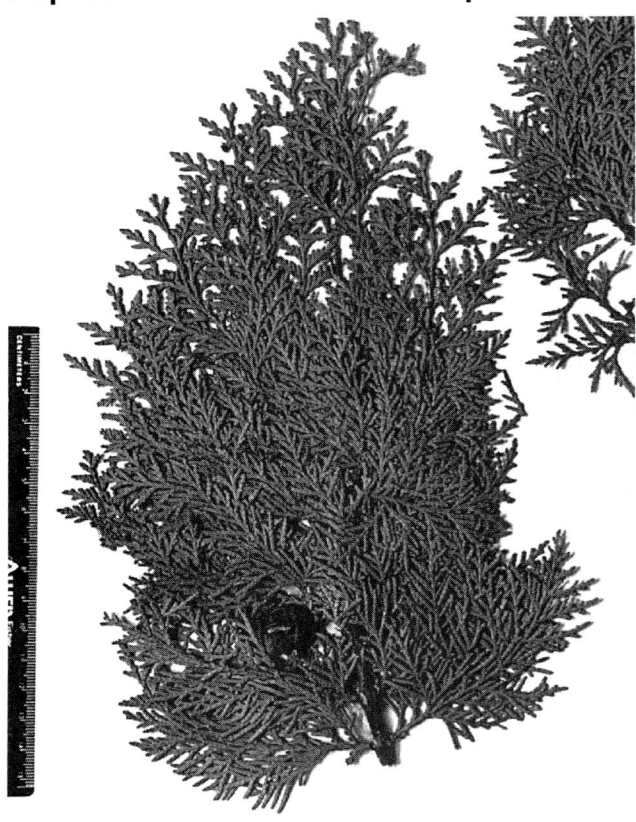

Figure 8.48 Herbarium specimen of *Thuja orientalis*. A cultivated tree of Old World origin, this tree provides seeds for many species of birds. *Thuja orientalis* is a member of family Cupressaceae.

Figure 8.49 Herbarium specimen of a juniper, *Juniperus*. Within the family Cupressaceae, this tree along with the pinyon pine, comprises the dominant vegetation type within the pinyon-juniper forest in semiarid environments of the western United States. Juniper berries are used to flavor gin.

Figure 8.50 Herbarium specimen of the pond cypress, *Taxodium ascendens*. Within the family Taxodiaceae, this species is a swamp tree that can tolerate brackish water and grow to 40 feet.

Representative Herbarium Specimens of Conifers

Figure 8.51 Herbarium specimen of the giant sequoia (or Sierra redwood), *Sequoiadendron giganteum*. The giant sequoia is within the family Taxodiaceae. Once widespread throughout temperate climates of North America, the giant sequoia occurs now in isolated groves in the Sierra Nevada mountain range of Northern California. Many trees of the Sequoia National Park are over two thousand years old.

Figure 8.52 Herbarium specimen of *Pinus flexilis*. This species is a high altitude tree within the family Pinaceae.

Figure 8.53 Herbarium specimen of the Douglas fir, *Pseudotsuga menziesii*. The Douglas fir is a valuable lumber tree within the family Pinaceae.

Representative Herbarium Specimens of Conifers

Figure 8.54 Herbarium specimen of a fir tree, *Abies lasiocarpa*. Within the family Pinaceae, fir species often inhabit high elevations near the timberline.

Figure 8.55 Herbarium specimen of a larch, *Larix dahurica*, family Pinaceae. This larch occurs in high latitudes near the Arctic Circle. The leaves of the larch are deciduous and spirally arranged.

DIVISION GNETOPHYTA (gnetophytes)

Figure 8.56 *Ephedra* is one of three genera of shrubs within the division Gnetophyta. Although found throughout most arid or semiarid regions of the world, *Ephedra* is the only one of the three genera of gnetophytes found in the United States. It is a highly branched shrub with very small leaves.

Figure 8.57 Herbarium specimen of Mormon tea, *Ephedra fasciculata*. Species of *Ephedra* in the Old World are used in the treatment of colds and asthma.

DIVISION GNETOPHYTA (gnetophytes)

Figure 8.58 Mormon tea, *Ephedra*, is a small shrub within the family Ephedraceae. Its common name comes from its use by pioneers in the American West to make a hot beverage.

Figure 8.59 A stem of *Ephedra* with several microsporangiate cones attached.
1. Stem
2. Cone

Figure 8.60 Stems and scale-like leaves of *Ephedra*. (X4)
1. Stems 2. Leaves

Figure 8.61 A transverse section through a stem of *Ephedra*. Note that unlike most other gymnosperms, *Ephedra* has vessel elements in the xylem similar to those found in angiosperms. (X10)
1. Epidermis 2. Vessel elements 3. Pith

DIVISION GNETOPHYTA (gnetophytes)

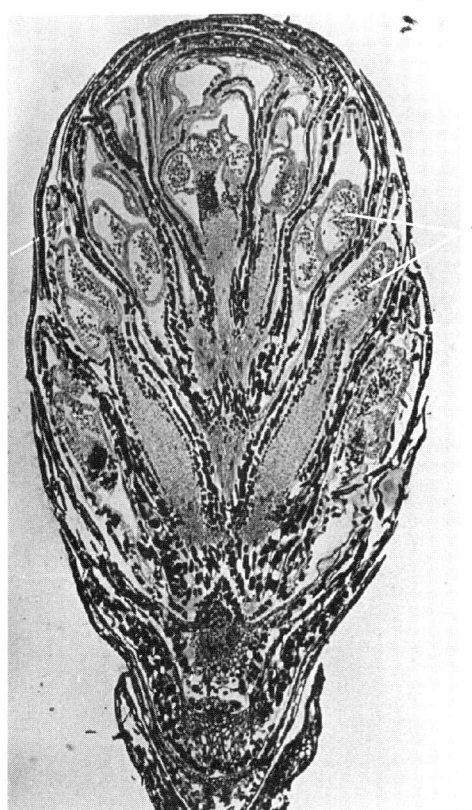

Figure 8.62 A longitudinal section through a microsporangiate cone of *Ephedra*. (X20)

1. Microsporangia

Figure 8.63 A stem of *Ephedra* with four attached megasporangiate cones.

1. Cones
2. Stem

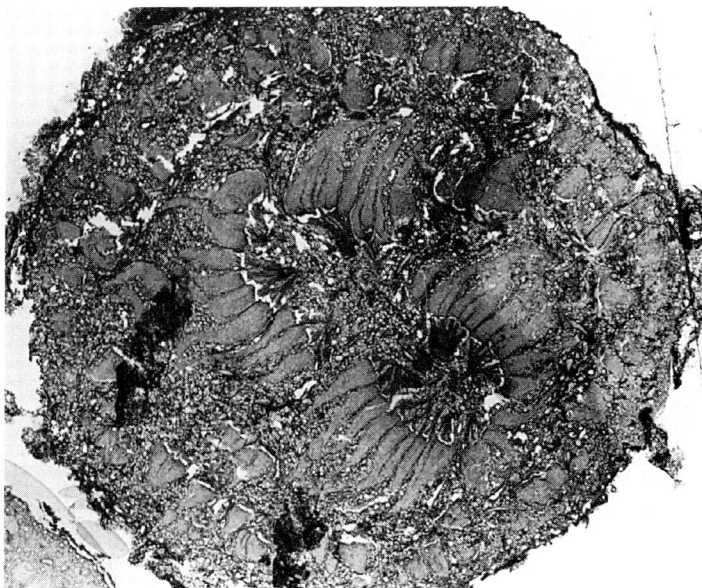

Figure 8.65 A transverse section through a young stem of *Welwitschia*, a gnetophyte that grows in the coastal desert of southwestern Africa. Most of this unusual plant is buried in sandy soil. The exposed portion consists of a woody disk that produces two strap-shaped leaves. Cone-bearing branches arise from meristematic tissue on the margin of the disk.

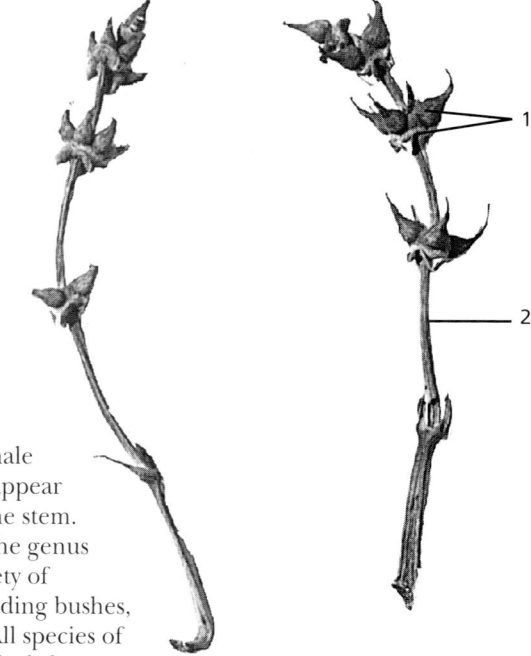

Figure 8.64 Female cones of *Gnetum* appear in whorls along the stem. Included within the genus *Gnetum* are a variety of gnetophytes including bushes, trees, and vines. All species of *Gnetum* have relatively large, leathery leaves.

1. Cones
2. Stem

Kingdom Plantae: Division Anthophyta — Angiosperms (Enclosed Seed Plants — Flowering Plants)

Angiosperms are plants producing true flowers and fruits. They range in size from the minute duckweed (1 mm in total size) to Eucalyptus trees (100 m tall). A few are saprophytic or parasitic, but most are free-living. Angiosperms, because of their rapid life cycle, proliferate in all major habitats, including terrestrial, marsh, fresh water, and marine.

Anthophyta is the most recent plant division to have evolved. Flowering plants have only existed since late Jurassic or early Cretaceous times, but currently dominate the Earth's flora in terms of variety and biomass. Most flowering plants produce a crop of mature seeds in a single year, and many species are able to pass from a seed to a mature seed-producing plant in a matter of a few weeks. This rapid cycle has allowed these plants to be efficient in occupying new territory. (Many gymnosperms require two years to complete their life cycle.)

Angiosperms are divided into the class *Monocotyledonae* (monocots), including the grasses, palms, lilies, and orchids; and class *Dicotyledonae* (dicots), including most other familiar flowers, shrubs, and trees (except conifers). An estimated 65,000 species of monocots and 170,000 species of dicots have been identified, although many botanists believe the actual number of species will turn out to be much higher.

Like most plants, angiosperms consist of a vegetative portion and a germinative portion. The vegetative portion includes the *roots*, *stems*, and *leaves*. It is concerned primarily with the manufacture of food and plant growth. The germinative portion includes the *flowers* and is concerned primarily with seed production.

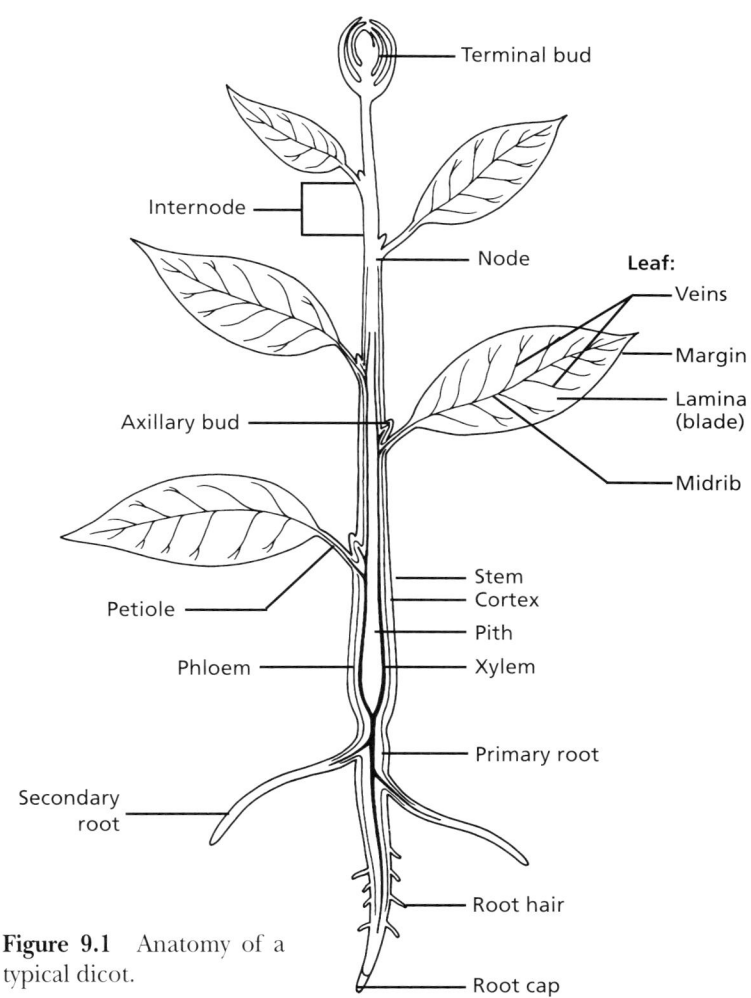

Figure 9.1 Anatomy of a typical dicot.

TABLE 9.1
Angiosperms within the Kingdom Plantae

Class and Representative Kinds	Characteristics
Class Monocotyledonae (monocots)	One cotyledon; leaf veins usually parallel, vascular bundles within stem are scattered; fibrous root system; floral parts usually in multiples of three
Class Dicotyledonae (dicots)	Two cotyledons; leaf veins usually netlike; vascular bundles within stem arranged in ring; taproot usually present; floral parts usually in multiples of four or five

DIVISION ANTHOPHYTA (angiosperms: monocots and dicots)

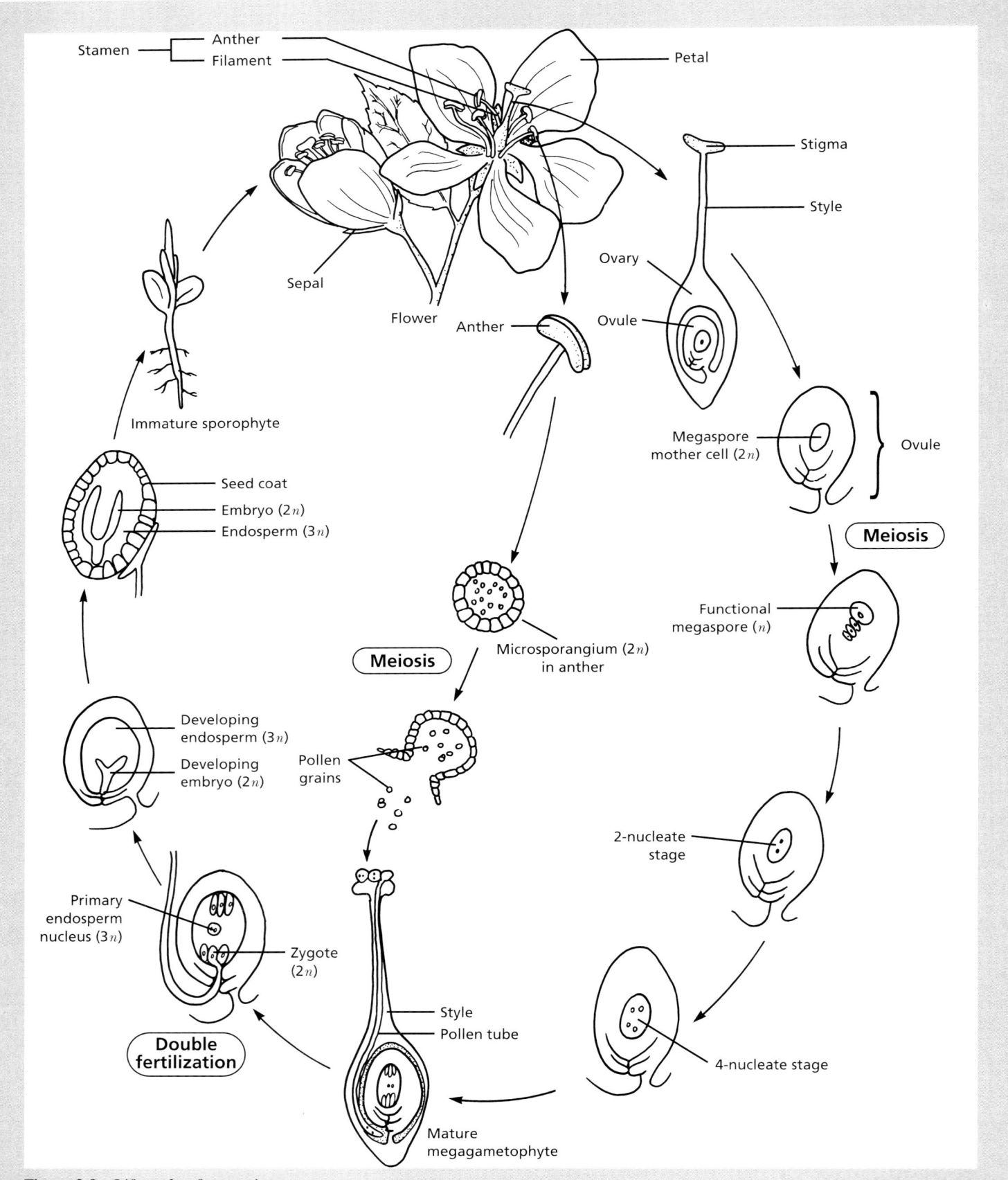

Figure 9.2 Life cycle of an angiosperm.

CHAPTER 9 Kingdom Plantae: Division Anthophyta — Angiosperms (Enclosed Seed Plants — Flowering Plants)

DIVISION ANTHOPHYTA (angiosperms: monocots and dicots)

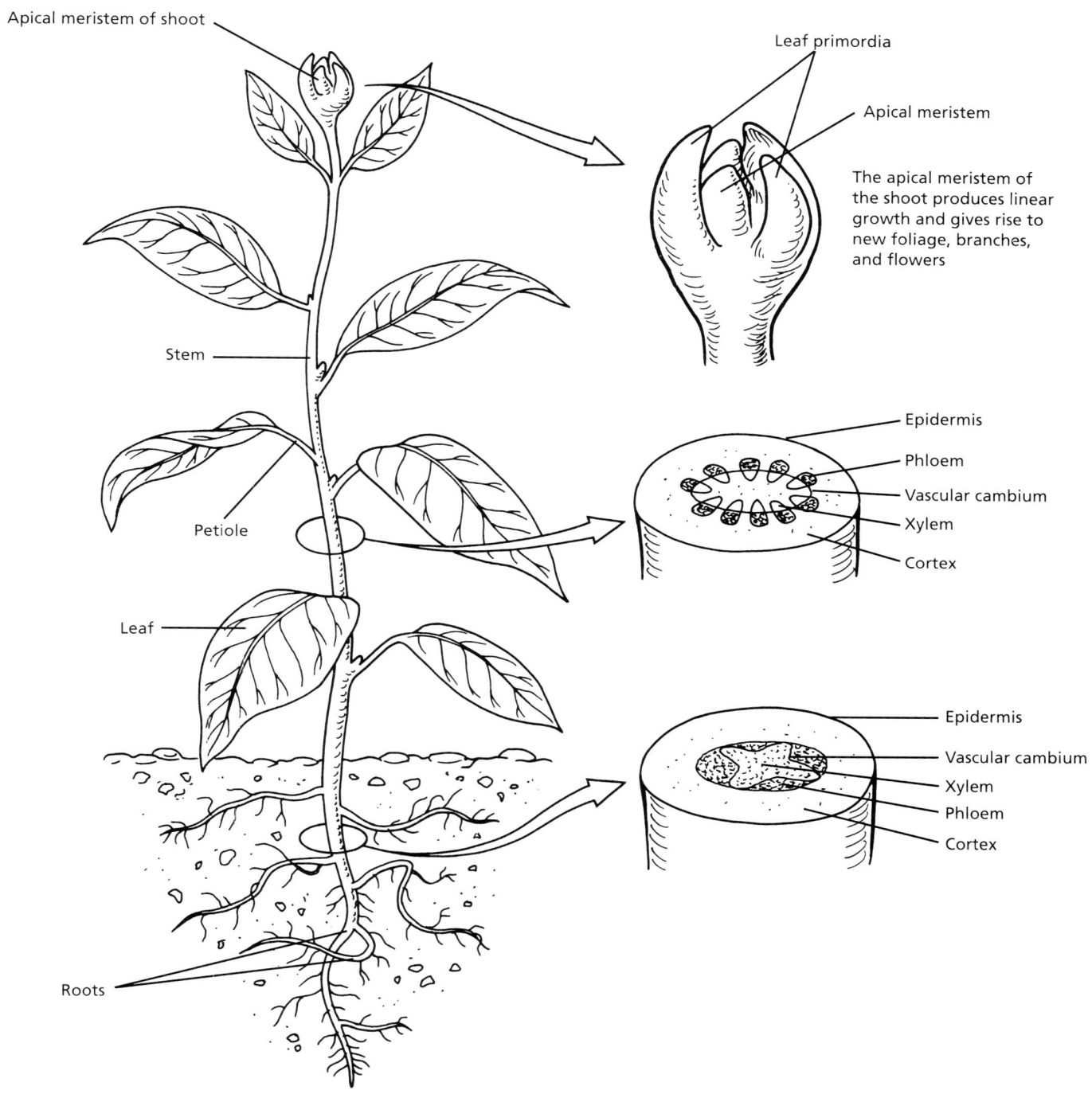

Figure 9.3 Diagram illustrating the principal organs and tissues of a typical dicot.

DIVISION ANTHOPHYTA (angiosperms: monocots and dicots)

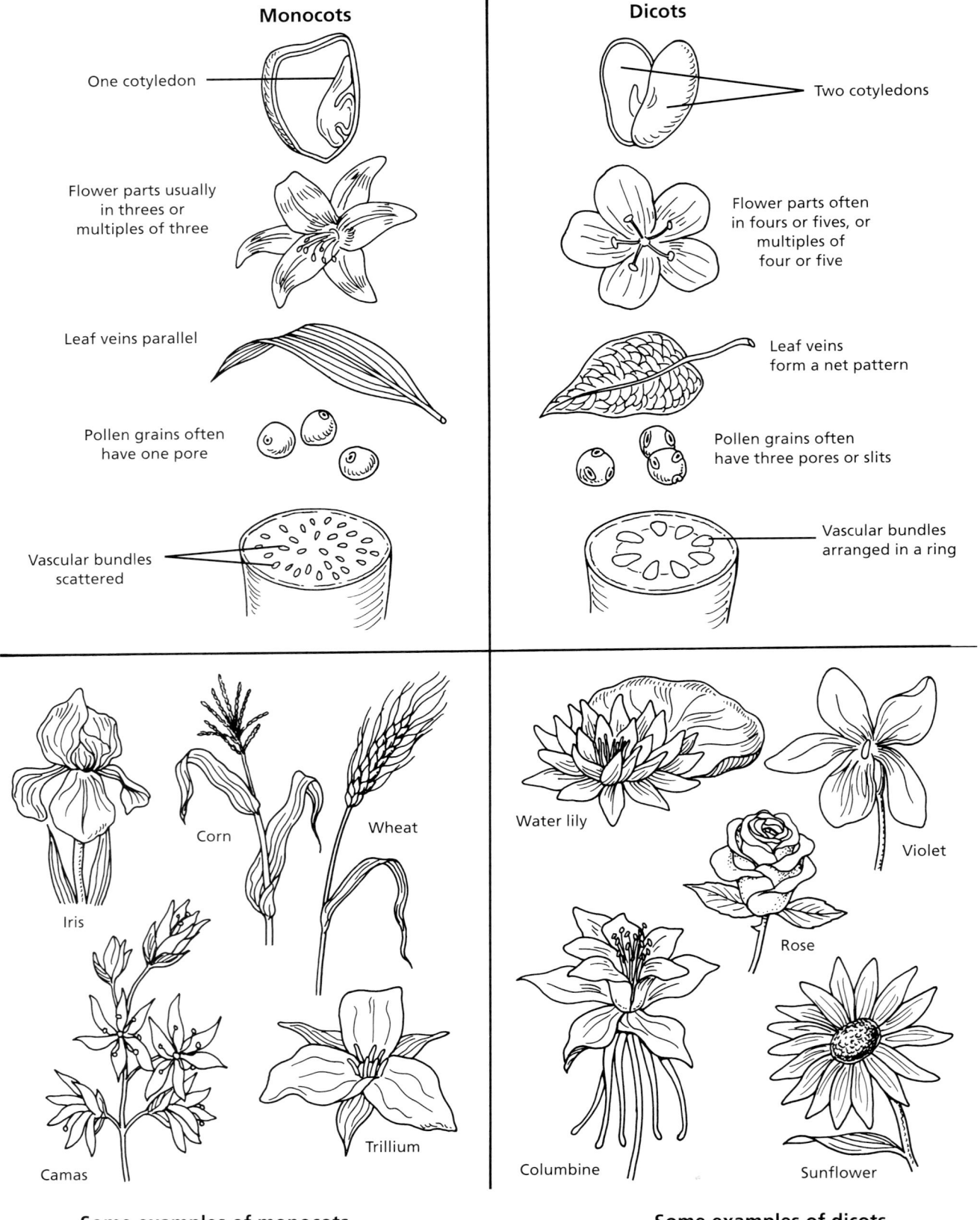

Figure 9.4 Comparison and examples of monocots and dicots.

Roots of Angiosperms

Plants are only as healthy as are their root systems. The root system of an angiosperm is the descending portion of the plant. The roots can make up more than half the plant body. Roots anchor the plant. Water and nutrients are absorbed, stored, and conducted by the roots. The root system of a plant is influenced by soil type and mineral content and the amount and timing of moisture.

Monocots, such as grasses, have *fibrous*, or *diffuse*, root systems. Dicots, such as shrubs and most woody plants, have *taproot systems*. Specialized supporting root systems include *aerial roots* and *prop roots*. Taproots, such as found in carrots and turnips, are capable of storing large amounts of food.

The active root system of most angiosperms consists of four main regions (see Figure 9.11):

Root cap — A root cap region is a cluster of cells at the tip of the root. It protects the root during growth.

Meristematic region — The meristematic region of the root is just behind the root cap, where new cells are added to the growing root by active cell division.

Elongation region — The elongation of the root is where newly added cells increase in size.

Maturation region — The maturation region of the root is where cells differentiate into epidermal and cortex layers and xylem and phloem tissues (vascular tissues). The *root hairs* formed in the epidermis of this region greatly increase the surface area for absorption. The cortex stores reserve food.

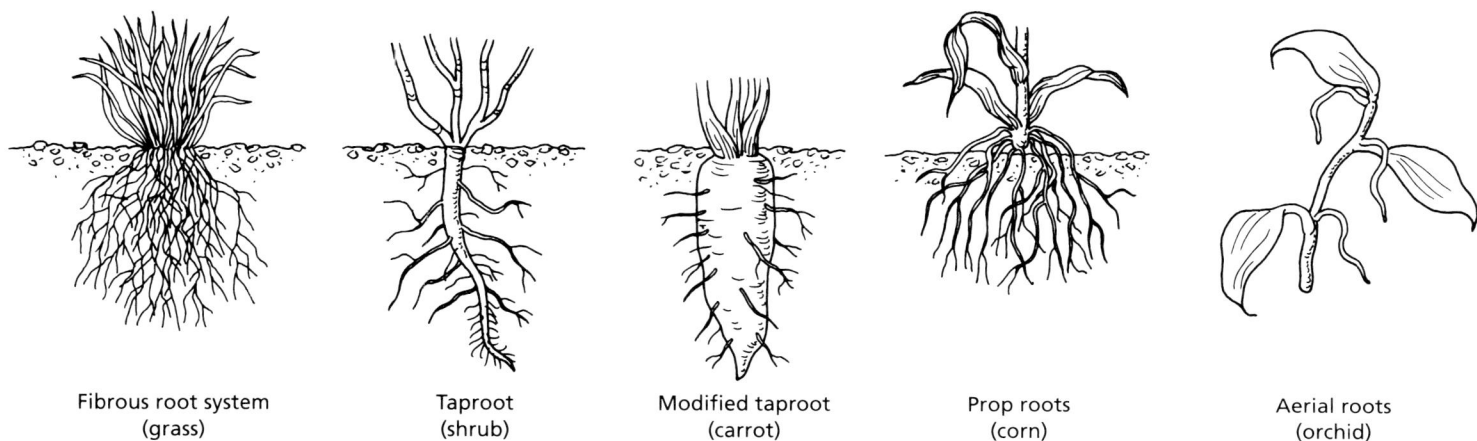

Figure 9.5 Root systems of angiosperms.

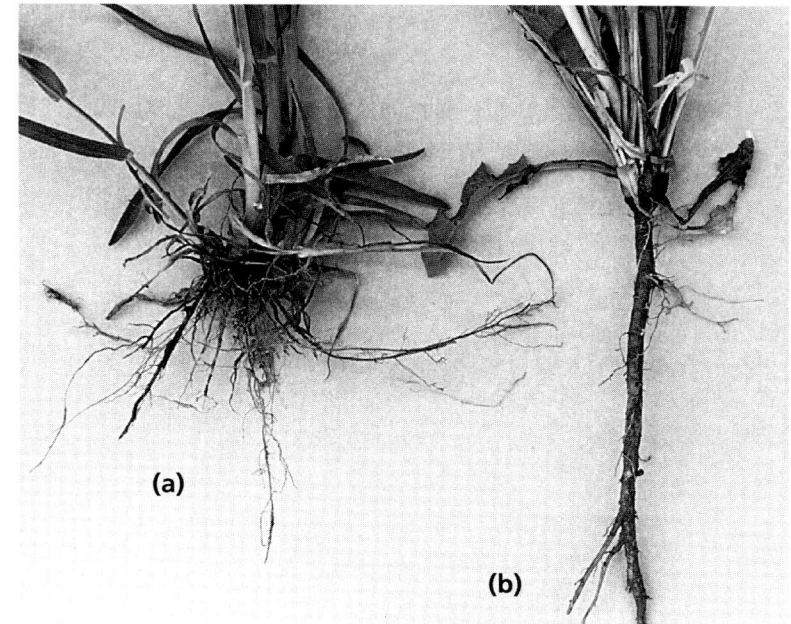

Figure 9.6 Typical root systems in (a) monocots and (b) dicots. Monocot roots are fibrous, with many roots of more or less equal size. Dicots usually have a taproot system, consisting of a long central root with many smaller, secondary roots branching from it.

Roots of Angiosperms

Figure 9.7 Prop roots of the avocado plant, *Persea*.

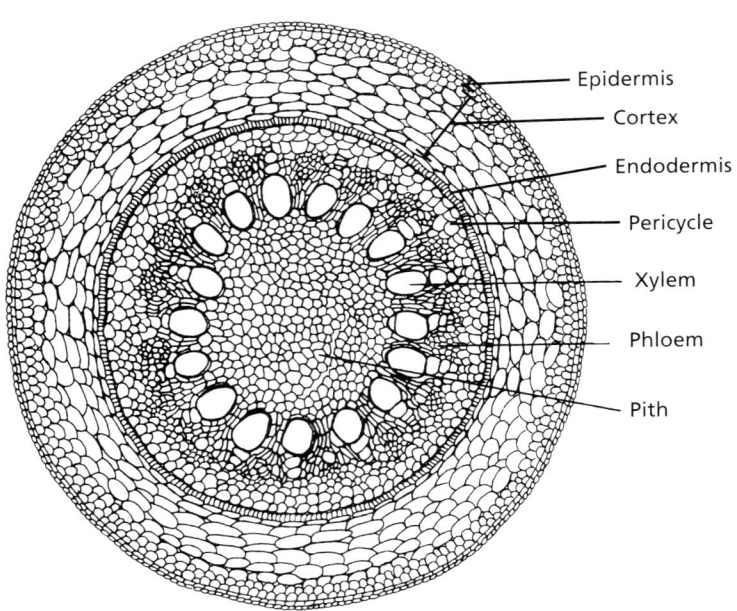

Figure 9.8 A transverse section through the monocot root of a greenbrier, *Smilax*.

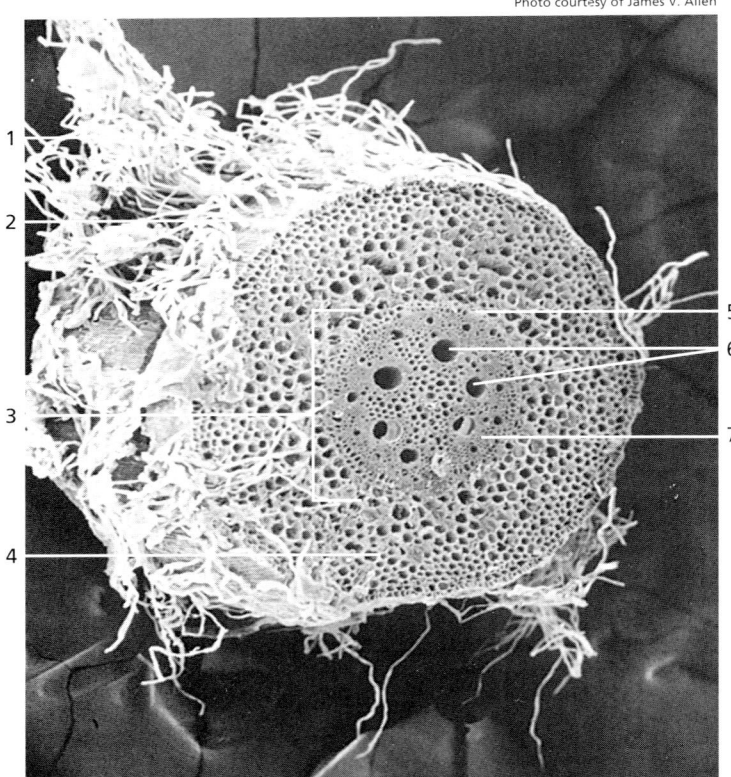

Figure 9.9 A scanning electron micrograph of a young root from a corn plant, *Zea mays*, in transverse view. (X30)

1. Root hairs 3. Stele 5. Endodermis 7. Phloem
2. Epidermis 4. Cortex 6. Xylem

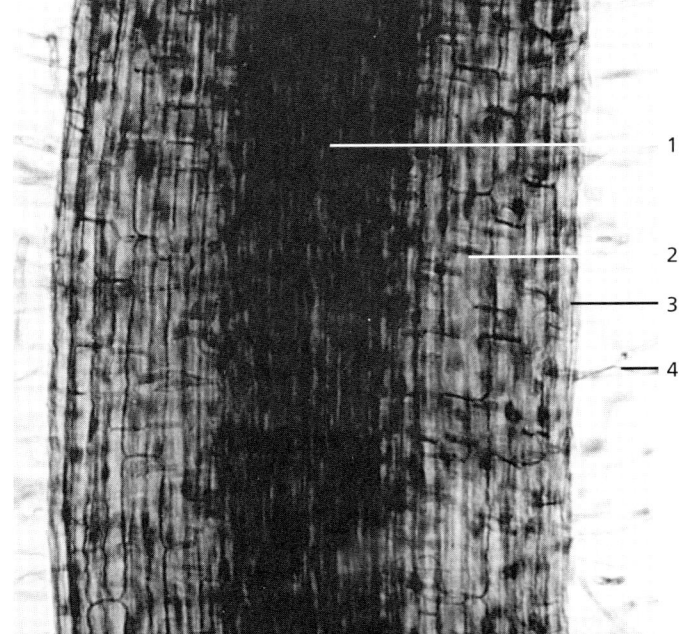

Figure 9.10 Photomicrograph of a young root of wheat, *Triticum*, showing root hairs. (X40)

1. Stele 2. Cortex 3. Epidermis 4. Root hair

CHAPTER 9 Kingdom Plantae: Division Anthophyta — Angiosperms (Enclosed Seed Plants — Flowering Plants)

Roots of Angiosperms

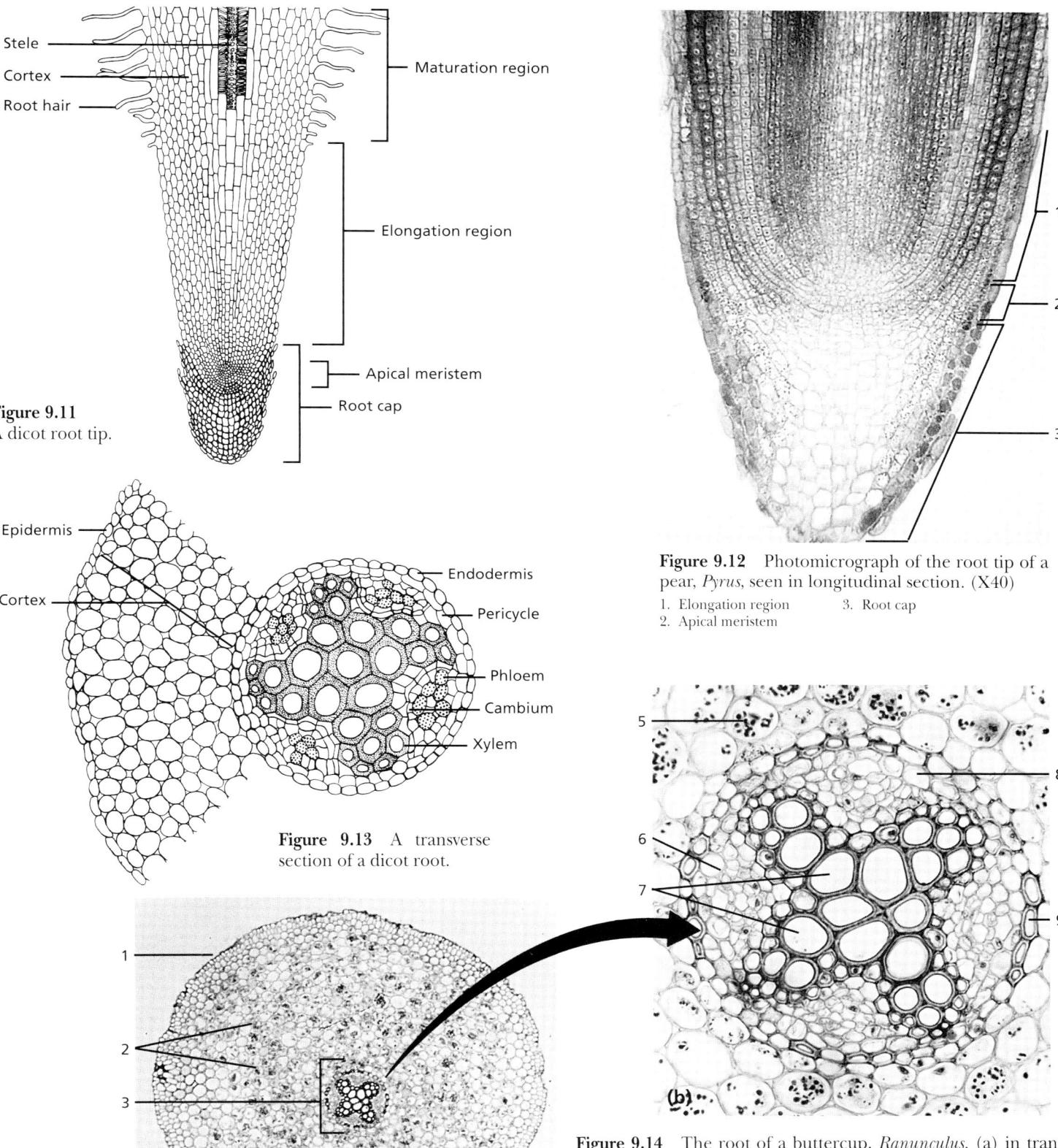

Figure 9.11 A dicot root tip.

Figure 9.12 Photomicrograph of the root tip of a pear, *Pyrus*, seen in longitudinal section. (X40)
1. Elongation region
2. Apical meristem
3. Root cap

Figure 9.13 A transverse section of a dicot root.

Figure 9.14 The root of a buttercup, *Ranunculus*, (a) in transverse section showing all of the root tissues (X40), and (b) in transverse section showing the stele. (X430)

1. Epidermis
2. Parenchyma cells of cortex
3. Stele
4. Cortex
5. Starch grains within parenchyma cells
6. Phloem
7. Xylem
8. Pericycle
9. Endodermis

Stems of Angiosperms

The *stem* of an angiosperm is the ascending portion of the plant. It helps produce and support leaves and flowers; transport and store water and nutrients; and provide growth through cell division. *Branches* and *twigs,* smaller extensions and/or branches of the stem of certain plants, directly support leaves and flowers.

Herbaceous stems are soft and succulent. The herbaceous stems of monocots have scattered vascular bundles, while herbaceous stems of dicots have vascular bundles arranged in a ring. Some species of monocots have stems reinforced with secondary fibrous strands and appear woody.

Woody stems are hardened and often larger in diameter (they usually increase in diameter by secondary growth). Woody stems of dicots have three parts:

1. *Bark,* which contains *periderm* and *phloem;*
2. *Wood,* which contains the annual rings of *xylem;*
3. *Pith,* composed of loosely packed parenchyma cells at the center of the vascular tissue. The pith will often be lost as the stem increases in size.

Linear growth of woody dicot stems occurs at *terminal buds,* where mitosis occurs at the *apical meristem.* Buds also contain developing *leaves* and, in certain locations, developing *flowers. Nodes* of branches or twigs are the points of leaf attachments, and *internodes* are the spaces between the nodes. *Lenticels* are pores in the bark which facilitate gas exchange.

People use stems for products including paper, building materials, furniture, and fuel. In addition, food is obtained from the stems of potatoes, onions, celery, cabbage, and other plants.

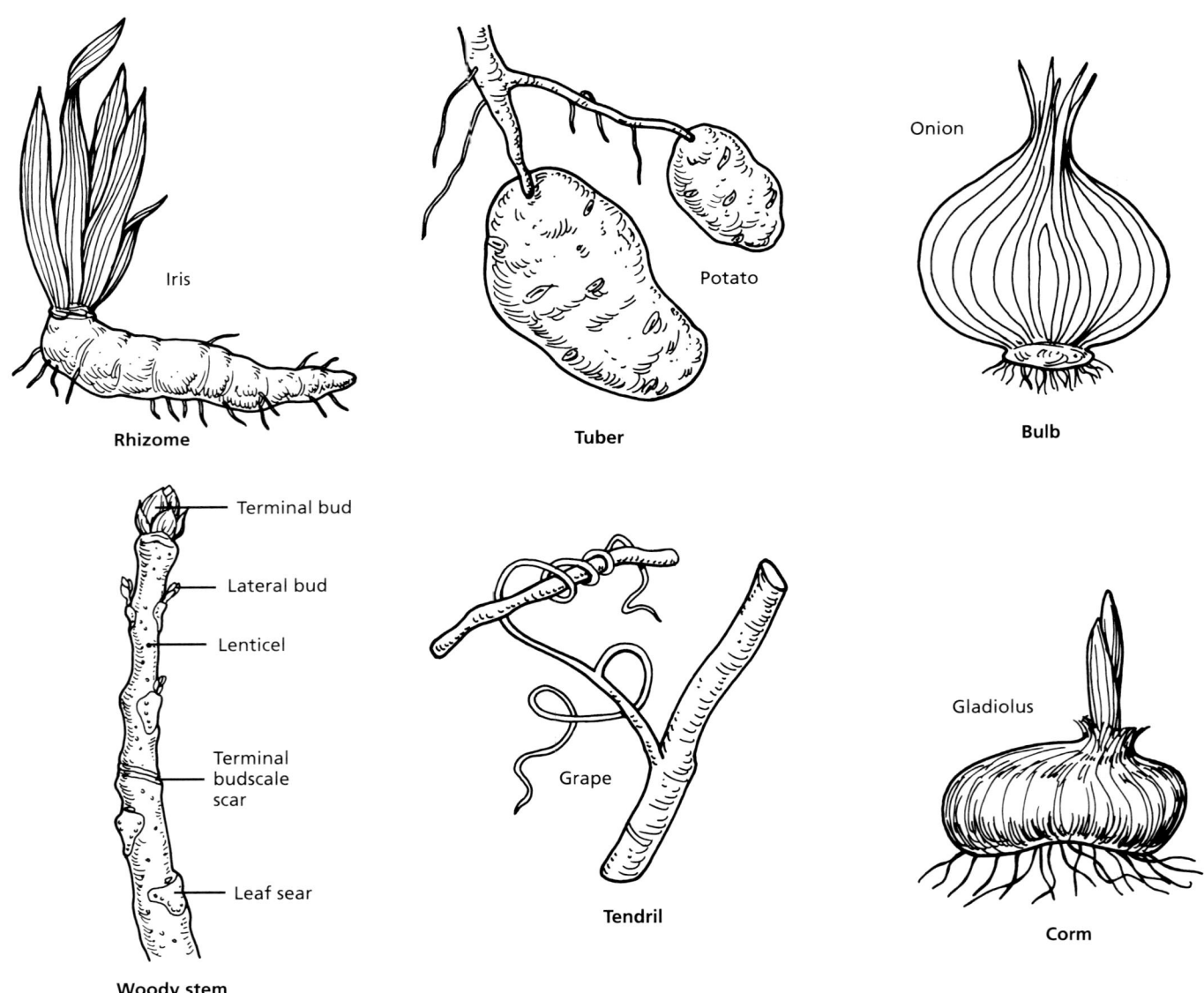

Figure 9.15 Examples of the variety and specialization of angiosperm stems.

CHAPTER 9   Kingdom Plantae: Division Anthophyta — Angiosperms (Enclosed Seed Plants — Flowering Plants)

Stems of Angiosperms

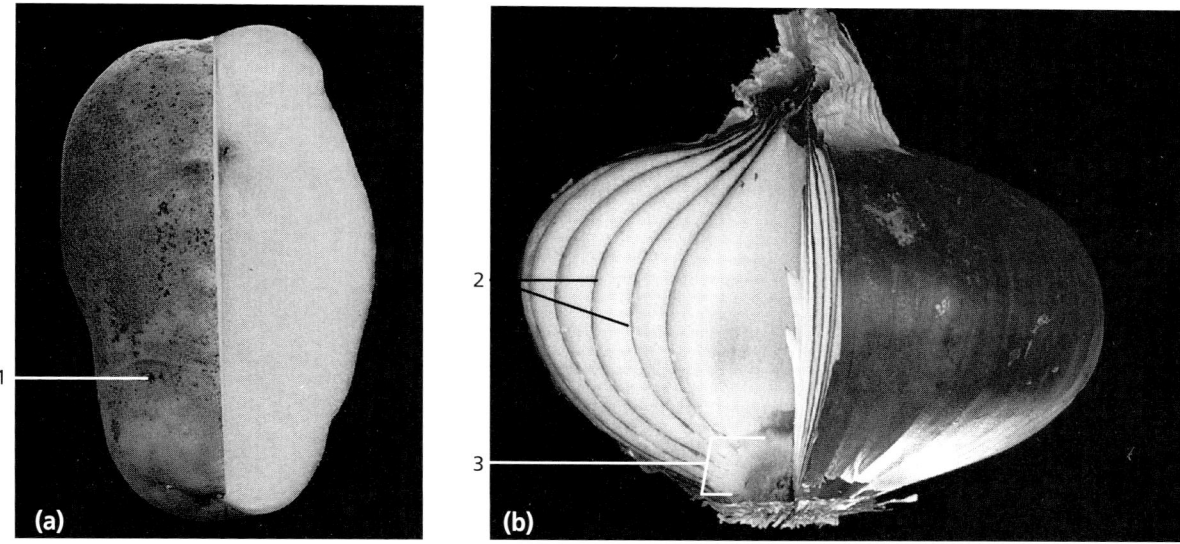

Figure 9.16 Specialized underground stems, (a) potato, and (b) onion.
1. Eye (bud) in the axil of a minute scale leaf
2. Fleshy leaf bases
3. Short stem

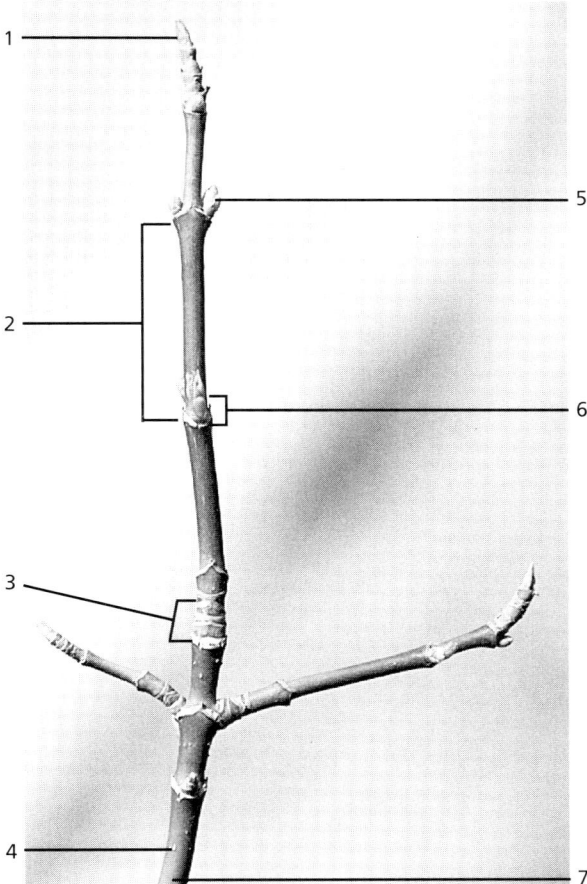

Figure 9.17 A woody branch of a dicot, seen in early spring just as the buds are beginning to swell. Branches and twigs are small extensions of the stems of certain angiosperms, directly supporting leaves and flowers.

1. Terminal bud
2. Internode
3. Terminal budscale scars
4. Lenticel
5. Lateral bud
6. Node area
7. Stem

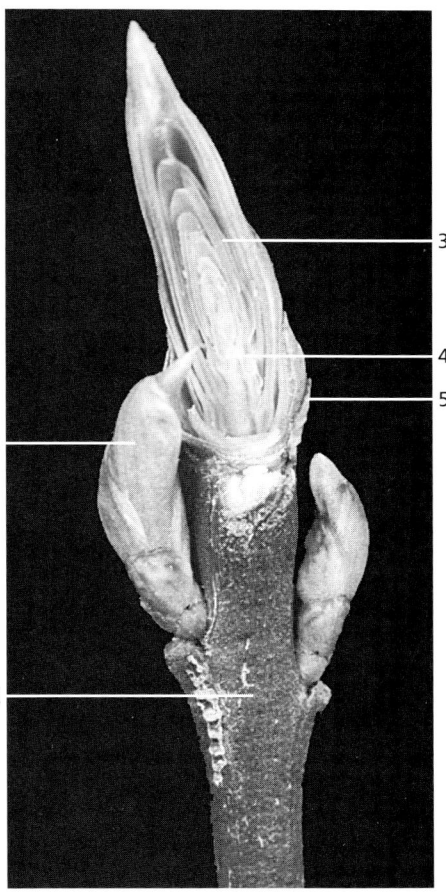

Figure 9.18 A terminal bud of a woody stem that has been longitudinally sectioned to show a developing leaf.

1. Lateral bud
2. Stem
3. Leaf primordium
4. Apical meristem
5. Bud scale

Stems of Angiosperms

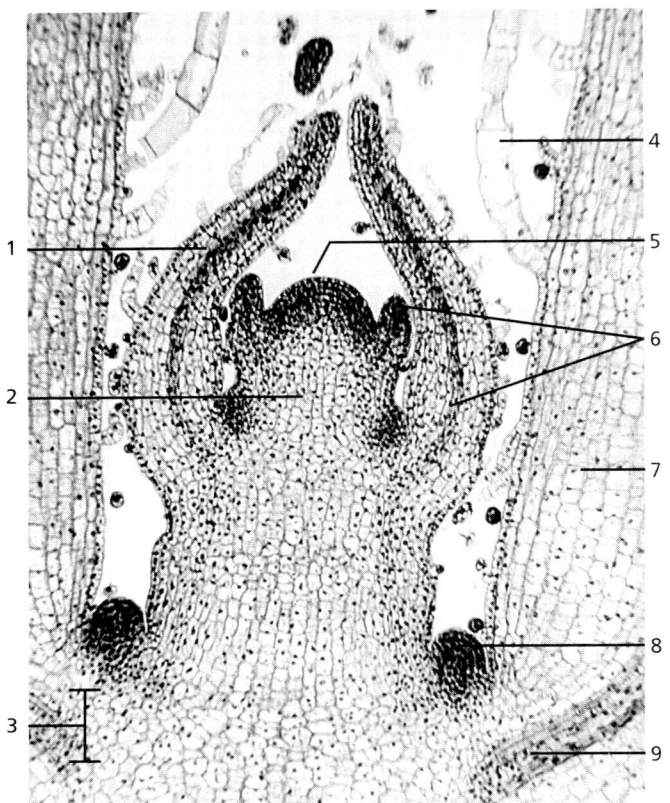

Figure 9.19 Longitudinal section of the stem tip of the common houseplant *Coleus*. (X40)

1. Procambium
2. Ground meristem
3. Leaf gap
4. Trichome
5. Apical meristem
6. Developing leaf primordia
7. Leaf primordium
8. Axillary bud
9. Developing vascular tissue

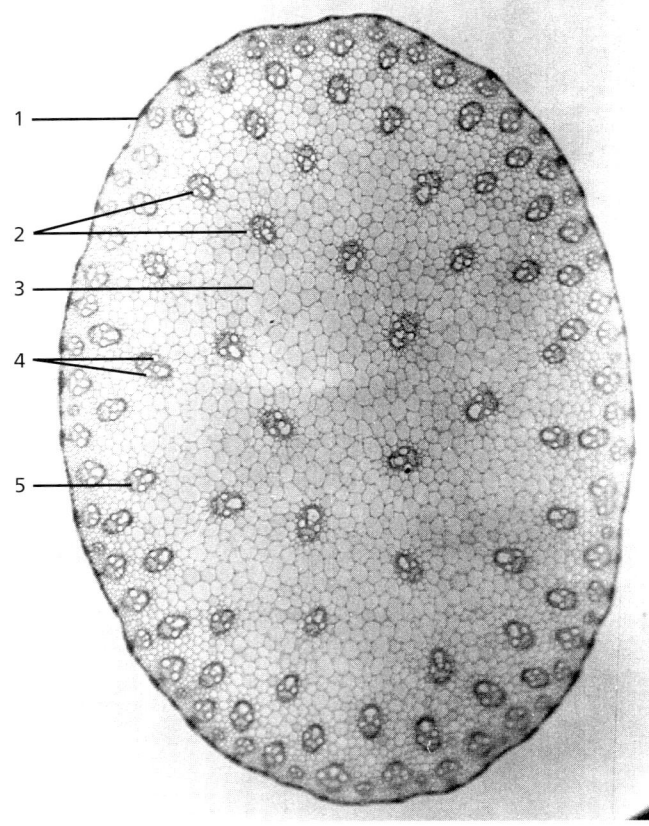

Figure 9.20 A transverse section from the stem of a monocot, *Zea mays* (corn). (X40)

1. Epidermis
2. Vascular bundles
3. Parenchyma cells (ground tissue)
4. Xylem
5. Phloem

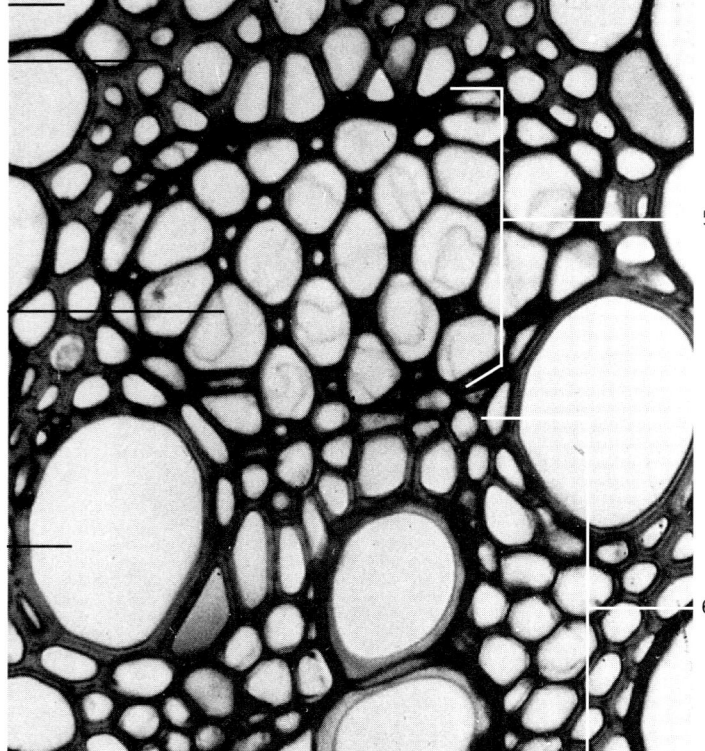

Figure 9.21 A transverse section through the stem of a corn plant, *Zea mays*, showing the vascular tissue. (X200)

1. Ground tissue
2. Sclerenchyma
3. Sieve tube cell in phloem
4. Vessel element in xylem
5. Phloem
6. Xylem

CHAPTER 9   *Kingdom Plantae: Division Anthophyta — Angiosperms (Enclosed Seed Plants — Flowering Plants)*   131

Stems of Angiosperms

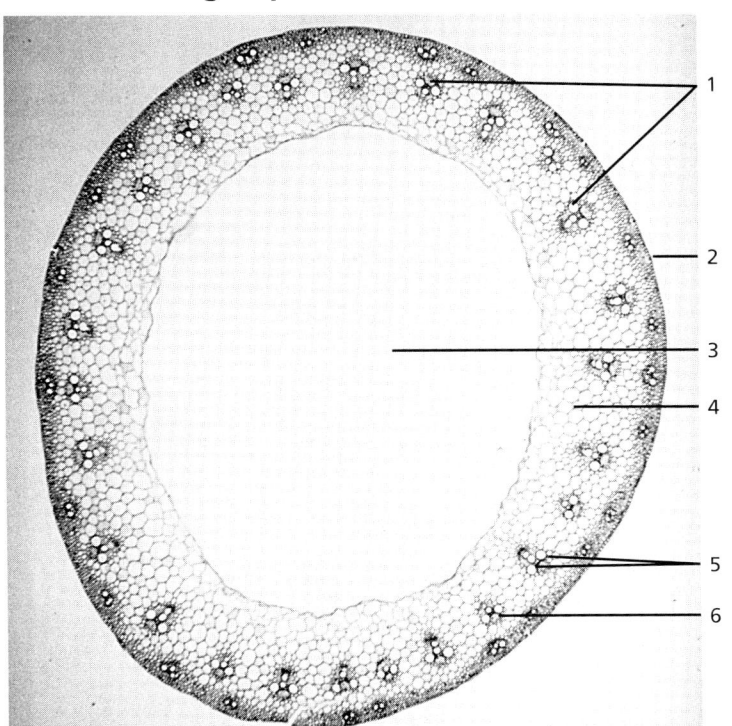

Figure 9.22 A transverse section through the stem of a monocot, *Triticum*, wheat. (X40)

1. Vascular bundles
2. Epidermis
3. Ground tissue cavity
4. Parenchyma cells of ground tissue
5. Xylem
6. Phloem

Figure 9.23 A transverse section through the secondary xylem (wood) of the stem of an oak, *Quercus*. (X100)

1. Summer wood
2. Spring wood
3. Vessel element

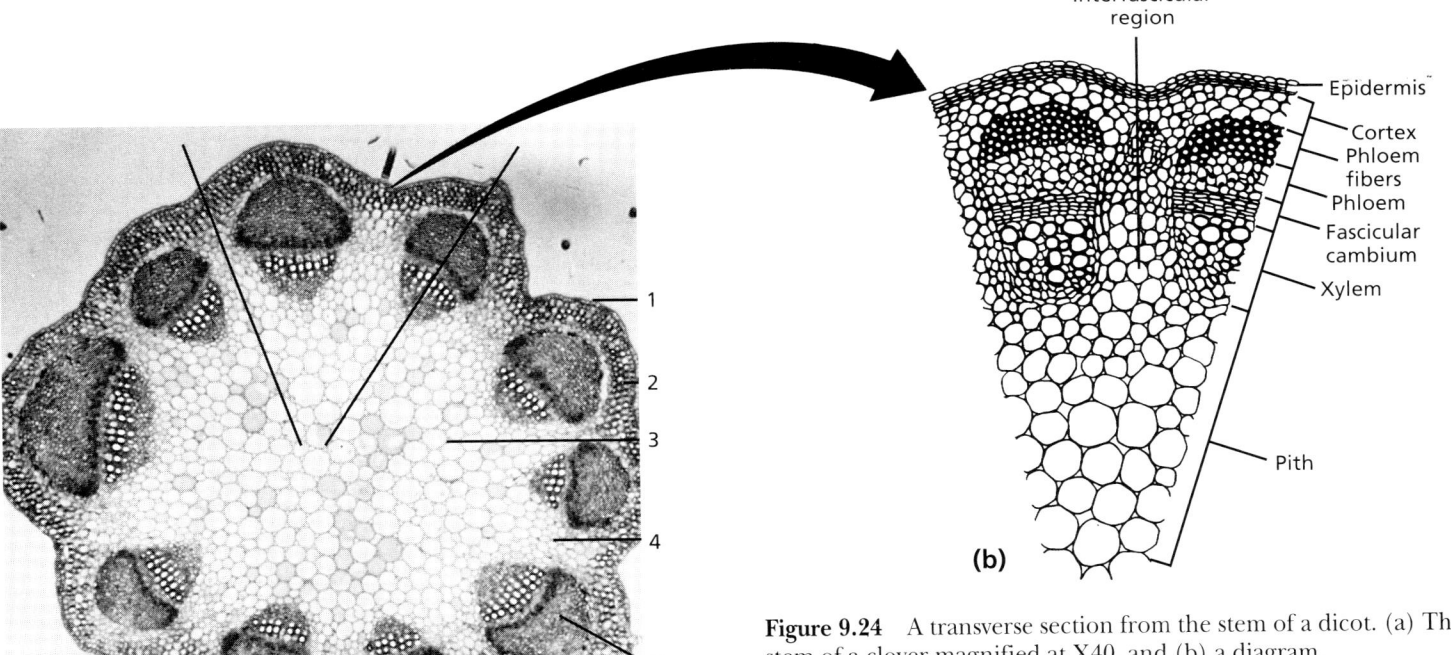

Figure 9.24 A transverse section from the stem of a dicot. (a) The stem of a clover magnified at X40, and (b) a diagram.

1. Epidermis
2. Cortex
3. Pith
4. Interfascicular region
5. Vascular bundles with caps

Stems of Angiosperms

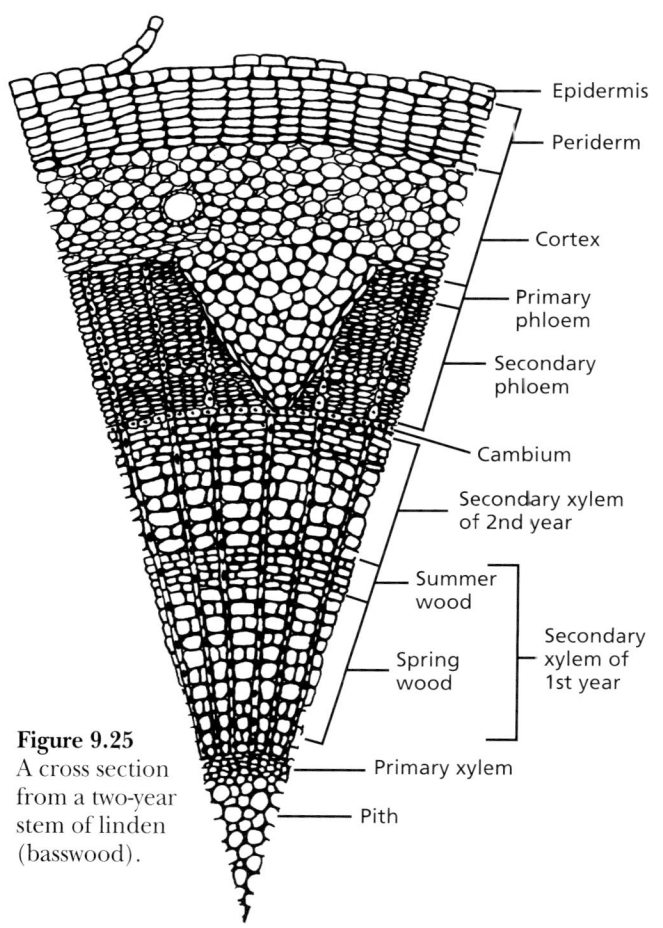

Figure 9.25
A cross section from a two-year stem of linden (basswood).

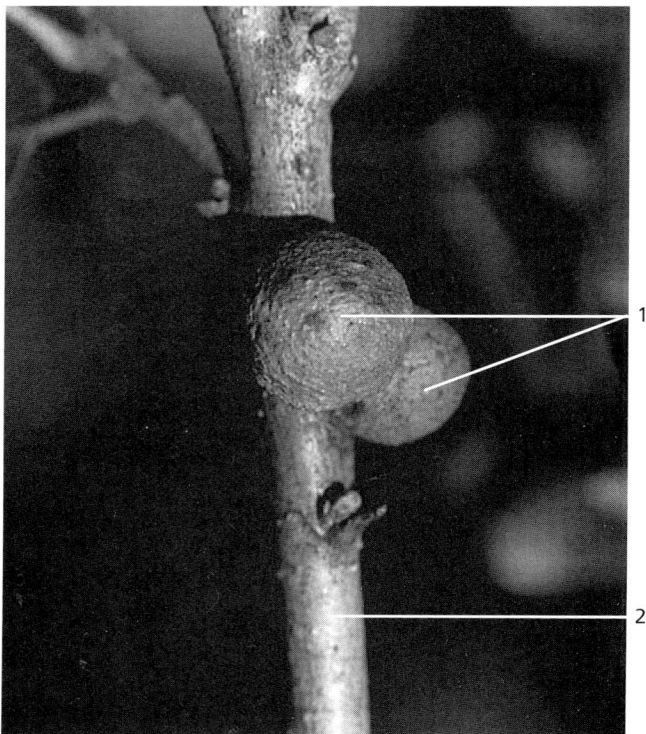

Figure 9.26 Galls on an oak, *Quercus*, stem. The feeding of a gall wasp larva causes abnormal growth and the formation of a gall. The wasp larva feeds upon the gall tissue, pupates within this enclosure, and then chews an exit to emerge.
1. Galls
2. Stem

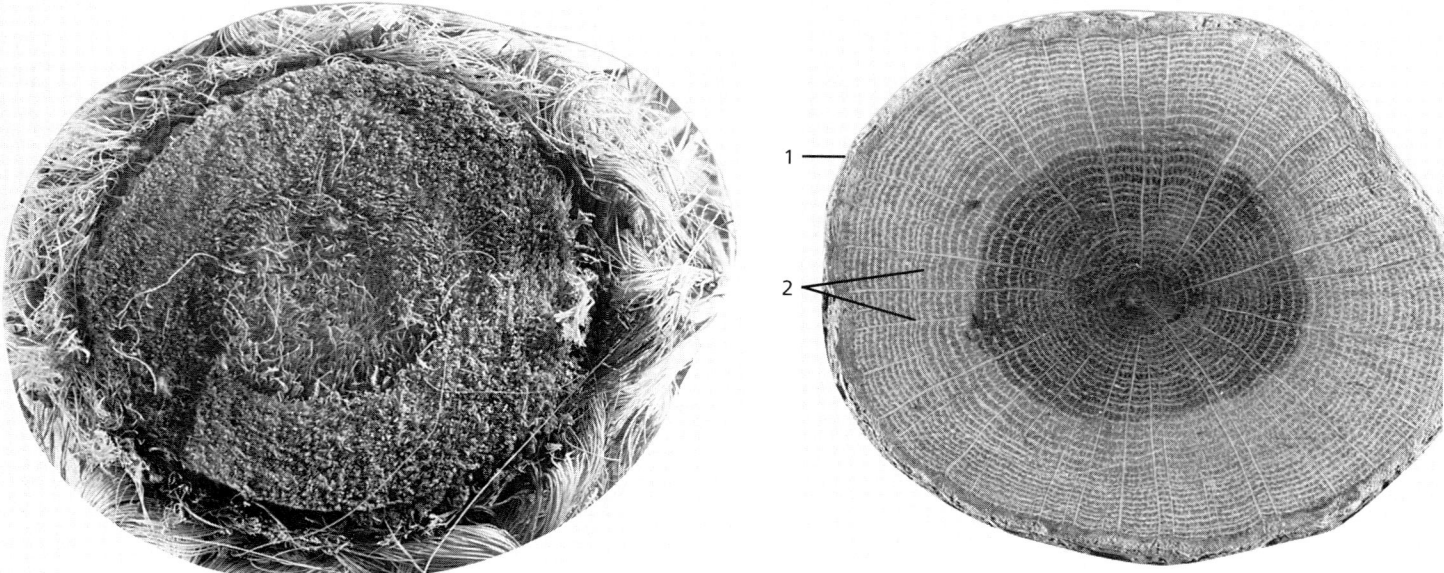

"Woody" monocot

Woody dicot

Figure 9.27 A comparison of the cross sections of stems of a "woody" monocot and a woody dicot. The stem of the "woody" monocot is rigid because of the fibrous nature of the numerous vascular bundles. The stem of the woody dicot is rigid because of the compact xylem cells at the center of the stem forming the dense, hardened wood, seen as annual rings.
1. Bark
2. Annual rings

Stems of Angiosperms

Figure 9.28 The angiosperm, *Ruscus aculeatus*, is characterized by stems (a) that resemble leaves in form and function. Note the true leaves (b) arising from the leaf-like stems.
1. Stem
2. Leaf

Leaves of Angiosperms

The leaf of an angiosperm manufactures food by *photosynthesis*, which is the production of sugar from carbon dioxide and water, in the presence of chlorophyll, with sunlight as the source of energy.

Leaves are attached to the *nodes* of a stem by *petioles*. The leaf *midrib* is vascular tissue continuous with the vascular tissue of the petiole through the *lamina* (blade) and gives rise to numerous branching *veins*. Leaves may be classified on the basis of arrangement on a petiole, the arrangement of the veins, and the appearance of the margins. A *deciduous leaf* is one that is shed during the autumn season as the petiole detaches from the stem.

The typical tissue arrangement of a leaf includes an *upper epidermis*, a *lower epidermis* and the centrally located *mesophyll*. The cells of the mesophyll contain *chloroplasts*, which are necessary for photosynthesis. Mesophyll is often divided into *palisade mesophyll* and *spongy mesophyll*. *Veins* within the mesophyll conduct material through the leaf. Atmospheric gasses containing carbon dioxide enter the leaf through *stomata*, the shape of which is regulated by *guard cells*.

Leaves comprise the foliage of plants. Leaves provide habitat and food source for many animals, including humans. They also provide protective ground cover and are the portion of the plant most responsible for oxygen replenishment into the atmosphere.

Figure 9.29 Leaf compression fossils of four angiosperms.

Kingdom Plantae: Division Anthophyta — Angiosperms (Enclosed Seed Plants — Flowering Plants)

Leaves of Angiosperms

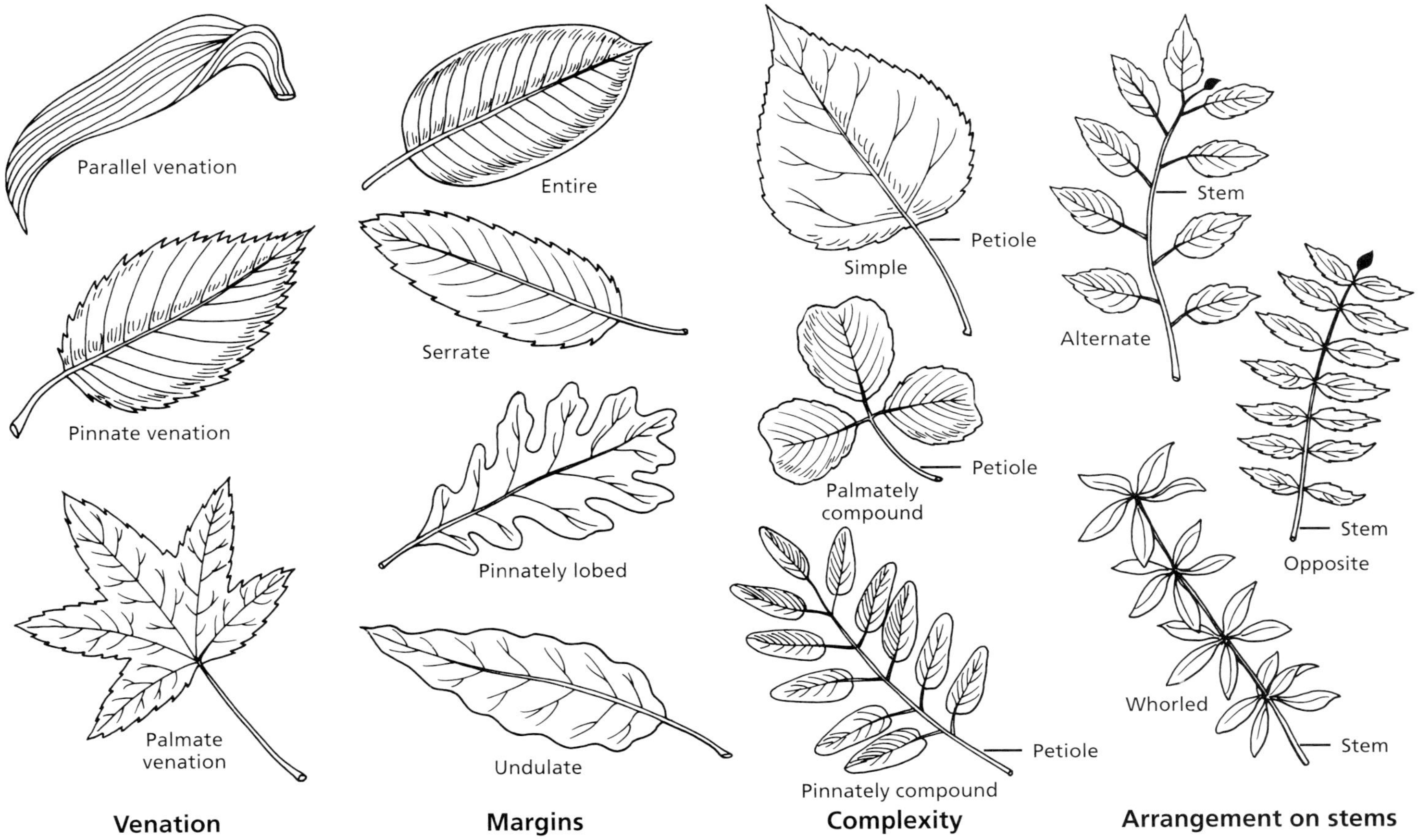

Figure 9.30 Several representative angiosperm leaf types.

Venation — Margins — Complexity — Arrangement on stems

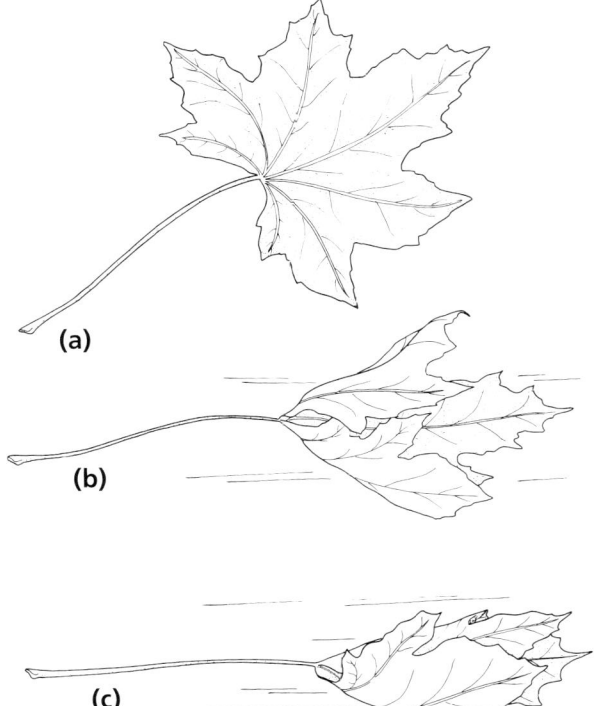

Figure 9.31 The shape of a leaf (a) is of adaptive value to withstand wind. As the speed of the wind increases (b and c), the leaf rolls into a tight cone shape, avoiding damage.

Figure 9.32 Typical angiosperm leaf showing characteristic surface features.

1. Lamina
2. Petiole
3. Serrate margin
4. Midrib
5. Veins

Leaves of Angiosperms

Figure 9.33 The undersurface of an angiosperm leaf showing the vascular tissue lacing through the lamina, or blade, of the leaf.
1. Midrib
2. Veins

Figure 9.34 The organic decomposition of a leaf is a gradual process beginning with the softer tissues of the lamina, leaving only the vascular tissues of the midrib and the veins, as seen in this photograph. With time, these will also decompose.

Figure 9.35 The Panama hat plant, *Carludovica palmata*, is a monocot with leaves that have parallel venation.

Figure 9.36 Oleander, *Nerium oleander*, is a xerophyte (adapted to arid conditions), as reflected by rather thick, waxy leaves. Commonly, oleander plants in southern California and Texas have brilliantly colored flowers. Oleander is native to Old World subtropics.

Leaves of Angiosperms

Figure 9.37 Water hyacinths, *Eichhornia*, have modified leaves that buoy the plants on the water surface. Water hyacinths are common in New World tropical fresh-water habitats, where they may become so thick that they choke out bottom-dwelling plants.

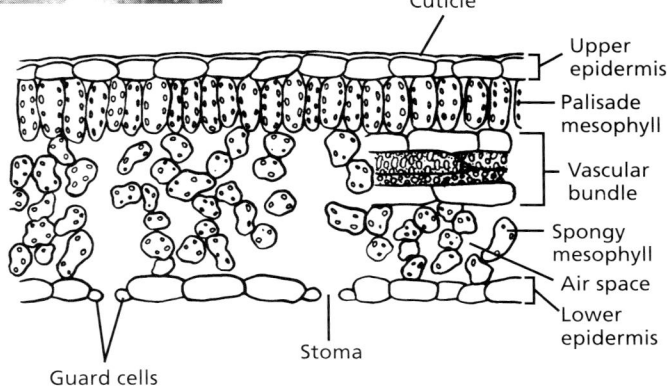

Figure 9.38 A dicot leaf in transverse section.

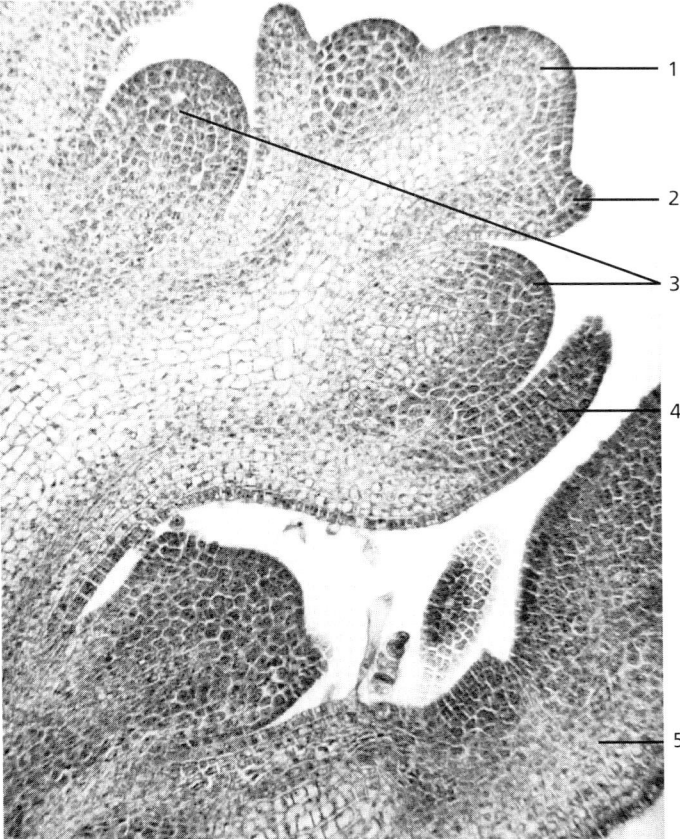

Figure 9.39 A longitudinal section of the stem tip of a garden bean, *Phaseolus*, showing the development of alternate leaves. (X40)

1. Apical meristem
2. Early leaf primordium
3. Axillary buds
4. Developing leaf primordium
5. Leaf primordium

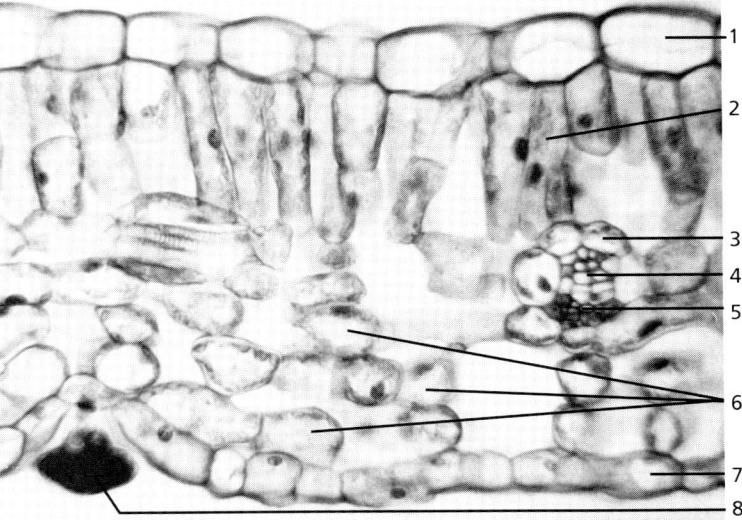

Figure 9.40 A transverse section through the leaf of the common hedge privet, *Ligustrum*. (X100)

1. Upper epidermis
2. Palisade mesophyll
3. Bundle sheath
4. Xylem
5. Phloem
6. Spongy mesophyll
7. Lower epidermis
8. Trichome (leaf hair)

Leaves of Angiosperms

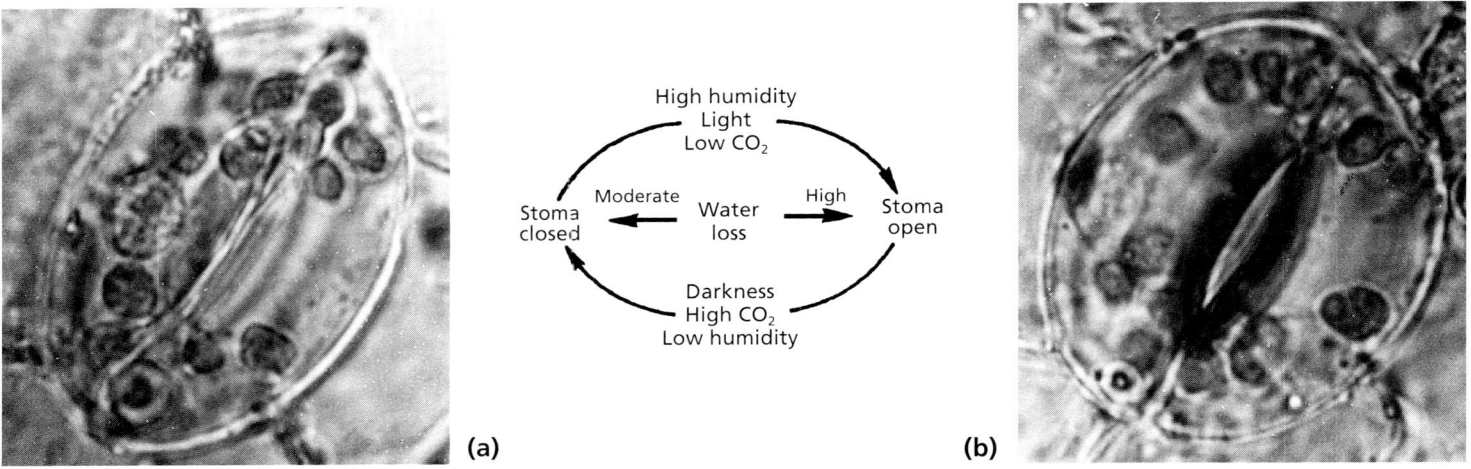

Figure 9.41 Guard cells in many plants regulate the opening of the stomata according to the environmental factors, as indicated in this diagram. (a) A face view of a closed stoma of a geranium, and (b) an open stoma. (X1500)

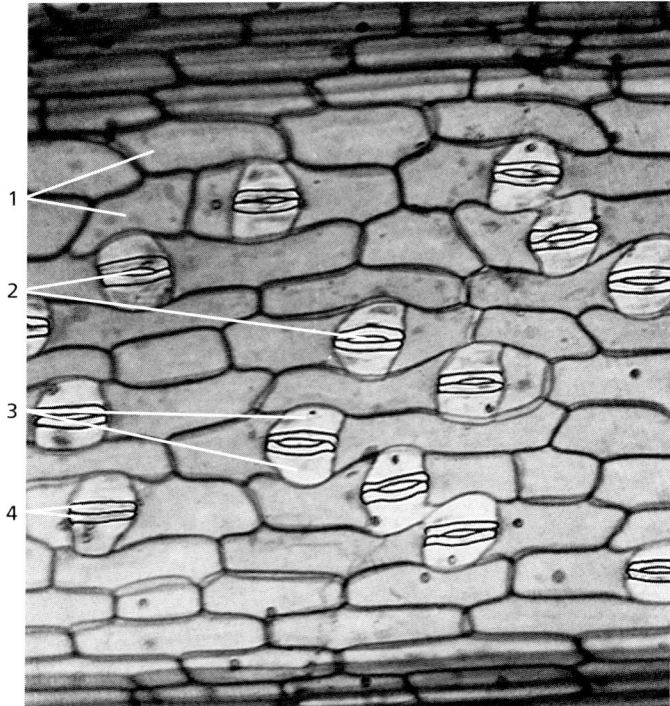

Figure 9.42 Face view of the leaf epidermis of *Tradescantia*. (X430)

1. Epidermal cells
2. Stomata
3. Subsidiary cells
4. Guard cells

Figure 9.43 As seen on the leaflets in the upper left of this photograph, the leaves of the sensitive plant, *Mimosa pudica*, droop upon being touched. The drooping results from differential changes in turgor of the leaf cells in the pulvinus, a thickened area at the base of the leaflet.

Leaves of Angiosperms

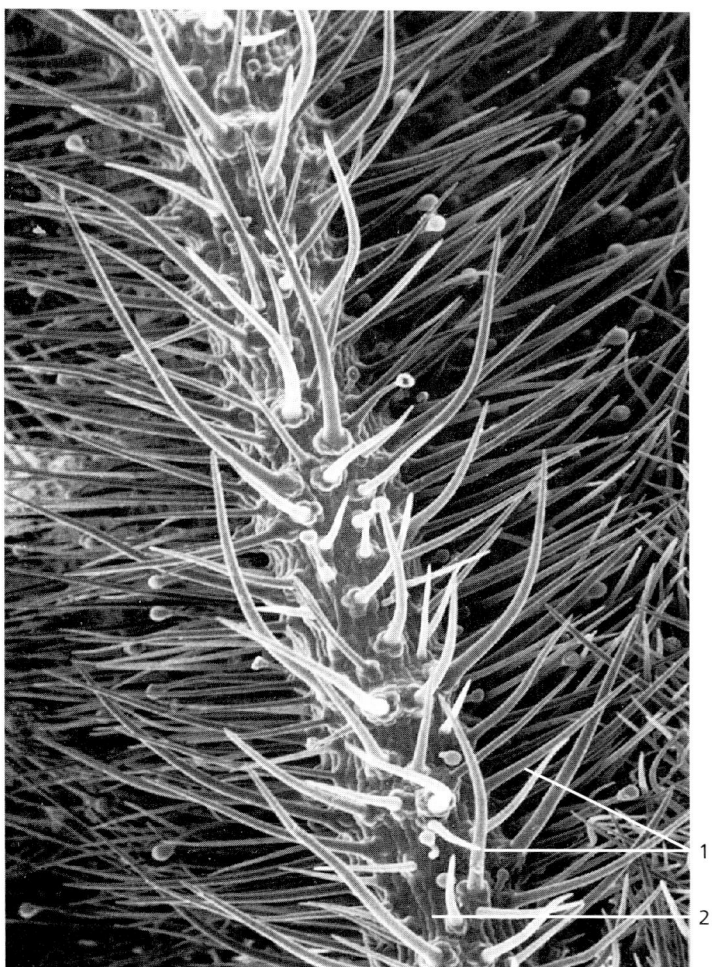

Figure 9.44 A scanning electron micrograph of a geranium leaf showing the prominent and abundant epidermal hairs. (X300)

1. Epidermal hairs
2. Epidermis

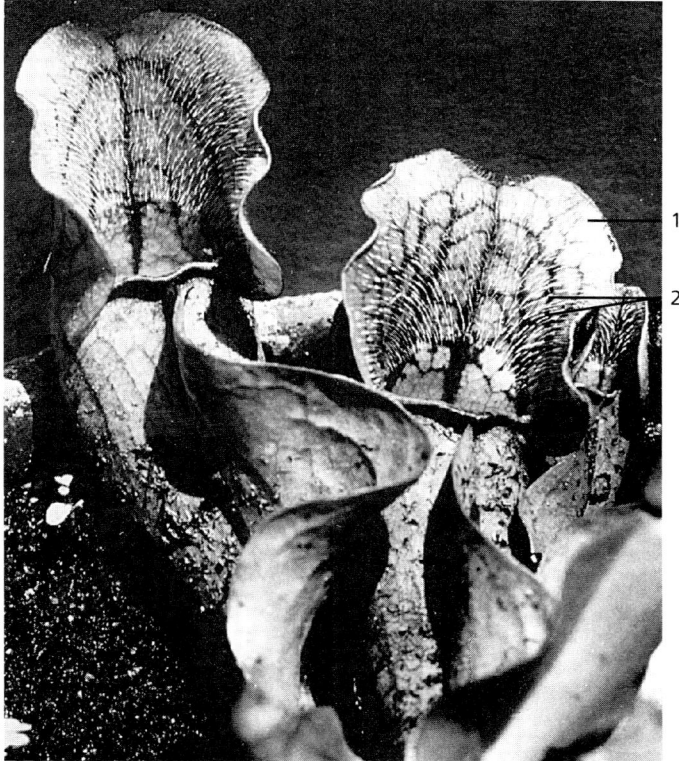

Figure 9.45 The leaves of the pitcher plant are adapted to entrap insects. The leaves are funnel-shaped and have epidermal hairs pointed toward the base of the leaf. Insects are attracted to the funnel where they are entrapped, die, and are digested by the plant.

1. Leaf
2. Epidermal hairs

Figure 9.46 The leaves of the venus flytrap, *Dionaea muscipula*, are adapted to entrap insects. An insect is attracted by nectar secreted on the surface of the leaf. The movement of the insect upon the leaves stimulates the sensitive trichomes on the upper surface of the leaves, triggering the leaves to close, entrapping the insect. (X3)

Leaves of Angiosperms

Figure 9.47 Prickly pear, *Opuntia*. Cacti have several modifications to withstand drought. They have spine-like leaves to prevent water loss through transpiration; they have developed tissue that stores water after rain; and their surface is coated with a waxy substance to aid in water retention.

Figure 9.48 The leaf of *Yucca* shows a thick cuticle covering the epidermis of the leaf. The cuticle protects against excessive water loss. (X100)
1. Cuticle
2. Epidermis

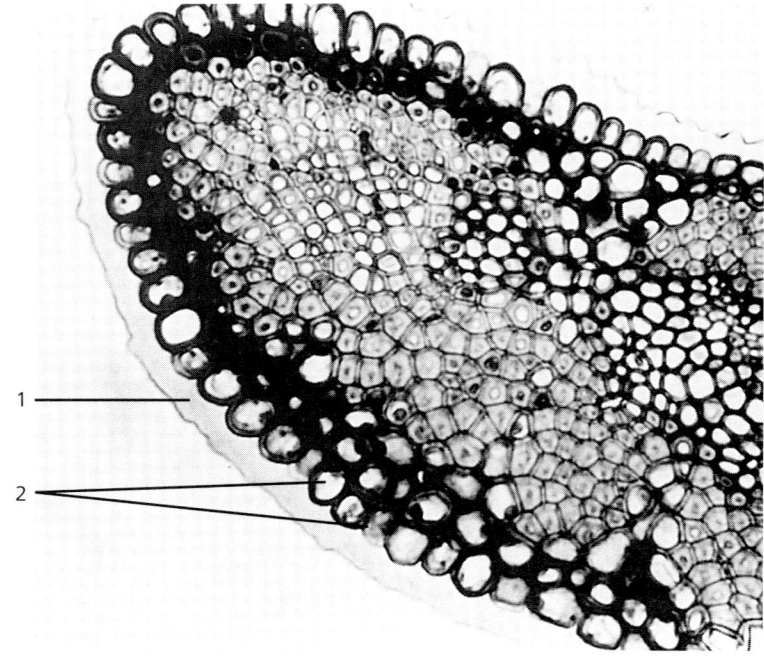

Flowers of Angiosperms

The angiosperm flower is composed of sepals, petals, stamens, and a carpel or carpels (gynecium). The *sepals* are the outermost circle of protective leaf-like structures. They are usually green and are collectively called the *calyx*. The *petals* generally form a whorl to the inside of the calyx and are collectively called the *corolla*. Petals are often brightly colored and may secrete aromatic substances and nectar to attract pollinating insects. The *stamens* and the *carpels* are the reproductive parts of a flower. A stamen consists of the *filament* (stalk) and the *anther*, where *pollen* is produced. The centrally positioned *pistil* consists of a *stigma* at the tip that receives pollen and a *style* that leads to the ovary. The ovary is composed of one or more modified leaves known as carpels. A *carpel* is a megasporophyll upon which ovules are produced. The carpel encloses the ovules so that seeds are produced within a protective layer that matures to form a fruit. Most flowers contain both stamens and a pistil, although some species produce unisexual flowers.

On the basis of position of the ovaries, flowers are classified as *hypogynous* (with flower parts below the ovary), *epigynous* (with flower parts above the ovary), or *perigynous* (centrally-positioned ovary). Regardless of the position of the ovary, most angiosperms rely on wind or animals for pollination. *Pollination* is the placement of pollen from the anther onto the stigma of the pistil by wind or animal vectors and is a prerequisite to fertilization. Wind is the primary pollinating agent for grasses and many trees. Because of this random dispersal, enormous quantities of pollen grains are released by the anthers of the flowers. Many angiosperms are pollinated by bees and other insects. The flowers of these plants are generally brightly colored and sweet smelling. Flowers pollinated by hummingbirds often have nectar deep in slender floral tubes where most other animals cannot reach.

When a *pollen grain* adheres to the stigma of the same species of plant, it swells and splits its outer coat. A tube cell grows and digests a tube down the style toward an ovule in the ovary. Directed by chemical attraction, the tip of the pollen tube enters the ovule and discharges two sperm nuclei into the embryo sac. One sperm fertilizes the *egg*, and the other combines with the two polar nuclei to form a triploid ($3n$) nucleus. The fusion of the egg with one sperm and the polar nuclei cell with the second sperm nucleus is called *double fertilization* and is unique to flowering plants. After fertilization, the other cells of the embryo sac degenerate and the ovule begins developing into a seed.

Flowers have contributed greatly to the success of angiosperms because they enhance the efficiency of plant reproduction by attracting and rewarding pollen-carrying animals. The beauty and fragrance of flowers have always appealed to humans. Even many perfumes have chemicals extracted from flowers as an important ingredient.

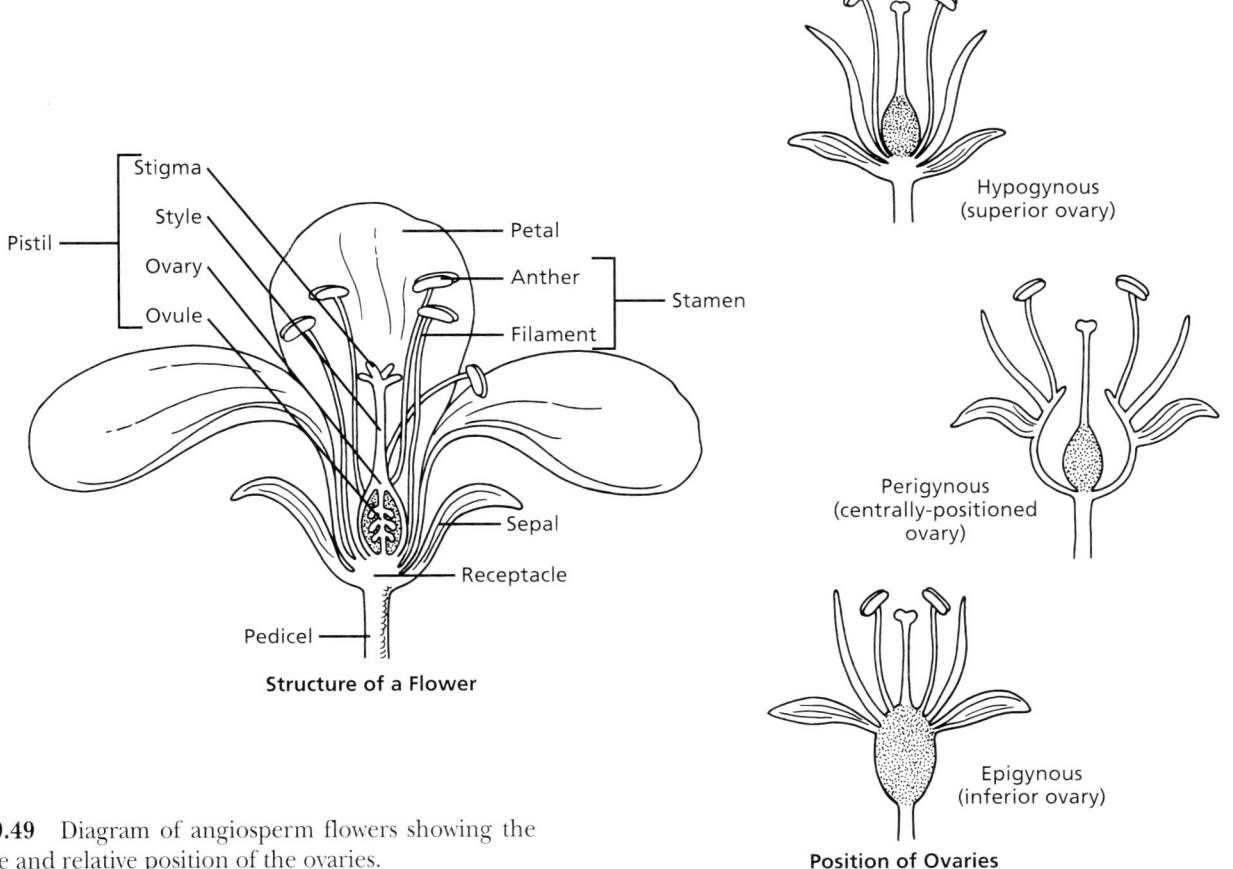

Figure 9.49 Diagram of angiosperm flowers showing the structure and relative position of the ovaries.

Flowers of Angiosperms

Rose, *Rosa woodsii*

Amaryllis, *Amaryllis belladonna*

Lily, *Lilium columbianum*

Morning Glory, *Ipomoea arvensis*

Chicory, *Cichorium intybus*

Gladiolus, *Gladiolus grandiflorus*

Figure 9.50 A variety of angiosperm flowers.

Figure 9.51 *Hibiscus* flower. *Hibiscus*, the state flower of Hawaii, is characterized by the stigma and stamens being extended well outside of the petals and sepals of the flower.

Figure 9.52 Floral parts of a tulip.
1. Stigma
2. Anther
3. Petal
4. Ovary

Flowers of Angiosperms

Figure 9.53 Rose flower.
1. Petals
2. Stigma
3. Anther
4. Sepal

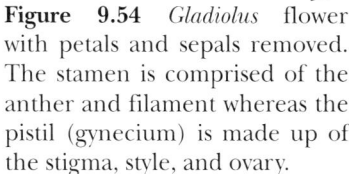

Figure 9.54 *Gladiolus* flower with petals and sepals removed. The stamen is comprised of the anther and filament whereas the pistil (gynecium) is made up of the stigma, style, and ovary.

1. Anther
2. Filament
3. Ovules
4. Receptacle
5. Stigma
6. Style
7. Ovary

Figure 9.55 *Gladiolus* anthers and stigma.
1. Anther
2. Pollen grains
3. Stigma
4. Style
5. Filament

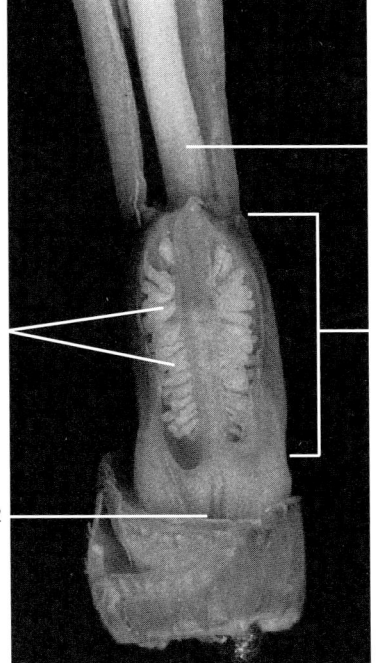

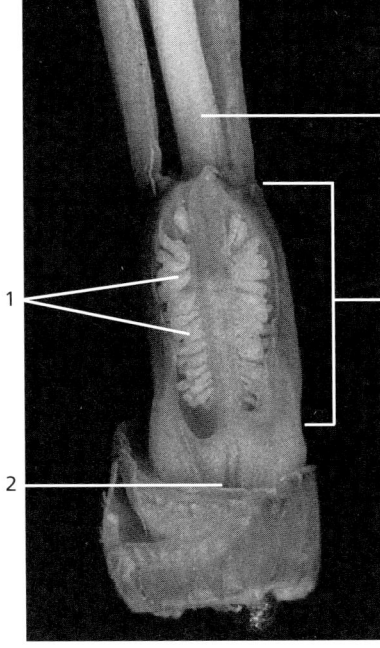

Figure 9.56 *Gladiolus* ovary.
1. Ovules
2. Receptacle
3. Style
4. Ovary

Flowers of Angiosperms

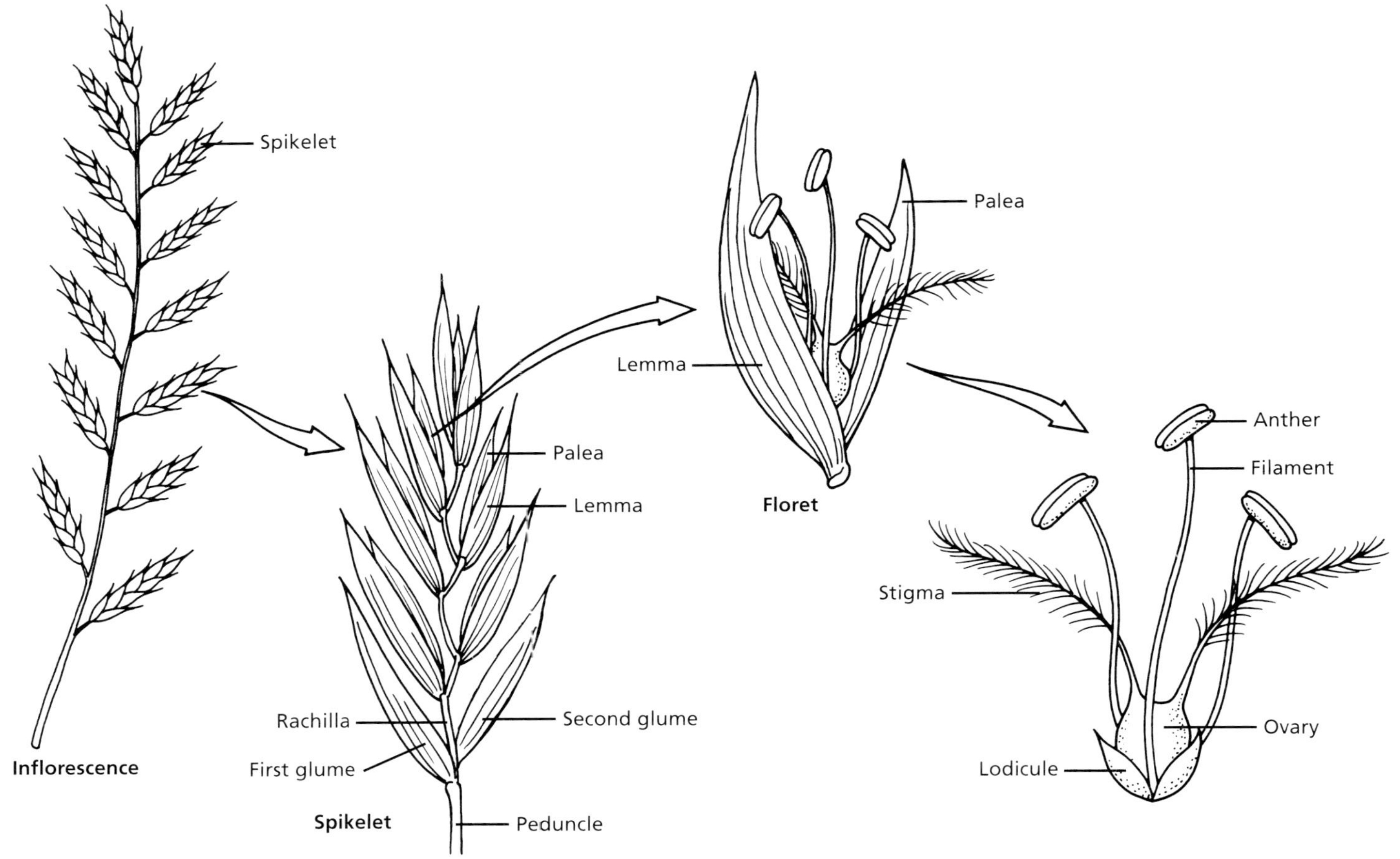

Figure 9.57 Flower structure in grasses.

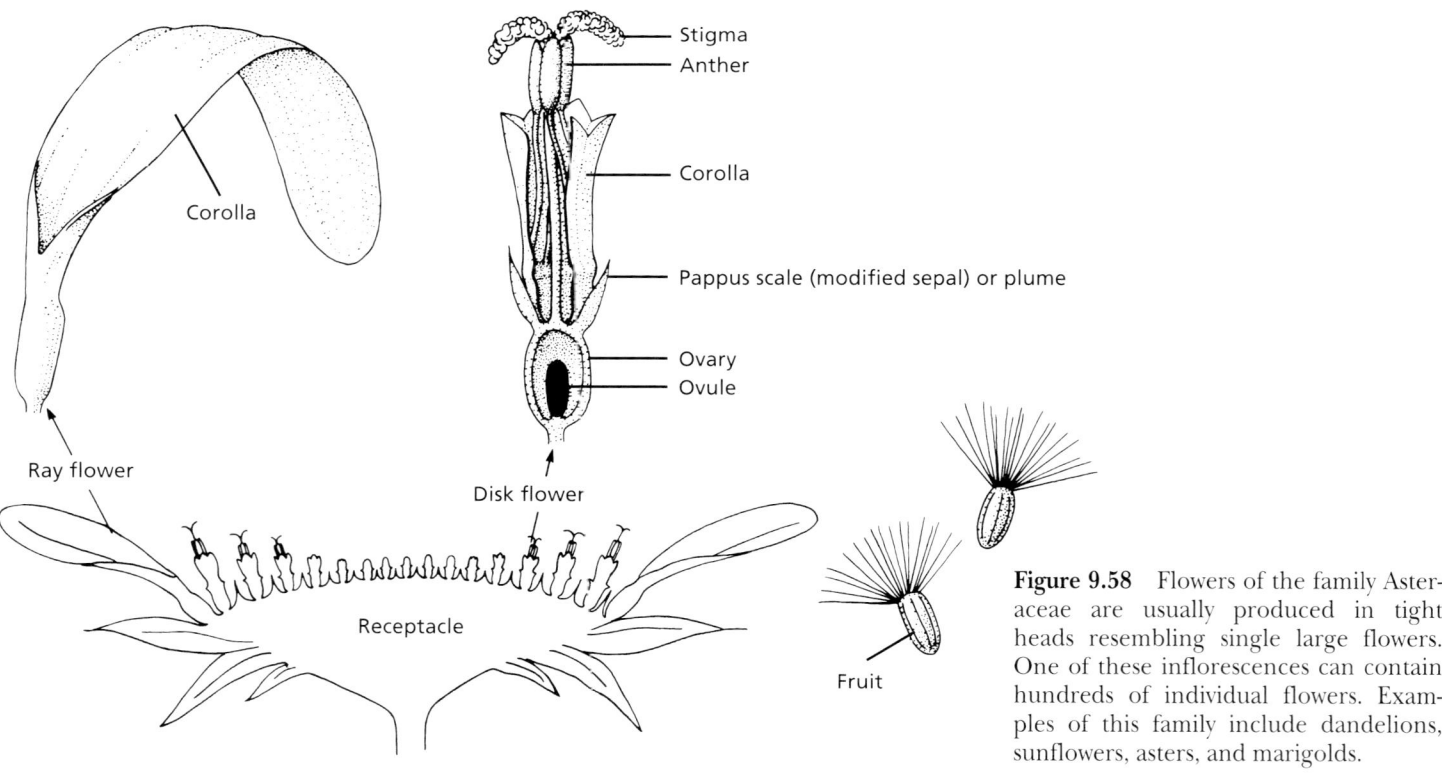

Figure 9.58 Flowers of the family Asteraceae are usually produced in tight heads resembling single large flowers. One of these inflorescences can contain hundreds of individual flowers. Examples of this family include dandelions, sunflowers, asters, and marigolds.

Flowers of Angiosperms

Figure 9.59 A scanning electron micrograph of the stigma of an angiosperm pistil. The stigma is the location where pollen grains adhere and germinate to produce a pollen tube. (X100)

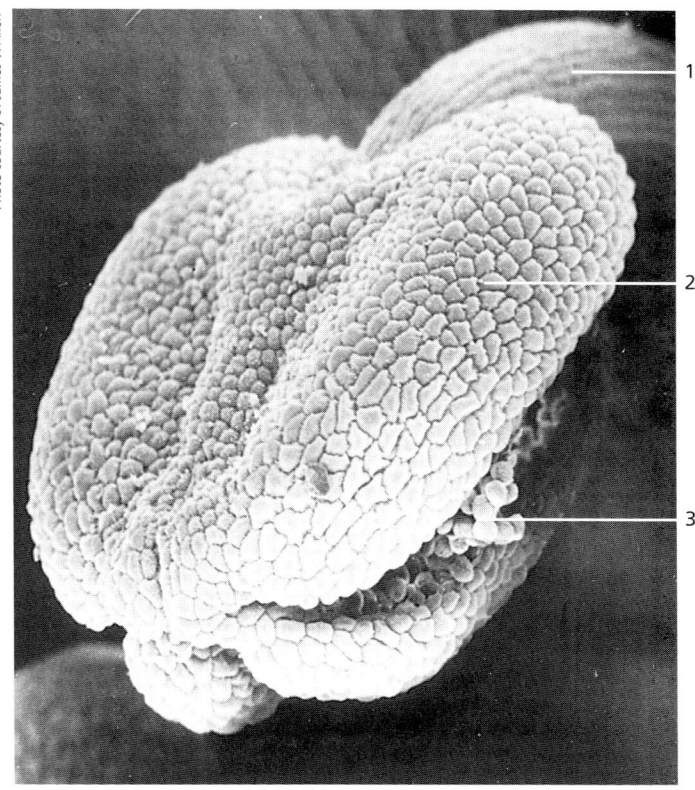

Figure 9.60 A scanning electron micrograph of the anther of candy tuft, *Lobularia*. The anther has ruptured, resulting in the release of pollen grains. (X10)

1. Filament
2. Anther
3. Pollen grains

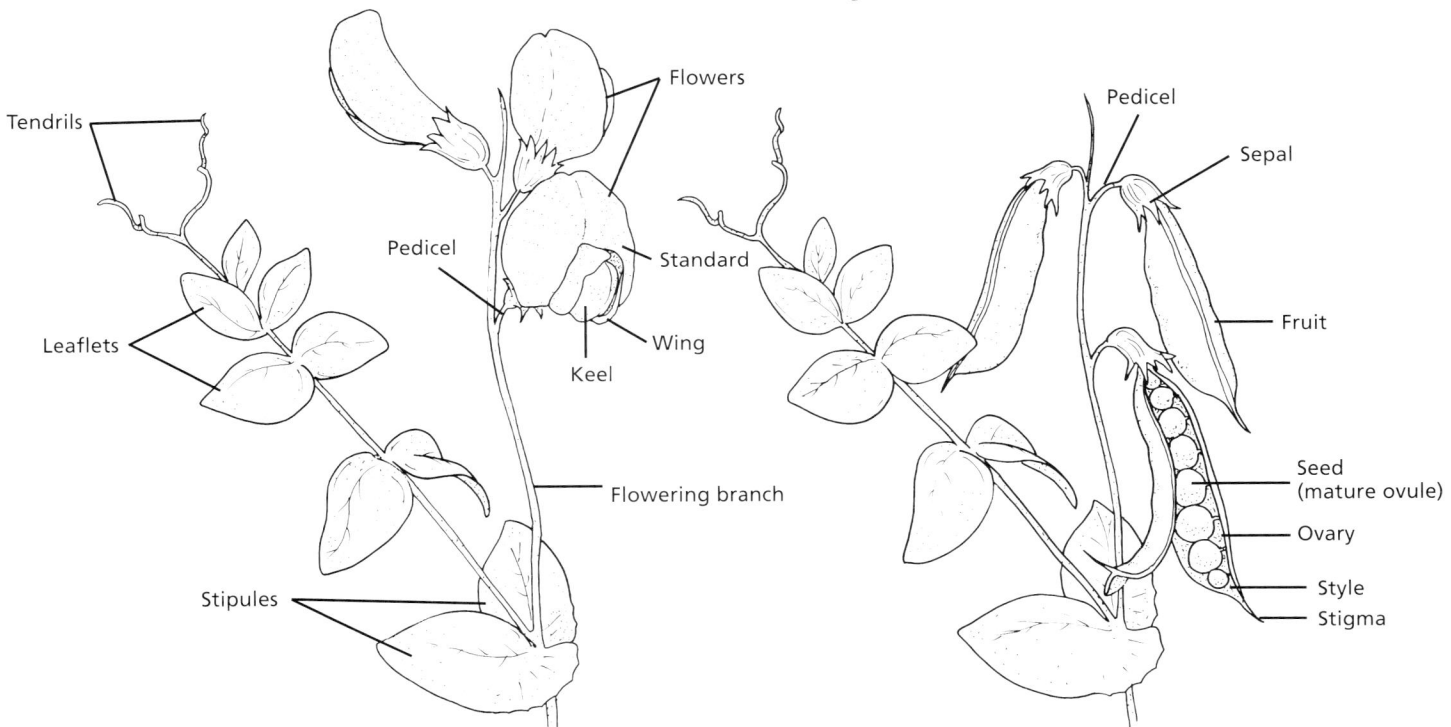

Figure 9.61 Flower and fruit of the pea, *Pisum*.

Flowers of Angiosperms

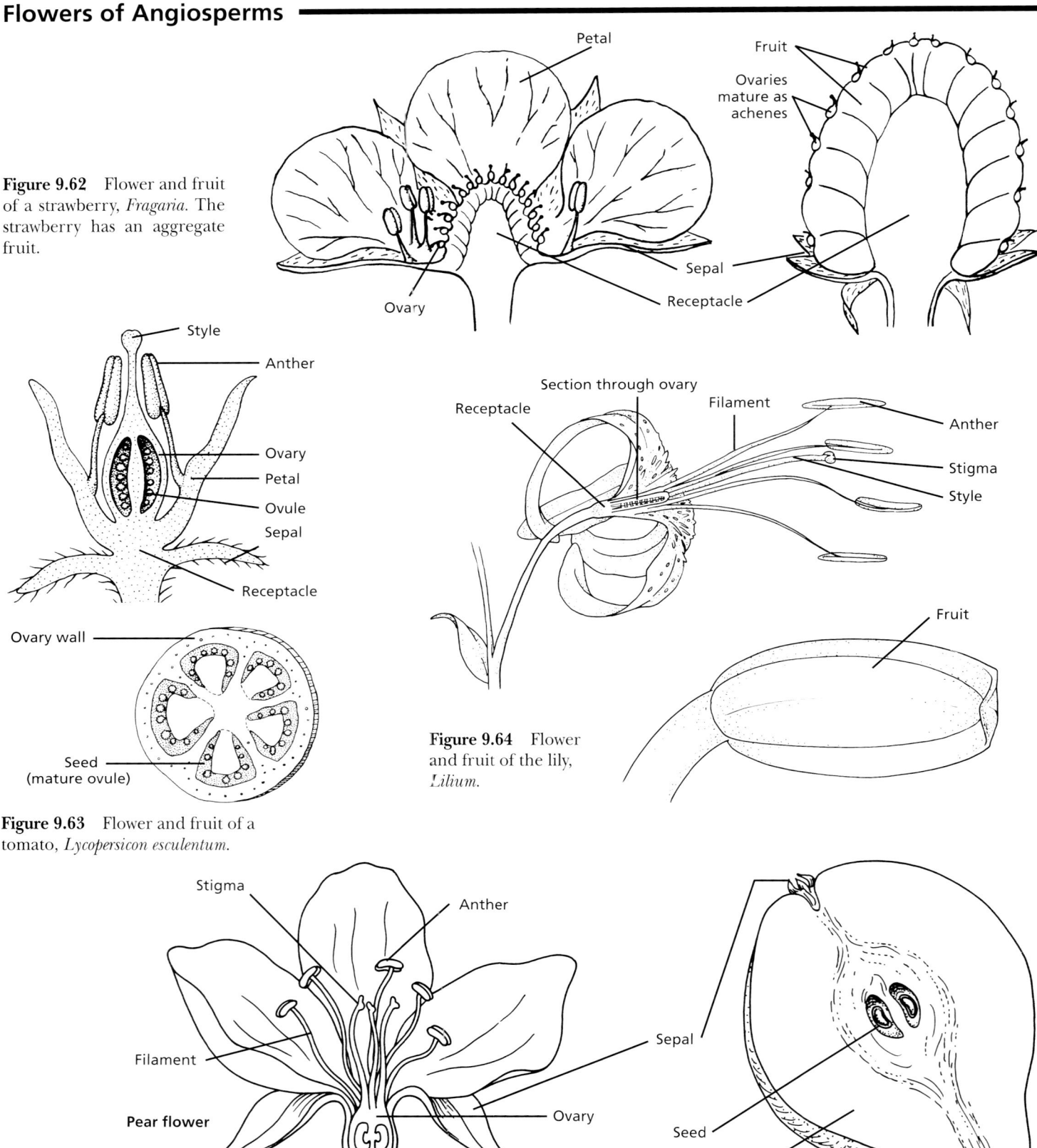

Figure 9.62 Flower and fruit of a strawberry, *Fragaria*. The strawberry has an aggregate fruit.

Figure 9.63 Flower and fruit of a tomato, *Lycopersicon esculentum*.

Figure 9.64 Flower and fruit of the lily, *Lilium*.

Figure 9.65 Structures of the flower and fruit of an angiosperm, the pear. The pear fruit develops from the floral tube (fused perianth) as well as the ovary.

Seeds, Fruits, and Seed Germination of Angiosperms

Seeds are the reproductive structures of gymnosperms and angiosperms. The seeds of gymnosperms (see Chapter 8) develop on the exposed surface of the scales of cones, whereas the seeds of angiosperms usually develop within a fruit produced from the ovary of a flower.

A typical seed of an angiosperm consists of a protective *seed coat*, a sporophyte *embryo*, and a thick layer of nutritive tissue called the *endosperm*. The endosperm, consisting of cells rich in proteins, fats, oils, and starch, is absorbed by the seed embryo during development, or germination. The embryo consists of *cotyledons, epicotyl, hypocotyl,* and *radicle*. During germination, the cotyledons become the embryonic leaves, the epicotyl becomes the shoot from which derives the first plant foliage, the hypocotyl becomes the point of attachment of the epicotyl, and the radicle becomes the primary root. When fully developed and prior to germination, some seeds dry out and become dormant with only 5 to 20% water content. A protective *fruit* develops around the angiosperm seed as it matures. Seeds and fruit are a source of food for many kinds of animals, including humans.

Although each species of angiosperm has evolved specific mechanisms for seed dispersal, three basic methods are used (see Figure 9.76).

1. *Animal-dispersed seeds* include the majority of the fleshy fruits (berries, grapes, cherries, apples) eaten by vertebrates. Seeds are dispersed unharmed as they are passed through the digestive tract. The enticing flavor and color of fruits are examples of coevolution of animals and flowering plants. Many other plants have fruits or seeds that have hooks, spines, or sticky coverings and are dispersed by adhering to fur or feathers.

2. *Water-dispersed seeds* include those from plants that grow near or in water and have seeds or fruit adapted for floating. In these species, either the seed or the fruit is buoyant. Nearly all Pacific islands have coconut trees that are seeded by buoyant coconuts. Rainfall is important in seed dispersal of some species.

3. *Wind-dispersed seeds* include those that are lightweight and buoyant in the air. The fruits of dandelions, for example, have dry, plumelike structures attached that carry the wind-borne fruits great distances. Each dandelion fruit contains one seed. Other plants, such as maples, develop fruits that dry into winglike structures. Tumbleweeds scatter their seeds as the detached plant blows along the ground. Other plants, such as the poppy, disperse their seeds aloft into the wind.

Fruits are classified on the basis of development into three principal groups (see Figure 9.75): simple, aggregate, and multiple. *Simple fruits* develop from single ovaries and may be fleshy, such as cherries, or dry, such as legumes (beans and peas). *Aggregate fruits* develop from single flowers that have several separate carpels, such as strawberries and blackberries. *Multiple fruits* develop from groups of separate flowers clustered tightly, such as pineapples.

Seed germination occurs when appropriate environmental conditions are present. Some seeds must be exposed to extended cold; others must undergo a drying period followed by adequate moisture. Many seeds with hardened coats must be physically or chemically scarified before they can germinate. *Imbibition*, or the absorption of water, is the first step in the germination of most seeds. This *hydration* causes a seed to expand and rupture its coat. Once the germination process is initiated, the *radicle*, or root of the embryo, emerges from the seed and grows downward into the soil. In monocots, the shoot of the seedling grows upward through the tube of the *coleoptile*. In dicots, the hypocotyl of the seed grows upward, pulling the shoot and cotyledons from the soil. As leaves emerge from the seedling, the cotyledons die and wither, or may persist for weeks or months.

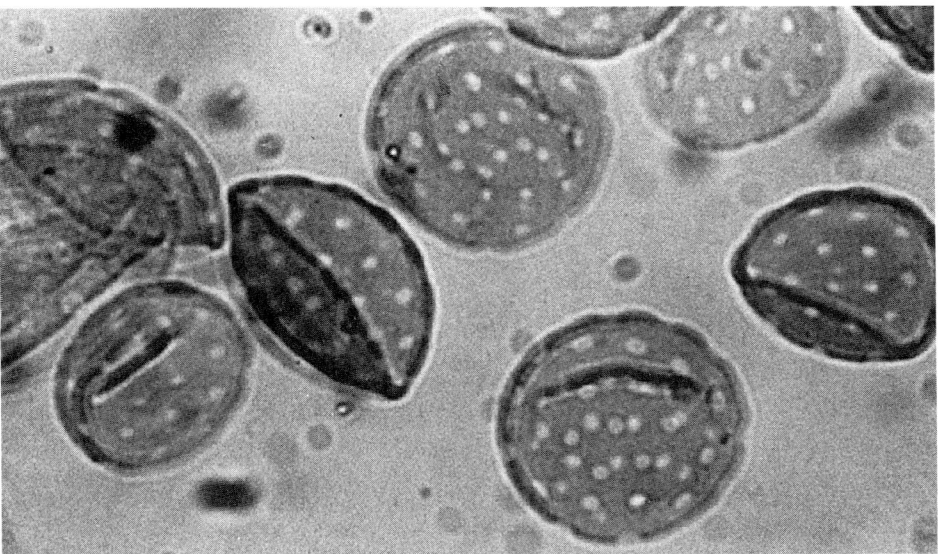

Figure 9.66 Pollen grains of the dicot pigweed, *Amaranthus*. (X430)

Seeds, Fruits, and Seed Germination of Angiosperms

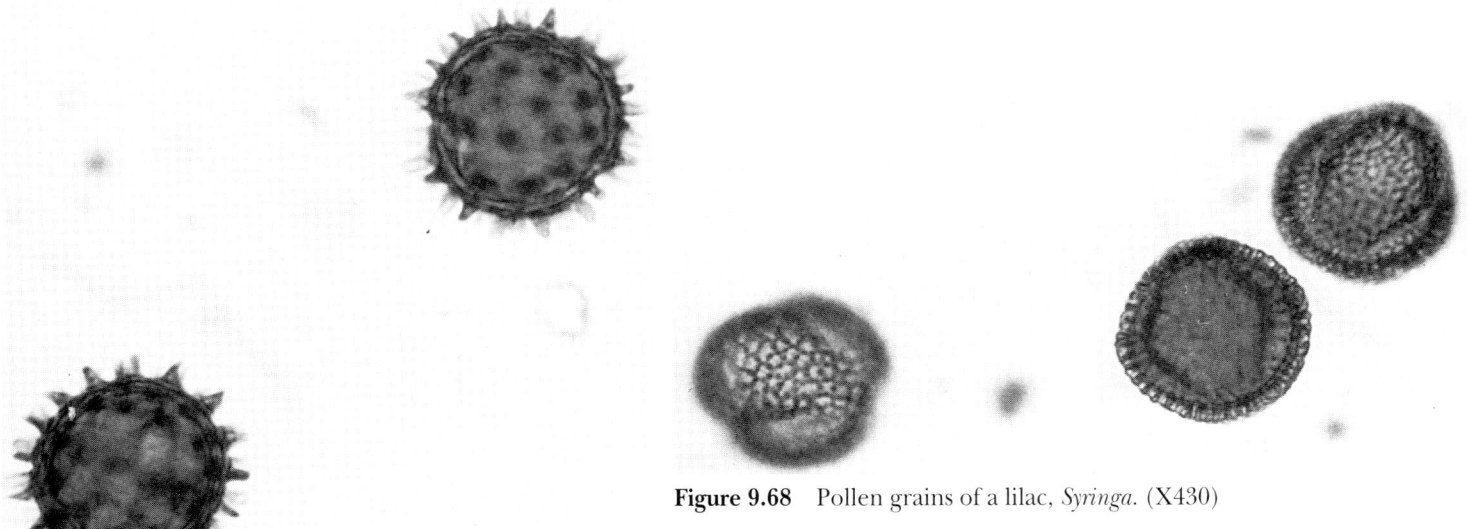

Figure 9.68 Pollen grains of a lilac, *Syringa*. (X430)

Figure 9.67 Pollen grains of a common dicot, arrowroot. (X430)

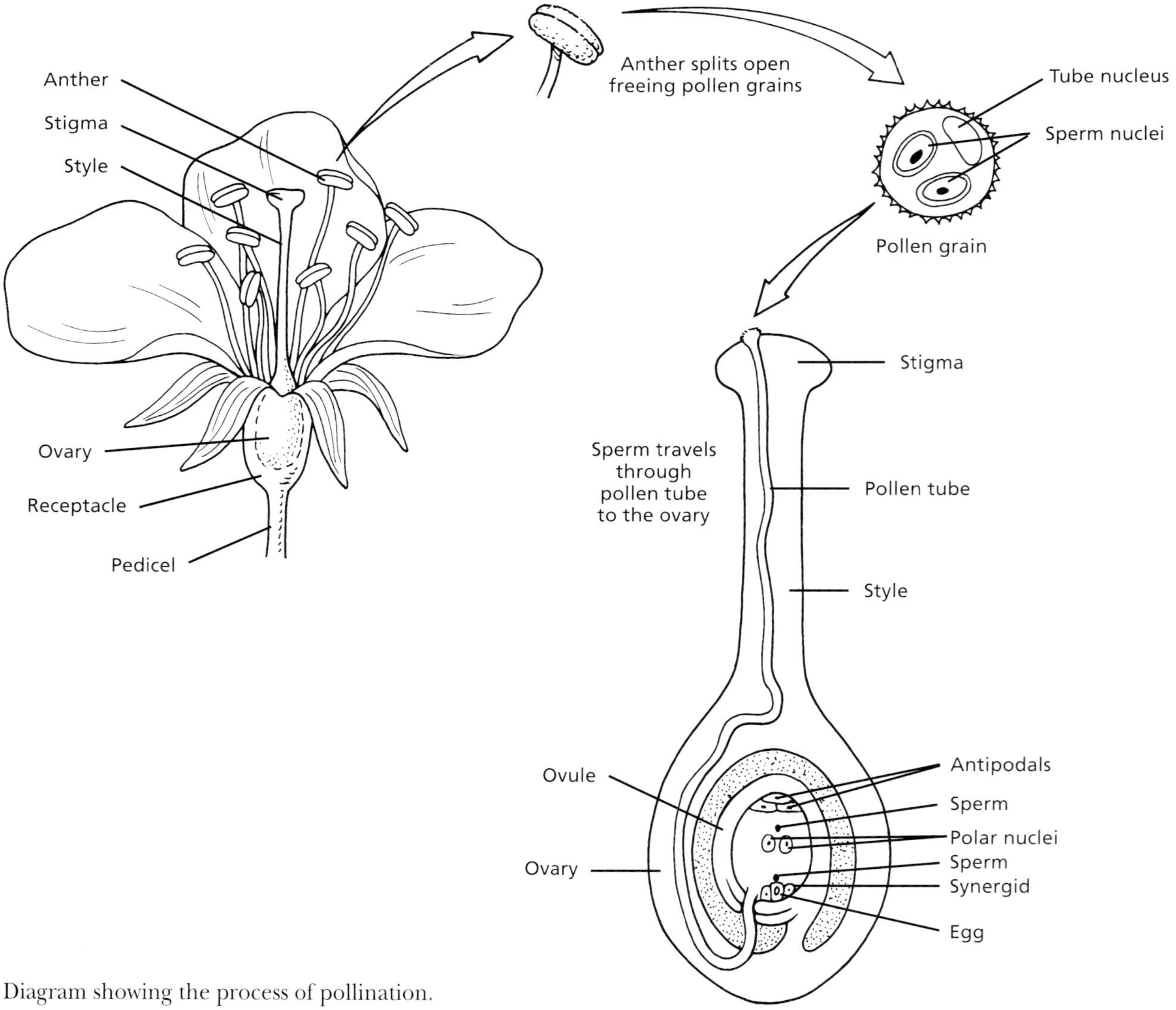

Figure 9.69 Diagram showing the process of pollination.

CHAPTER 9 Kingdom Plantae: Division Anthophyta — Angiosperms (Enclosed Seed Plants — Flowering Plants)

Seeds, Fruits, and Seed Germination of Angiosperms

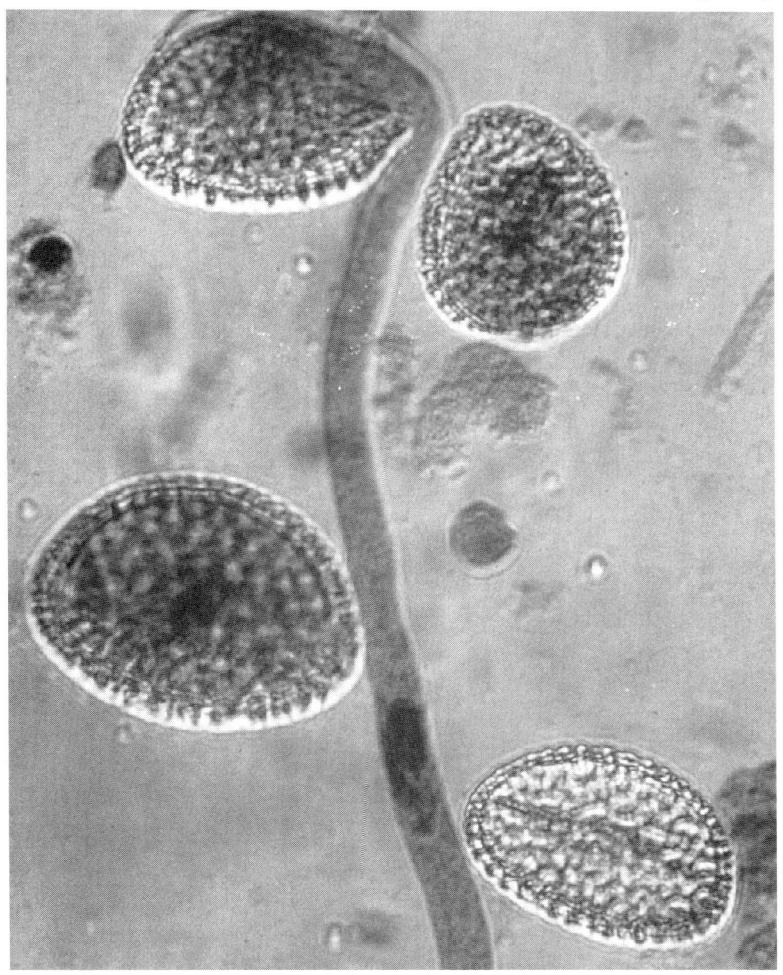

Figure 9.70 Pollen grains of a lily. The pollen grain at the top of the photo has germinated to produce a pollen tube. (about X500)

Figure 9.71 Lady slipper orchid, *Paphiopedilum venustii*. The flower of the lady slipper orchid fills with rain and drops to the ground, allowing ants to enter and fertilize the flower.

Figure 9.72 The flowers of many angiosperms are adapted for insect pollination.

Photos courtesy of Armand T. Whitehead

Seeds, Fruits, and Seed Germination of Angiosperms

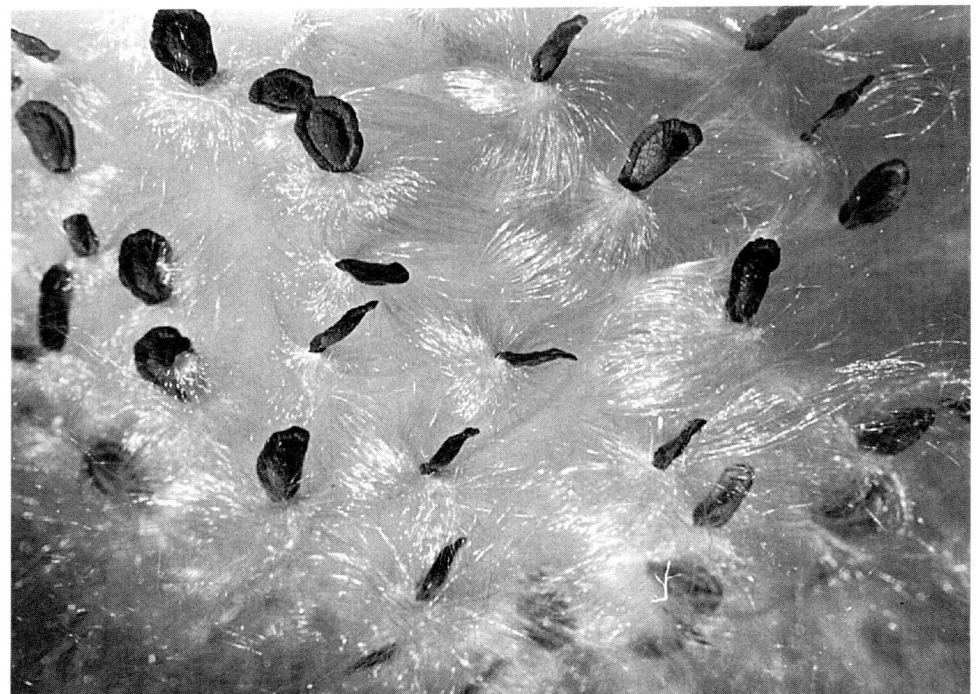

Figure 9.73 Seeds of the milkweed, *Asclepias*.

(a)

(b)

Figure 9.74 The flower (a) and the fruits (b and c) of the dandelion, *Taraxacum*. The dandelion has a composite flower. The wind-borne fruit (containing one seed) of a dandelion, and many other members of the family Asteraceae, develop a plumelike pappus, which enables the light fruit to float in the air.
1. Pappus
2. Ovary wall, with one seed inside

(c)

CHAPTER 9   *Kingdom Plantae: Division Anthophyta — Angiosperms (Enclosed Seed Plants — Flowering Plants)*   151

Seeds, Fruits, and Seed Germination of Angiosperms

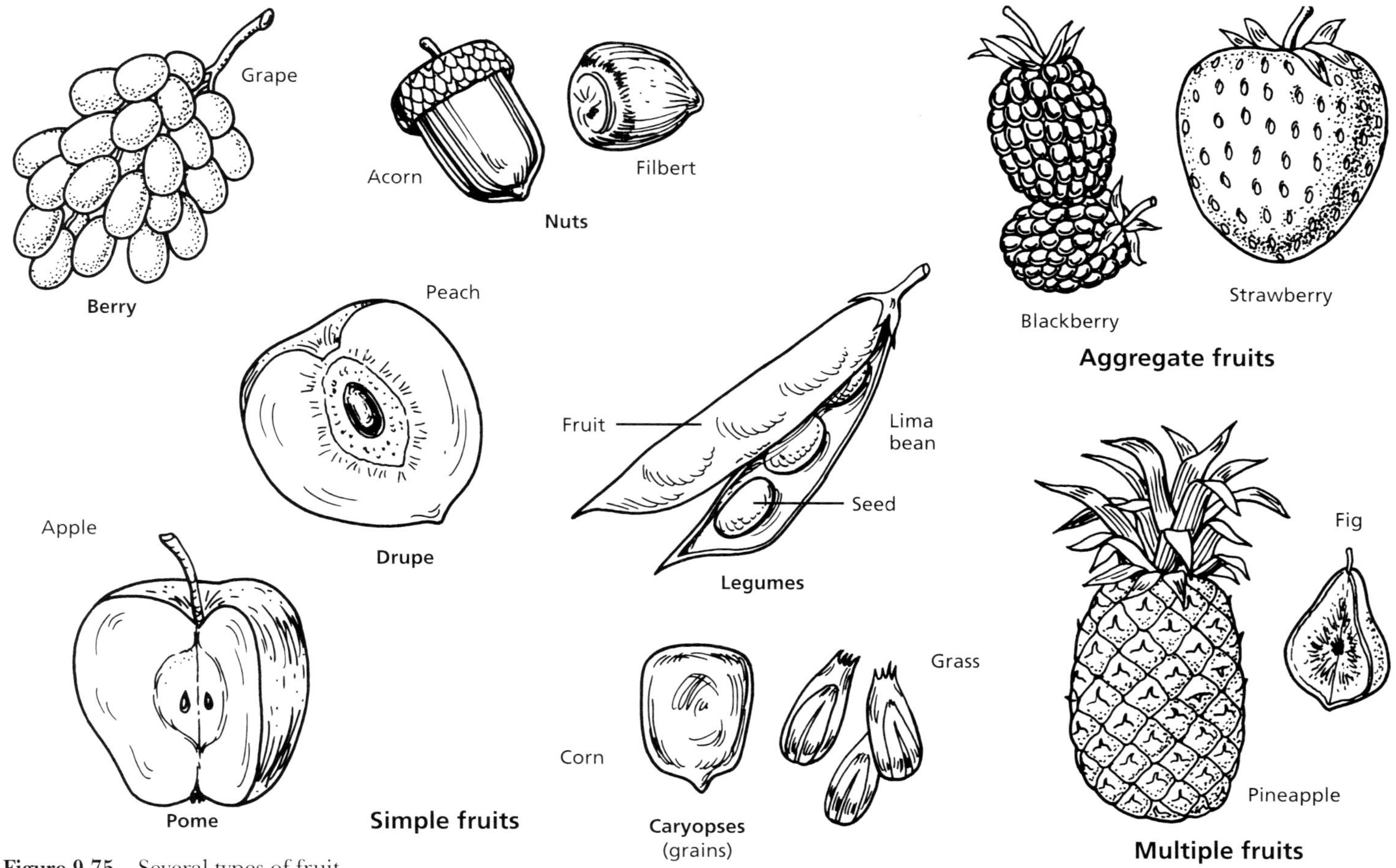

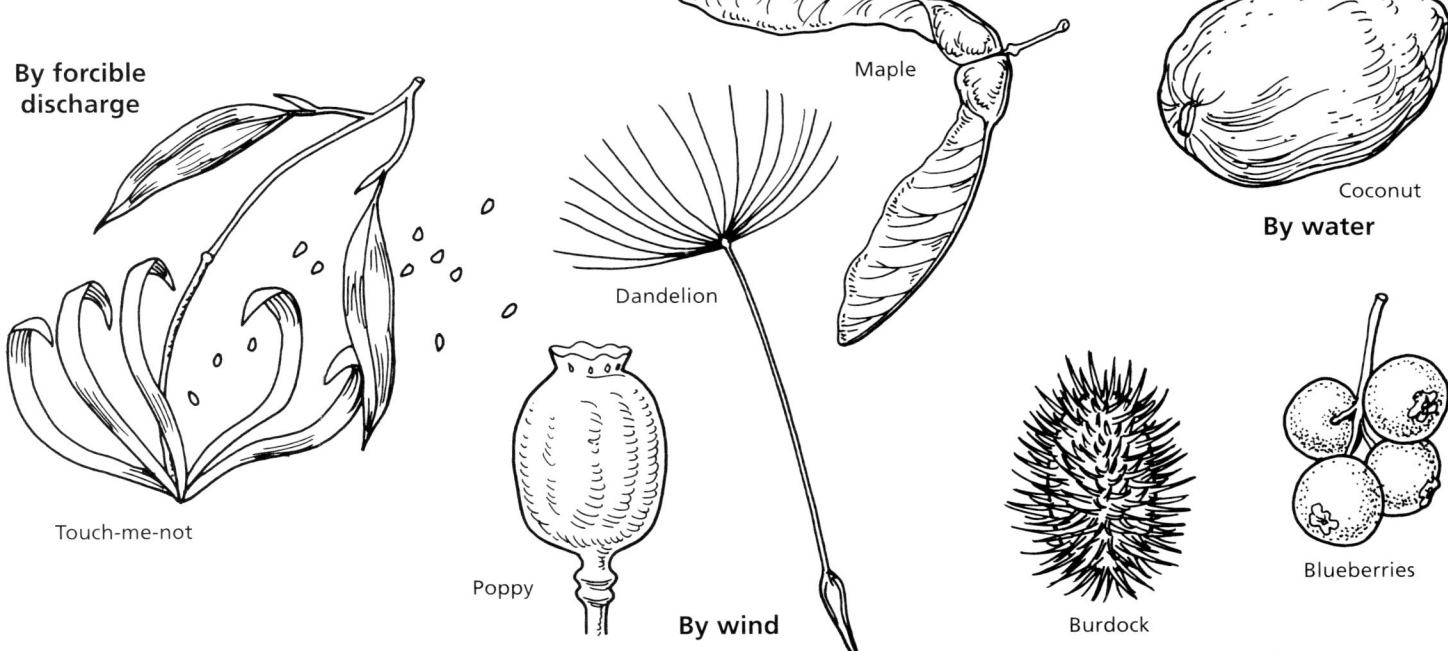

Figure 9.75   Several types of fruit.

Figure 9.76   Several fruits and seeds to illustrate seed dispersal.

Seeds, Fruits, and Seed Germination of Angiosperms

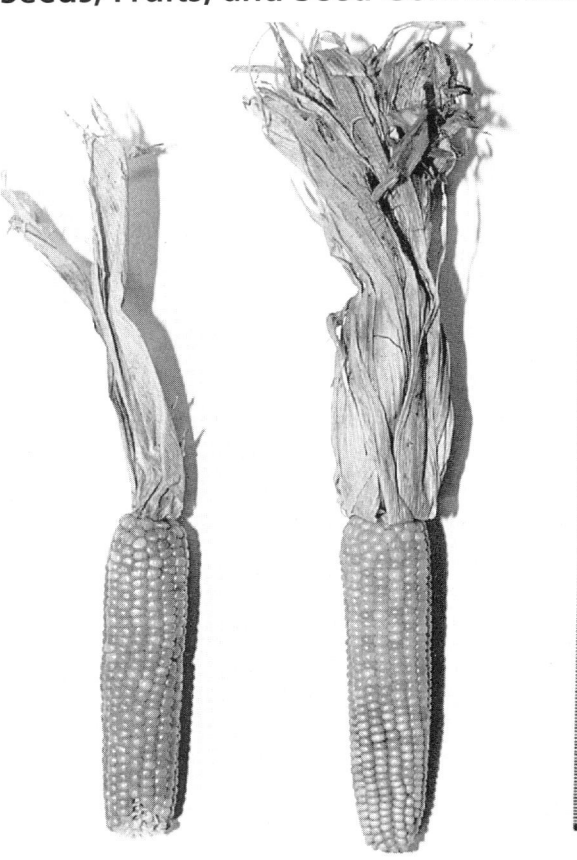

Figure 9.77 Herbarium specimen of cobs of corn from *Zea mays*. Corn was domesticated approximately 7,000 years ago from a Mexican grass, family Poaceae.

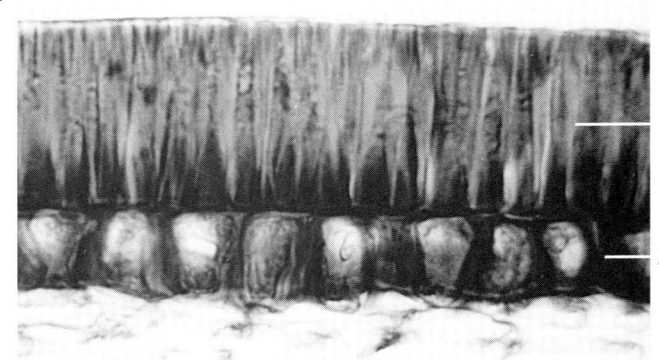

Figure 9.78 Closeup of the seed coat of the garden bean, *Phaseolus*, showing sclerified epidermis. (X100)
1. Macrosclerids
2. Subepidermal sclerids

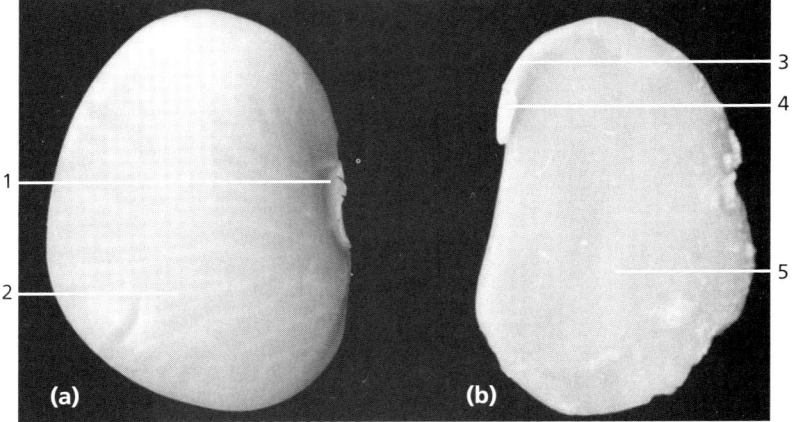

Figure 9.79 Lima bean. (a) Entire bean seed; (b) longitudinally sectioned seed.
1. Hilum
2. Integument (seed coat) (2n)
3. Hypocotyl (2n)
4. Radicle (2n)
5. Cotyledon (2n)

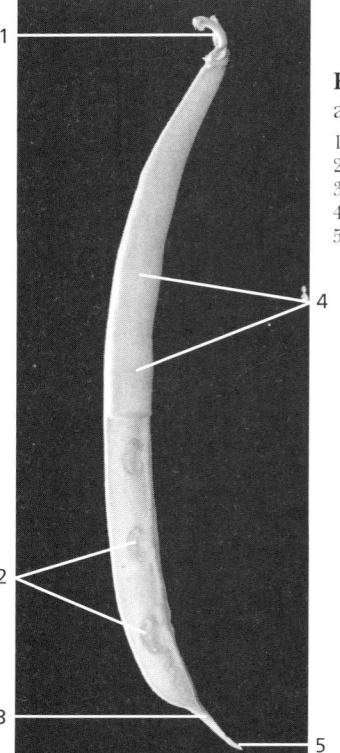

Figure 9.80 Photograph of a legume, string bean.
1. Pedicel
2. Seeds
3. Style
4. Fruit
5. Stigma

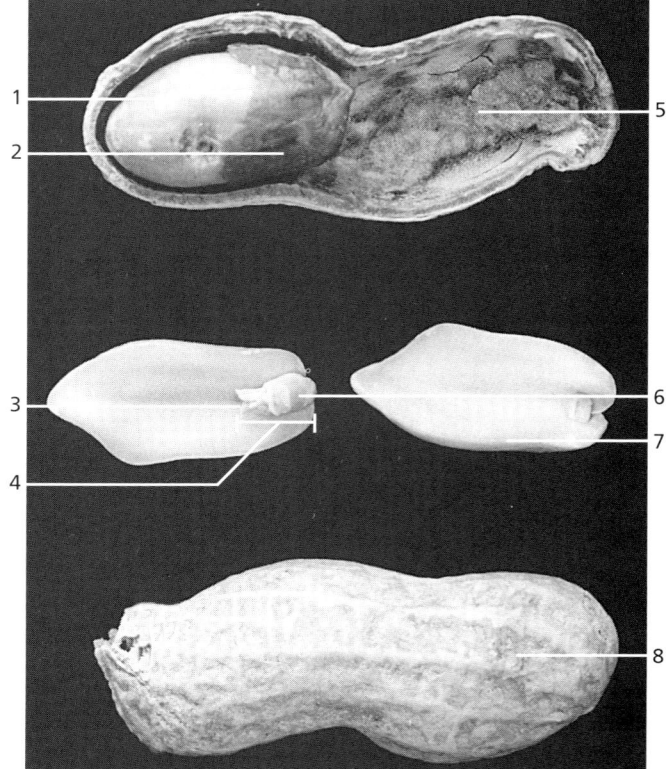

Figure 9.81 The fruit and seeds of a peanut plant.
1. Cotyledon
2. Seed coat
3. Plumule
4. Embryo
5. Interior of fruit
6. Radicle
7. Cotyledon
8. Fruit wall

CHAPTER 9 Kingdom Plantae: Division Anthophyta — Angiosperms (Enclosed Seed Plants — Flowering Plants) 153

Seeds, Fruits, and Seed Germination of Angiosperms

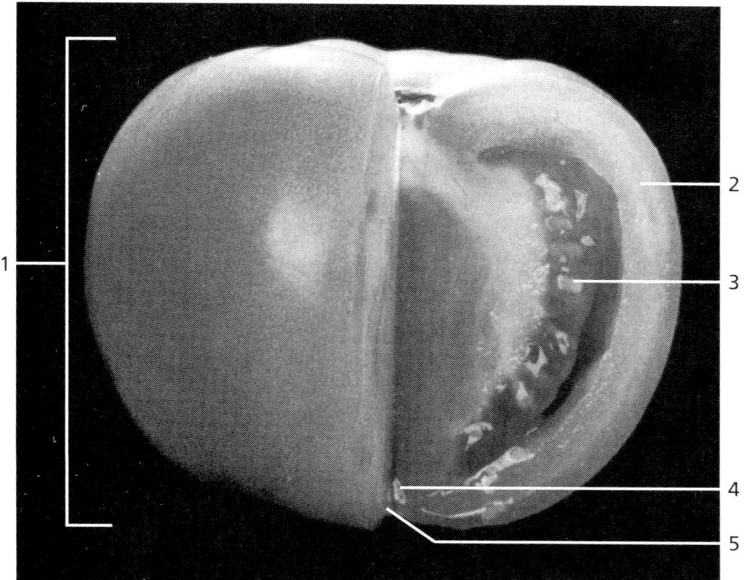

Figure 9.82 Longitudinal section of a tomato fruit.
1. Fruit (mature ovary)
2. Fruit wall
3. Seed (mature ovule)
4. Remnant of style
5. Remnant of stigma

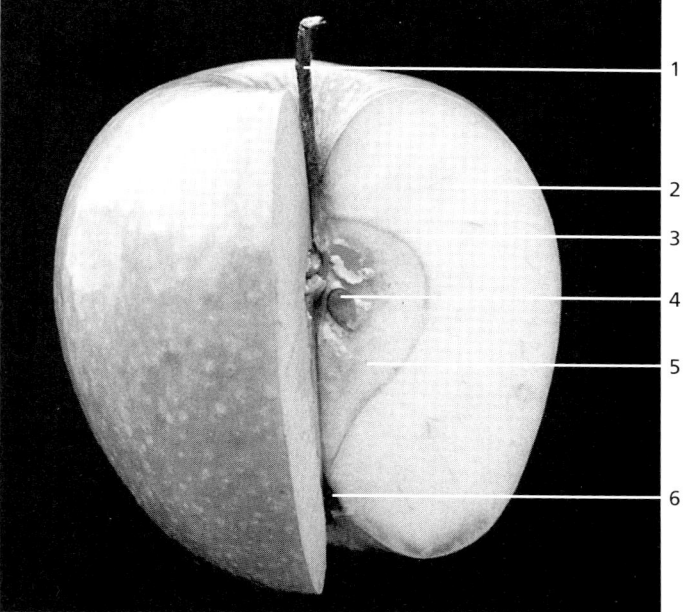

Figure 9.83 Longitudinal section of an apple fruit.
1. Pedicel
2. Receptacle
3. Ovary wall
4. Seed (mature ovule)
5. Mature ovary
6. Remnants of floral parts

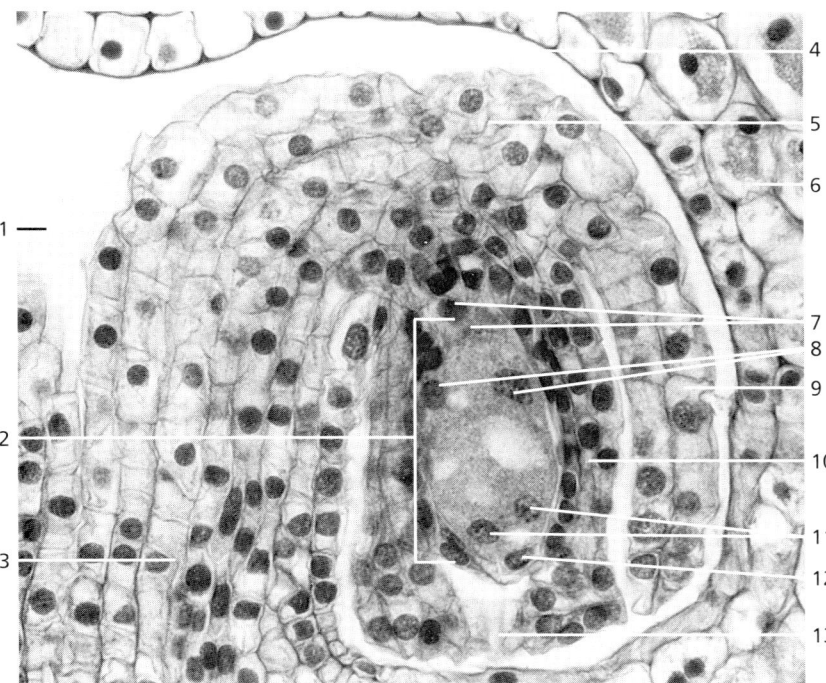

Figure 9.84 A longitudinal section of an eight-nucleate embryo sac of an ovule from a lily, *Lilium*. (X430)

1. Locule
2. Embryo sac
3. Funiculus
4. Chalaza
5. Ovule
6. Ovary
7. Antipodal cells
8. Polar nuclei (one is n, one is $3n$)
9. Outer integument ($2n$)
10. Inner integument ($2n$)
11. Synergid cells (n)
12. Egg (n)
13. Micropyle (pollen tube entrance)

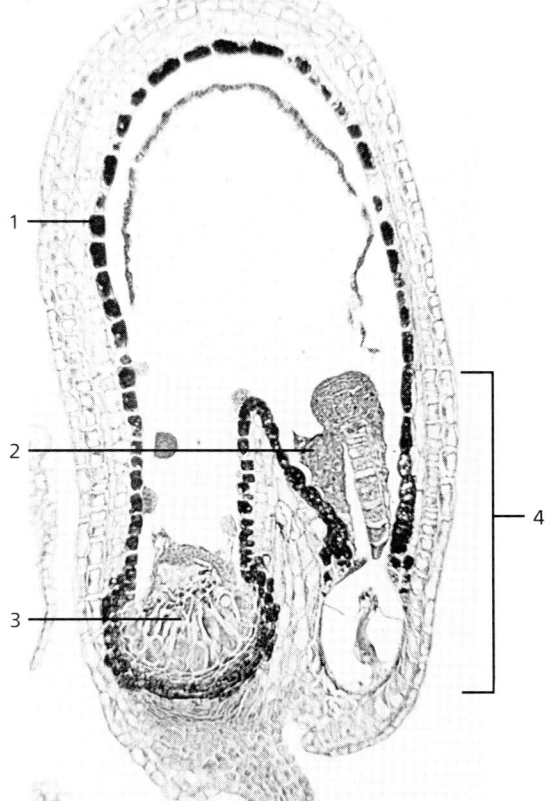

Figure 9.85 Photomicrograph of a developing dicot embryo from a shepherd's purse, *Capsella bursa-pastoris*. (X100)

1. Endothelium
2. Developing endosperm
3. Nucellar tissue
4. Developing embryo

Seeds, Fruits, and Seed Germination of Angiosperms

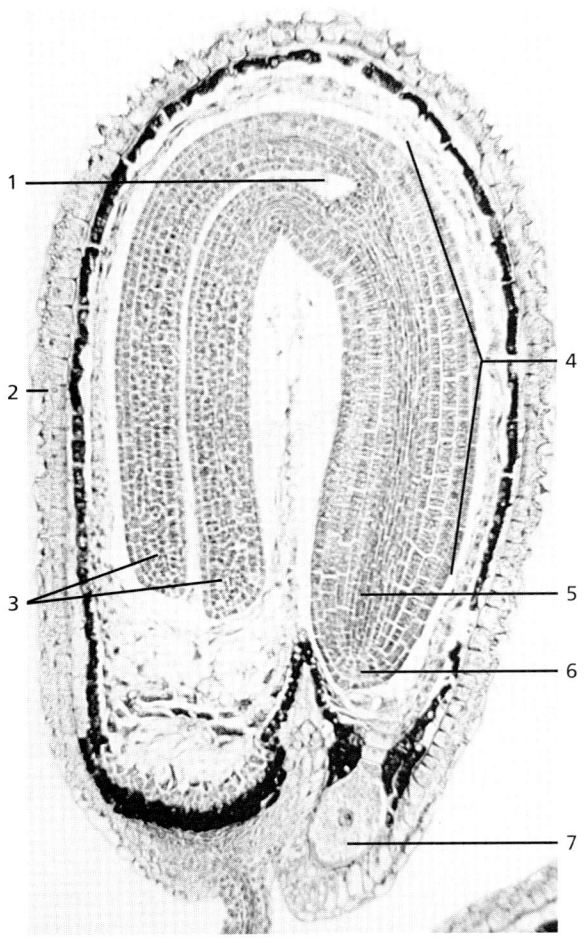

Figure 9.86 Photomicrograph of a mature dicot embryo from a shepherd's purse, *Capsella bursa-pastoris*. (X100)

1. Apical meristem
2. Seed coat
3. Cotyledons
4. Hypocotyl root axis
5. Developing vascular tissue
6. Root tip
7. Basal cell

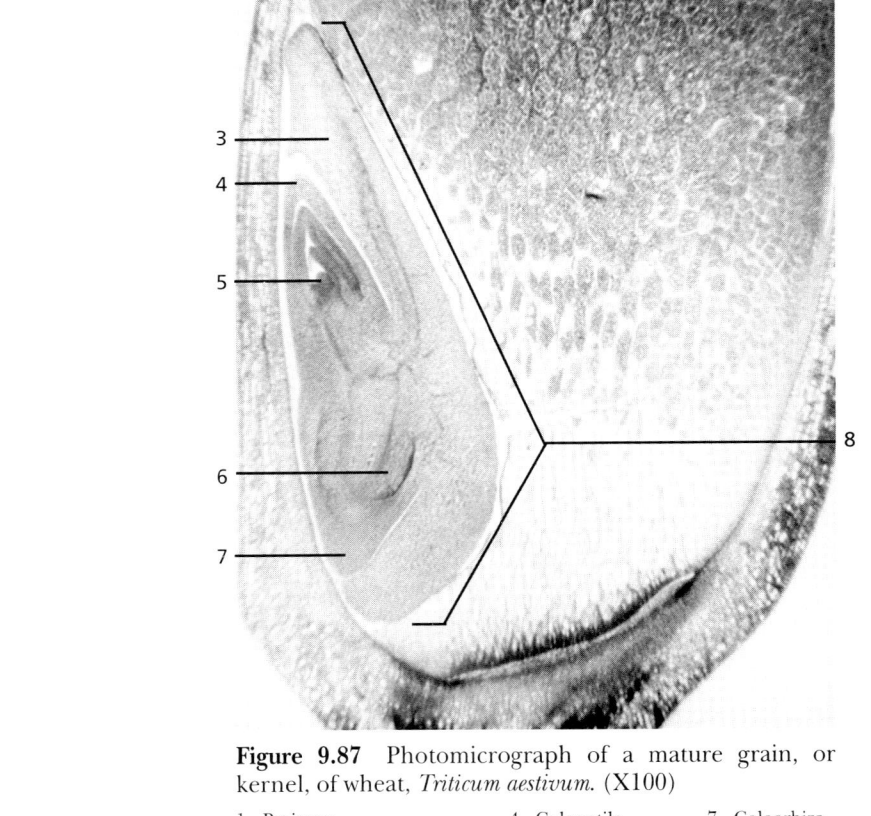

Figure 9.87 Photomicrograph of a mature grain, or kernel, of wheat, *Triticum aestivum*. (X100)

1. Pericarp
2. Starchy endosperm
3. Scutellum
4. Coleoptile
5. Shoot apex
6. Radicle
7. Coleorhiza
8. Embryo

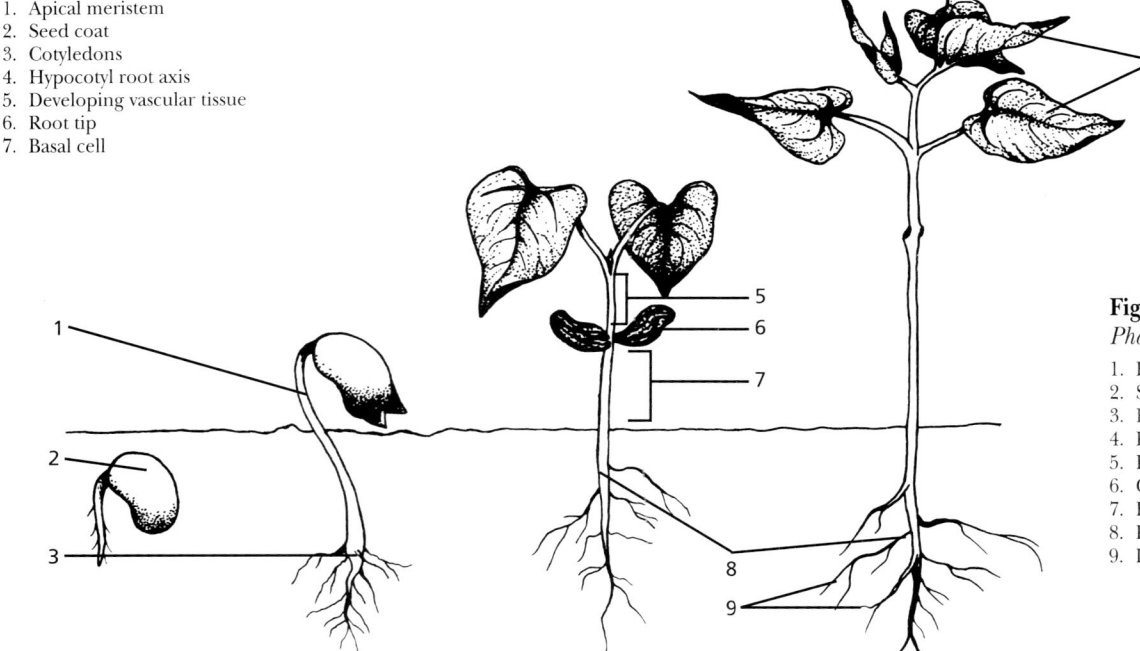

Figure 9.88 Diagram of bean, *Phaseolus*, germination.

1. Hypocotyl
2. Seed coat
3. Primary root
4. Foliage leaves
5. Epicotyl
6. Cotyledon
7. Hypocotyl
8. Primary roots
9. Lateral roots

CHAPTER 9   *Kingdom Plantae: Division Anthophyta — Angiosperms (Enclosed Seed Plants — Flowering Plants)*   **155**

Representative Specimens of Angiosperms

Figure 9.89   The duckweed, *Lemna*, is a small free-floating fresh-water plant found throughout the United States. The flowers are small and unisexual. Within the family Lemnaceae, *Lemna* is one of the smallest flowering plants in the world.

Figure 9.90   *Euphorbia canariensis*, a member of the spurge family, Euphorbiaceae. Living in African and Australian desert environments, species of *Euphorbia* provide examples of convergent evolution with specimens of the New World cacti of the family Cactaceae.

Figure 9.91   Herbarium specimen of a sage, *Salvia dorrii*, family Lamiaceae. *Salvia* lives in arid environments where it produces terpenes that inhibit the growth of other plants. This specialization tends to insure adequate moisture for an established plant.

Figure 9.92   Herbarium specimen of loco weed, *Astragalus oophorus*, family Leguminosae. Loco weed is toxic to livestock on the semiarid open ranges in the Western United States.

Representative Specimens of Angiosperms

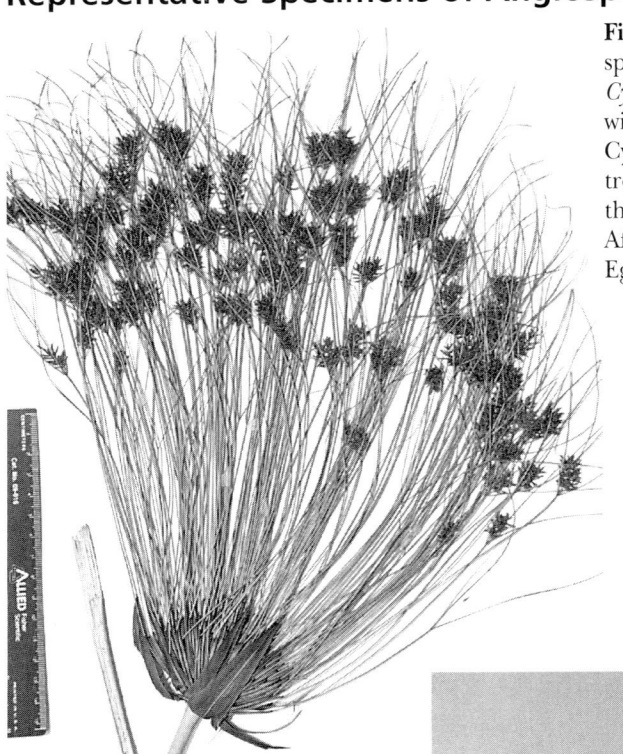

Figure 9.93 Herbarium specimen of papyrus, *Cyperus papyrus*. *Cyperus* is within the family Cyperaceae. Papyrus, a tropical reed that grows in the waterways of Northern Africa, was used by the Egyptians to make paper.

Figure 9.94 Herbarium specimen of Indian rice grass, *Stipa hymenoides*. Indian rice grass is a member of the family Poaceae and was used by Native Americans to make a mush-like food, rich in protein. *Stipa hymenoides* is the state grass of both Utah and Nevada.

Figure 9.95 Herbarium specimen of the sedge, *Carex scirpoidea*, family Cyperaceae. This species of sedge is a high altitude plant that occurs above the timberline in North America. The higher the altitude, the smaller the plant specimens become.

Figure 9.96 The saguaro cactus, *Carnegiea gigantea*, is the largest of all North American cacti. Arms begin to develop on the saguaro when the plant is about 75 years old. A saguaro cactus may live over 250 years and reach a height of more than 50 feet.

Figure 9.97 Herbarium specimen of a wild rose, *Rosa woodsii*. With only five petals, *Rosa woodsii* is an ancestral form of cultivated roses within the rose family, Rosaceae.

Representative Specimens of Angiosperms

Figure 9.98 Herbarium specimen of holly atriplex, *Atriplex hymenelytra*. Holly atriplex is a salt-tolerant plant within the family Chenopodiaceae. This species of atriplex has small, felt-like leaves. Its range is arid to semiarid regions in the Western United States.

Figure 9.99 Herbarium specimen of a lady slipper, *Cypripedium calceolus*, family Orchidiaceae. The lady slipper orchid is found in wet climates in the eastern United States and in mountainous regions of other parts of the United States.

Figure 9.100 Herbarium specimen of *Verbascum thapsus*. An Old World plant within the family Scrophulariaceae, *Verbascum thapsus* was used in making fire torches. Portions of the plant were also used for medicinal purposes.

Figure 9.101 Herbarium specimen of breadroot, *Cymopterus purpurascens*. Breadroot is a member of the family Umbelliferae. Developing foliage in the spring, this plant was used as a food by many Native Americans.

Representative Specimens of Angiosperms

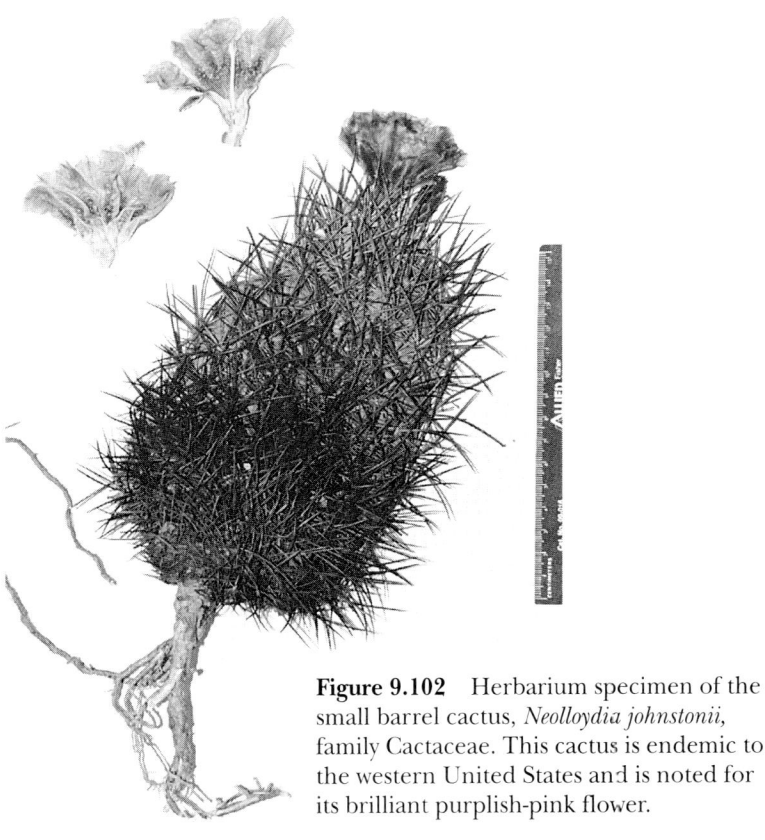

Figure 9.102 Herbarium specimen of the small barrel cactus, *Neolloydia johnstonii*, family Cactaceae. This cactus is endemic to the western United States and is noted for its brilliant purplish-pink flower.

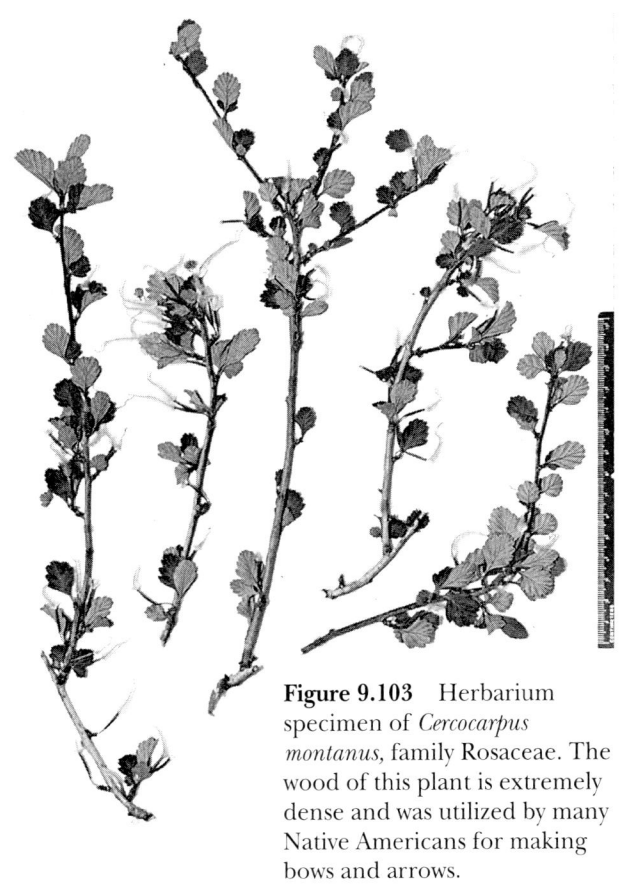

Figure 9.103 Herbarium specimen of *Cercocarpus montanus*, family Rosaceae. The wood of this plant is extremely dense and was utilized by many Native Americans for making bows and arrows.

Figure 9.104 Herbarium specimen of a sunflower, *Helianthus annuus*. Sunflowers produce flowers in a composite inflorescence and are members of the family Asteraceae.

Figure 9.105 Herbarium specimen of rice, *Oryza sativa*. A species within the grass family, Poaceae, *Oryza sativa* is the major food crop grown in Asia. Known to have been cultivated for over 7,000 years, rice requires warm temperatures and abundant moisture. Rice is one of the twelve most important human food plants.

Representative Specimens of Angiosperms

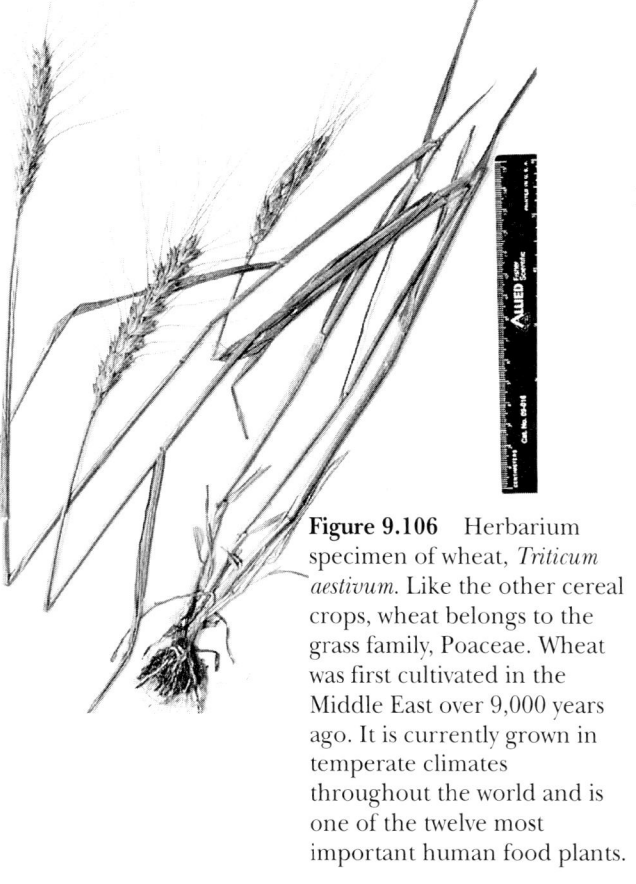

Figure 9.106 Herbarium specimen of wheat, *Triticum aestivum*. Like the other cereal crops, wheat belongs to the grass family, Poaceae. Wheat was first cultivated in the Middle East over 9,000 years ago. It is currently grown in temperate climates throughout the world and is one of the twelve most important human food plants.

Figure 9.107 Herbarium specimen of barley, *Hordeum vulgare*. Like wheat, rye is a member of the grass family, Poaceae. Rye has been cultivated as a grain crop since the time of ancient Rome. It is currently grown in cool climates of northern Europe, Asia, North America, and South America.

Figure 9.108 Herbarium specimen of maize (corn), *Zea mays*. Domesticated nearly 7,000 years ago from a Mexican grass in the family Poaceae, maize is currently grown throughout the world but more extensively in North America. During pre-Columbian times, it was cultivated by Native Americans in societies throughout North and South America. Although more than half of the cultivated maize in the United States is used for animal feed, it is the basis of many important food items ranging from cornbread to cereals to tortillas. Maize is one of the twelve most important human food plants.

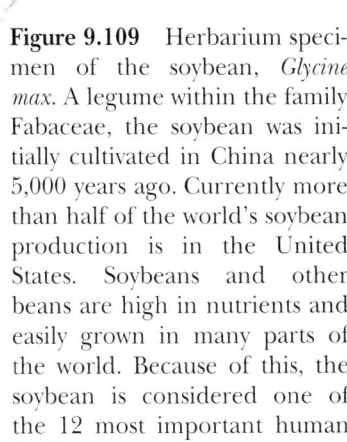

Figure 9.109 Herbarium specimen of the soybean, *Glycine max*. A legume within the family Fabaceae, the soybean was initially cultivated in China nearly 5,000 years ago. Currently more than half of the world's soybean production is in the United States. Soybeans and other beans are high in nutrients and easily grown in many parts of the world. Because of this, the soybean is considered one of the 12 most important human foods.

Representative Specimens of Angiosperms

Figure 9.111 Herbarium specimen of cassava, or manioc, *Manihot esculenta*. Within the spurge family Euphorbiaceae, cassava is an important starch root crop in South America, West Indies, Africa, and Indonesia. Cassava is a shrub that has large, starch-filled roots. Following the preparation of the root, the residue may be utilized in many ways, including baking as thin cake bread or drying as a meal called farinha. It is also the source of tapioca, used in puddings. Cassava is considered one of the twelve most important human food plants.

Figure 9.110 Herbarium specimen of a garden bean plant, *Dolichos lablab*. The garden bean is easily grown and the pod and seeds of the fruit provide a nutritious vegetable.

Figure 9.112 Herbarium specimen of the white potato, *Solanum tuberosum*. The potato is a member of the family Solanaceae. The potato tuber is a modified stem that is rich in nutrients and is a stable crop for millions of people. Cultivated by native South Americans, potatoes were introduced to Europe about 1570. Potatoes are currently grown worldwide, especially in temperate regions and at higher tropical elevations. The potato is one of the twelve most important human food plants.

Figure 9.113 Herbarium specimen of the coconut, *Cocos nucifera*. *Cocos nucifera* is a species within the palm family Arecaceae (or Palmae). Distributed along tropical shorelines, the coconut "meat" is a nutritious source of food for millions of people. Palm leaves are used in making shelters and the fibrous portion of the coconut fruit is used for making mats and rope. The coconut is one of the twelve most important human food plants.

Representative Specimens of Angiosperms

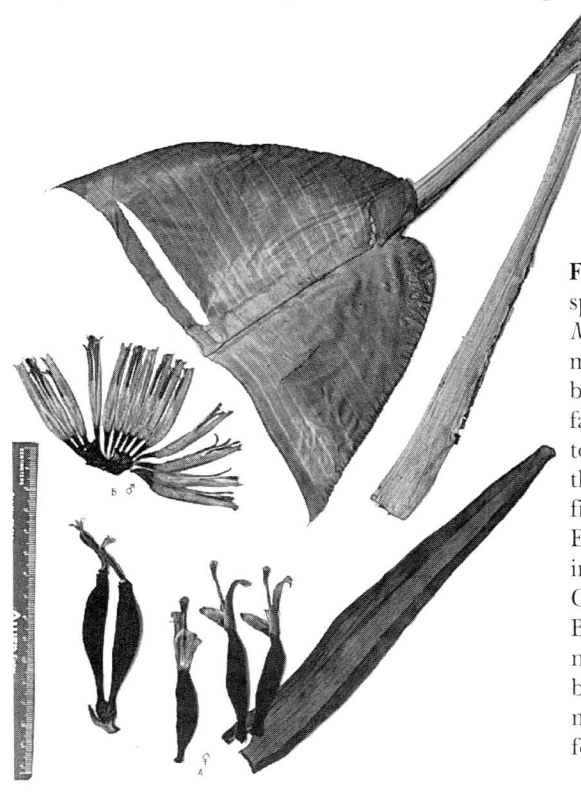

Figure 9.114 Herbarium specimen of the banana, *Musa balbiziana*. There are more than 300 varieties of bananas, all within the family Musaceae. Thought to have been domesticated thousands of years ago, the first became known to Europeans following the incursion of Alexander the Great into India (327 B.C.). Bananas are high in nutritional value. The banana is one of the twelve most important human food plants.

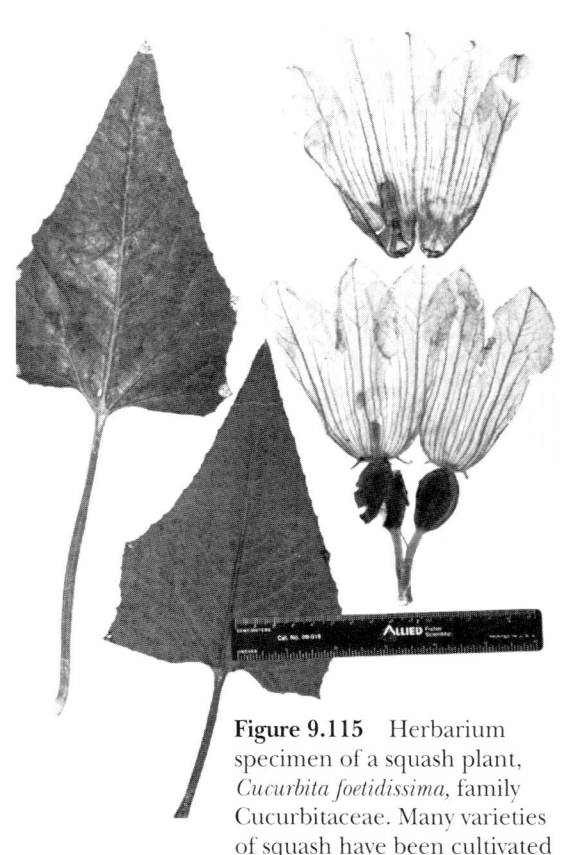

Figure 9.115 Herbarium specimen of a squash plant, *Cucurbita foetidissima*, family Cucurbitaceae. Many varieties of squash have been cultivated and are important food crops.

Figure 9.116 Herbarium specimen of tobacco, *Nicotiana tabacum*. The tobacco plant is in the nightshade family Solanaceae. Native to the American tropics, tobacco was first encountered by Columbus and his men in the West Indies. It is currently cultivated in countries throughout the world and as many as half the world's population chew, sniff, or smoke the products from the tobacco plant for the effects of the nicotine.

Figure 9.117 Herbarium specimen of sugarcane, *Saccharum ravennae*. There are several species of cultivated sugarcane grown in moist tropical regions throughout the world. Initially used in India nearly 5,000 years ago, sugarcane is a sturdy perennial grass with broad leaves. The sucrose content within the stems of certain varieties may reach 20% of the crop biomass. Sugarcane is one of the twelve most important human food plants.

Representative Specimens of Angiosperms

Figure 9.118 Herbarium specimen of the hot pepper, *Capsicum frutescens*. Native to North America, the hot pepper was introduced to Europe by Columbus. Because of its value in spicing foods, it is an important cultivated plant.

Figure 9.119 Herbarium specimen of the hemp plant, *Cannabis sativa*. The hemp plant is the source of marijuana and hashish. It is a dioecious annual that was initially cultivated in China as early as 3,000 B.C. The seeds of hemp are used for industrial oil and the plant body is a source of valuable fiber. In spite of its cultivation being illegal in many countries, products from the hemp plant are used by an estimated 200 million people throughout the world, making it an important cultivated crop.

Figure 9.120 Herbarium specimen of the opium poppy, *Papaver somniferum*. The latex of opium contains many alkaloids, of which morphine and codeine are the most important because of their medicinal benefits in making pain-killers. As a cultivated plant for the narcotics it produces, the opium poppy was first grown in Asia Minor as early as 2,500 B.C. Opium and its derivatives are commonly used narcotics, with an estimated 900 million users, mostly in Asia.

Figure 9.121 Herbarium specimen of the foxglove, *Digitalis purpurea*, within the family Scrophulariaceae. Foxglove is a native European wild flower that is an important medicinal plant to alleviate cardiac insufficiency and related problems.

Glossary of Terms

abiotic: that portion of the environment without living organisms; the non-living portion of the environment.

abscisic acid: a plant hormone that inhibits growth and promotes dormancy; helps the plant conserve energy and withstand stressful conditions.

abscission: the shedding of leaves, flowers, fruits, or other plant parts, usually following the formation of an abscission zone.

absorption: movement of a substance into a cell or an organism, or through a surface within an organism.

accessory bud: a bud developing on a stem or twig located above or on either side of the axillary bud.

acid: a substance that releases hydrogen ions (H^+) in a solution.

adsorption: movement of a substance onto a cell surface.

adhesion: an attachment between unlike substances or cells. No chemical bonds are formed between the two.

adventitious root: root developing from the stem of a plant; often functioning in support.

aeciospore: dikaryotic spore of rust fungi.

aecium: a cuplike structure in a rust fungus where aeciospores are produced.

aerobic: requiring free O_2 for growth and metabolism.

aggregate fruit: a fruit produced from a single flower with several separate carpels.

alga (*pl.* **algae**): any of a diverse group of aquatic photosynthesizing organisms that are either unicellular or multicellular; algae comprise the phytoplankton and seaweeds of the Earth.

alkaline: a substance having a pH greater than 7.0; basic.

allele: an alternative form of gene occurring at a given chromosome site, or locus; several alleles may exist for a single gene.

alternation of generations: two-phased life cycle characteristic of many plants in which sporophyte and gametophyte generations alternate.

amoeba: protozoans that move by means of pseudopodia.

anaerobic: metabolizing and growing in the absence of oxygen.

anatomy: the internal structure of an organism.

angiosperm: flowering plant, having double fertilization resulting in development of specialized seeds within fruits.

annual plant: a plant that completes its entire life cycle in a single year or growing season.

annual ring: yearly growth demarcation in woody plants formed by production of "spring wood" and "summer wood" in the secondary xylem.

anther: the terminal pollen sac of a stamen in an angiosperm flower where pollen grains with male gametes develop.

antheridium (*pl.* **antheridia**): a sperm-producing structure in an alga or non-seed plant.

apical meristem: embryonic plant tissue in the tip of a root, bud, or shoot where continual cell divisions cause growth in length.

archaebacteria: organisms of the kingdom Monera that represent an early group of simple life forms.

archegonium: multicellular female reproductive organ in certain nonseed plants; a gametangium where eggs are produced.

ascospore: a haploid spore produced within an ascus of a sac fungus (ascomycete).

asymmetry: a non-symmetrical morphology.

autosome: a chromosome other than a sex chromosome.

autotroph: an organism capable of synthesizing organic molecules (food) from inorganic molecules.

auxin: a category of plant hormone that stimulates cell differentiation and plant growth, such as phototropic response through cell elongation, stimulation of secondary growth, and development of leaf traces and fruit.

axillary bud: a group of meristematic cells at the junction of a leaf and stem which may develop a branch or flower(s); also called *lateral bud*.

bacillus (*pl.* **bacilli**): a rod-shaped bacterium.

bacteria: prokaryotes within the kingdom Monera, lacking the organelles of eukaryotic cells.

bark: outer tissue layers of a woody plant consisting of periderm, cortex, and phloem.

basal: at or near the base or point of attachment.

basidium (*pl.* **basidia**): a reproductive cell of basidiomycetes, where nuclear fusion to form a diploid cell followed by meiosis occurs to produce basidiospores.

berry: a simple fleshy fruit that develops from an ovary wall.

biennial plant: a plant that lives through two growing seasons; generally, these plants often have vegetative growth during the first season, and flower and set seed during the second.

bilateral symmetry: the morphologic condition of having similar halves.

binomial nomenclature: assignment of two names to an organism, the first of which is the genus and the second the species; together constituting the scientific name.

biome: a major climax community delineated by a characteristic group of plants and animals.

biosphere: the portion of the Earth's atmosphere and surface where living organisms exist.

biotic: pertaining to the living part of the environment.

bisexual flower: a flower that contains both male and female sexual organs.

blade: the broad expanded portion of a leaf.

brackish: water that is intermediate in saltiness between fresh water and seawater.

bryophyte: a plant within the division Bryophyta; a moss, liverwort, or hornwort; nonvascular plant that inhabits terrestrial environments but lacks many of the adaptations of most vascular plants.

budding: a type of asexual reproduction in which outgrowths from the parent plant pinch off to live independently or may remain attached to form colonies.

bulb: a thickened underground stem often enclosed by enlarged, fleshy leaves containing stored food.

callus: a mass of undifferentiated plant tissue often growing in a wound during the healing process.

carpel: the megasporophyll of an angiosperm.

carrying capacity: the maximum number of organisms of a species that can be maintained indefinitely in an ecosystem without causing damage.

catalyst: a chemical, such as an enzyme, that accelerates the rate of a reaction of a chemical process but is not used up in the process.

cell: the structural and functional unit of an organism; the smallest structure capable of performing all the functions necessary for life.

cell wall: a rigid protective structure of a plant cell surrounding the cell (plasma) membrane; often composed of cellulose fibers embedded in a polysaccharide/protein matrix.

cellular respiration: the reactions of glycolysis, Krebs cycle, and electron transport system that provided cellular energy and accompanying reactions to produce ATP.

cellulose: a polysaccharide produced as fibers that forms a major part of the rigid cell wall around a plant cell.

chlorophyll: green pigment in photosynthesizing organisms that absorbs energy from the sun.

chloroplast: a membrane-enclosed organelle that contains chlorophyll and is the site of photosynthesis.

chromosome: structure in the nucleus of a cell that contains the genes; comprised of a molecule of DNA and associated proteins.

climax community: a mature biological community that is the relatively stable terminal stage reached in ecological succession.

colony: an aggregation of organisms of the same species living together in close proximity.

community: an ecological unit composed of all the populations of organisms living and interacting in a given area.

competition: interaction between individuals of the same or different species striving to obtain a mutually necessary resource.

complete flower: a flower that has four whorls of floral components including sepals, petals, stamens, and carpels.

compound leaf: a leaf with a blade deeply divided into distinct leaflets.

conifer: a cone-bearing woody seed plant, such as pine, fir, and spruce.

convergent evolution: the evolution of similar structures in different groups of organisms occurring in similar environments.

cork: the protective outer layer of bark of many trees, composed of dead cells that may be sloughed off.

cortex: a primary tissue region of a plant root or stem, bounded externally by the epidermis and internally by the vascular system.

cotyledon: the leaves of a plant embryo, which in some plants enlarge and function as a storage site for nutrients to support early growth after seed germination.

crossing over: the exchange of corresponding chromatid segments (genetic material) of homologous chromosomes during synapsis in the first phase of meiosis.

cyanobacteria: photosynthetic prokaryotes that have chlorophyll and release oxygen; sometimes referred to as blue-green algae.

deciduous plant: a plant that seasonally sheds its leaves.

denitrifying bacteria: single-cellular organisms of the kingdom Monera that convert nitrate to atmospheric nitrogen.

detritus: non-living organic matter important in the nutrient cycle in soil formation.

diatoms: aquatic unicellular algae characterized by a cell wall composed of two silica impregnated valves.

dicot: a kind of angiosperm characterized by the presence of two cotyledons in the seed; also called *dicotyledon*.

diffusion: movement of molecules from an area of greater concentration to an area of lesser concentration.

dihybrid cross: a breeding experiment in which parental varieties differing in alleles for two traits are crossed.

dimorphism: two distinct forms within a species, with regard to size, color, organ structure, etc.

division: a major taxonomic grouping of plants that includes classes sharing certain features with close biological relationships.

dominant: a hereditary characteristic that is expressed when the genotype is homozygous or heterozygous.

dormancy: a period of suspended activity and growth.

double helix: a double spiral used to describe the three-dimensional shape of DNA.

ecology: the study of the relationship of organisms and the physical environment and their interactions.

ecosystem: a biological community and its associated abiotic environment.

embryo: a plant at an early stage of development; an embryo develops from a zygote and may begin growth immediately or become dormant.

endosperm: a plant tissue of angiosperm seeds that stores nutrients; the endosperm of an angiosperm is 3n in chromosome number.

epicotyl: portion of a plant embryo that develops to become part of the stem.

epidermis: the outermost protective layer of cells of a plant.

epiphyte: nonparasitic plant that grows on the surface of other plants.

estuary: the mixing zone between fresh water and seawater at the mouth of a river.

evolution: genetic and phenotypic changes occurring in populations of organisms through time, generally resulting in increased adaptation for continued survival.

fertilization: the fusion of two haploid gametes to form a diploid zygote.

fibrous root system: an intertwining mass of roots of about equal size.

filament: a long chain of cells.

filtration: the passage of a liquid through a filter or a membrane.

flora: a general term for the plant life of a region or area.

flower: the blossom of an angiosperm that contains the reproductive organs.

fossil: any preserved ancient remains or impressions of an organism.

frond: the leaf of a fern containing many leaflets.

fruit: a mature ovary enclosing a seed or seeds.

fruiting body: a reproductive structure of a fungus or slime mold in which spores are produced.

gamete: a haploid sex cell, often a sperm or egg.

gametophyte: the haploid, gamete-producing phase in a life cycle of a plant.

gemma (*pl.* **gemmae**): a small vegetative outgrowth of the thallus in liverworts or certain fungi that can develop into a new organism.

gene: part of the DNA molecule located in a definite position on a certain chromosome and coding for specific protein product.

gene pool: the total of all the alleles of the individuals in a population.

genetic drift: evolution by chance process; often due to the loss of parts of a population.

genetics: the study of genes, gene products and heredity.

genotype: the genetic makeup of an organism.

Glossary of Terms

genus: the taxonomic category above species and below family; the first name of a scientific binomial.

geotropism: plant growth oriented with respect to gravity; stems grow upward, roots grow downward.

germ cells: gametes or the cells that give rise to gametes.

germination: the process by which a spore or seed ends dormancy and resumes normal metabolism, development, and growth.

gibberellin: plant hormone producing increased stem growth by promoting cell division; also promotes seed germination and flowering.

girdling: removal of a strip of bark from around a tree down to the wood layer.

grana: a "stack" of membrane flattened disks within the chloroplast that contain chlorophyll.

growth ring: an annual growth layer of secondary xylem (wood) in gymnospems or angiosperms.

guard cells: epidermal cells at the side of a stomate that help to control the stoma size.

gymnosperm: a vascular plant producing naked (exposed) seeds, as in conifers.

gynecium: the carpel or carpels of an angiosperm flower.

habitat: the ecological abode of a particular organism.

herbaceous plant: a nonwoody plant.

herbaceous stem: stem of a non-woody plant; stem lacking wood.

heredity: the transmission of certain characteristics, or traits, from parents to offspring via the genes.

heterozygous: having two different alleles for a given trait.

holdfast: basal extension of a multicellular alga that attaches it to a solid object.

homothallic: species in which individuals produce both male and female reproductive structures and are self-fertile.

hybrid: an offspring from the crossing of genetically different strains or species.

hypha: a filament of cells that makes up the vegetative body of a fungus.

hypocotyl: portion of a plant embryo that contributes to the stem development; the hypocotyl is below the epicotyl.

indigenous: organisms that are native to a particular region; not introduced.

internode: region between stem nodes.

karyotype: the arrangement of chromosomes characteristic of the species or of a specific individual.

kingdom: a taxonomic category grouping related divisions (plants) or phyla (animals).

lateral root: a secondary root that arises by branching from an older root.

leaf veins: plant structures that contain the vascular tissues in a leaf.

legume: a member of the pea, or bean, family.

lenticel: spongy area in the bark of a stem or root that permits interchange of gases between internal tissues and the atmosphere.

lichen: an alga and fungus forming a single thallus and coexisting in a symbiotic relationship.

locus: the specific location or site of a gene within the chromosome.

marine: pertaining to the sea or ocean.

medulla: the center portion of an organ.

megaspore: a plant spore that will germinate to become a female gametophyte.

meiosis: cell division by which haploid cells are formed from a diploid cell; also referred to as *reduction division*.

meristem tissue: undifferentiated plant tissue that is capable of dividing and producing new cells.

mesophyll: the middle tissue layer of a leaf containing cells that are active in photosynthesis, gas exchange and sometimes storage.

microspore: a spore that develops to produce the male gametophyte.

migration: movement of organisms from one geographical site to another.

mitosis: the process of cell division, in which the two daughter cells are identical and contain the same number of chromosomes.

monocot: a type of angiosperm in which the seed has only a single cotyledon; also called *monocotyledon*.

mutation: a variation in heritable characteristic caused by a change in DNA; a permanent transmissible change in which the offspring differ from the parents.

mutualism: a beneficial relationship between two organisms of different species.

mycelium: the mass of hyphae that constitutes the body of a fungus.

natural selection: the evolutionary mechanism by which better adapted organisms are favored to reproduce and pass on their alleles to the next generation.

nitrogen fixation: a process carried out by certain organisms, such as some soil bacteria, whereby free atmospheric nitrogen is converted into ammonia compounds.

node: location on a stem where a leaf is attached.

nucleus: a spheroid body within the eukaryotic cell that contains the chromosomes of the cell.

nut: a hardened and dry single-seeded fruit.

oogonium: a unicellular female reproductive organ of some non-vascular plants that contains a single or several eggs.

organ: a structure consisting of two or more tissues, which performs a specific function.

organelle: a minute structure of the eukaryotic cell that performs a specific function.

organism: an individual living creature.

osmosis: the diffusion of water from a solution of lesser concentration to one of greater concentration through a semipermeable membrane.

ovule: the female reproductive structure in a seed plant that contains the megasporangium where meiosis occurs and the female gametophyte is produced. Ovules mature to become seeds.

palisade layer: the upper layer of the mesophyll of a leaf where abundant photosynthesis occurs.

parasite: an organism that resides in or on another from which it derives sustenance.

parenchyma: the principal structural cells of herbaceous plants; a relatively non-differentiated plant cell type.

pectin: an organic compound in the intercellular layer and primary wall of plant cell walls; the basis of fruit jellies.

pedicel: the stalk of a flower in an inflorescence.

perennial plant: a plant that lives throughout the year and grows during several to many growing seasons.

pericarp: the fruit wall that forms from the wall of a mature ovary; or female gametophyte tissue enclosing tetrasporophyte in some red algae.

pericycle: a tissue in the roots of plants, that is bounded externally by the endodermis and internally by the phloem.

petal: modified leaf occurring in a flower. Petals are often colored and functional in attracting pollinators; collectively called the *corolla*.

petiole: structure of a leaf that connects the blade to the stem.

phenotype: the appearance of an organism created by the genotype and environmental influences.

phloem: vascular tissue in plants that transports nutrients.

photosynthesis: the process of using the energy of the sun to make carbohydrates from carbon dioxide and water.

phototropism: plant growth or movement in response to a directional light source.

phytoplankton: microscopic, free-floating, photosynthetic organisms that are the major primary producers in fresh-water and marine ecosystems.

pistil: a reproductive structure of a flower comprised of the stigma, style, and ovary.

pith: a centrally located tissue within a dicot stem.

plankton: aquatic free-floating microscopic organisms.

plastid: an organelle of a plant where photosynthesis or food storage occurs.

pollen grain: mature microspore containing the male gametophyte generation of seed plants.

pollination: the delivery by wind, water, or animals of pollen to the ovule of a seed plant leading to fertilization.

population: all the organisms of the same species in a particular location.

primary producers: organisms within an ecosystem that synthesize organic compounds from inorganic constituents.

prokaryote: organism, such as a bacterium, that lacks the specialized organelles characteristic of complex cells.

prothallus: a heart-shaped structure that is the gametophyte generation of a fern.

protonema: the first stage of gametophyte development in mosses and liverworts.

radial symmetry: symmetry around a central axis so that any half of an organism is identical to the other.

receptacle: the tip of the axis of a flower stalk that bears the floral organs.

regeneration: regrowth of tissue or the formation of a complete organism from a portion.

renewable resource: a commodity that is not used up because it is continually produced in the environment.

replication: the process of producing a duplicate; DNA is replicated prior to cell division.

rhizoid: a minute hairlike extension of a fungus or plant that functions in nutrient and water absorption.

rhizome: an underground stem in some plants that stores photosynthetic products and gives rise to above-ground stems and leaves.

root: the anchoring subterranean portion of a plant that permits absorption and conduction of water, minerals, and nutrients.

root cap: end mass of parenchyma cells that protects the apical meristem of a root.

root hair: epidermal projection from the root of a plant that functions in absorption of water the nutrients.

salinity: saltiness in water or soil; a measure of the concentration of dissolved salts.

saprophyte: a heterotrophic bacterium, fungus, or plant that absorbs nutrients directly from dead and decaying organic matter.

savanna: open grassland with scattered trees.

sclerenchyma: supporting tissue in plants composed of cells with thickened walls.

secondary growth: plant growth in girth from secondary or lateral meristems.

seed: a plant embryo with a food reserve that is enclosed in a protective seed coat; seeds develop from matured ovules.

sepal: outermost whorl of flower structures beneath the petals; collectively called the *calyx*.

sessile: organisms that lack locomotion and remain stationary.

shoot: portion of a vascular plant that includes a stem with its branches and leaves.

sieve tube: a linear group of cells in the phloem functioning in translocation of dissolved photosynthetic products.

somatic cells: all the cells of the body of an organism except the germ cells.

sorus: a cluster of sporangia on the underside of fern pinnae.

species: a group of morphologically similar organisms that share a gene pool and are capable of interbreeding and producing fertile offspring and are generally reproductively isolated from other species.

spirillum (*pl.* **spirilla**): a spiral-shaped bacterium.

sporangium: any structure within which spores are produced.

spore: a reproductive cell capable of developing into an adult organism without fusion with another cell.

sporophyll: a sporangium-bearing leaf.

stamen: a reproductive structure of a flower, comprised of a filament and an anther, where pollen grains are produced.

starch: carbohydrate molecule synthesized from photosynthetic products; common food storage substance in many plants.

stele: the vascular tissue and pith or ground tissue at the central core of a root or stem.

stigma: the upper portion of the pistil of a flower. Pollen grains become attached to the stigma.

style: the long slender portion of the pistil of a flower.

succession: the sequence of ecological stages by which a particular biotic community gradually changes until replaced by another community.

sucrose: a disaccharide (double sugar) consisting of a linked glucose and fructose molecule; the principal transport sugar in plants.

symbiosis: a close association between two organisms where one or both species derive benefit.

syngamy: union of gametes in sexual reproduction; fertilization.

taproot: a plant root system in which a single root dominates the root system.

taxonomy: the science of describing, classifying, and naming organisms.

thallus: a flattened plant body often with little tissue specialization and lacking roots, stems, or leaves.

tissue: an aggregation of similar cells and their binding intercellular substance, joined to perform a specific function.

toxin: a poisonous compound.

trait: a distinguishing feature of an organism, often studied in heredity.

turgor pressure: osmotic pressure that provides rigidity to a cell.

vacuole: a membrane-bound, fluid-filled organelle.

vascular cambium: a layer of meristematic tissue in roots and stems of many vascular plants that continues to produce vascular tissue.

vascular tissue: plant tissue composed of xylem and phloem, functioning in transport of water, nutrients, and photosynthetic products throughout the plant.

vegetative: plant parts not specialized for reproduction; asexual reproduction.

wood: interior tissue of a tree composed of secondary xylem.

xylem: vascular tissue in plants that transports water and minerals.

zoospore: a flagellated motile plant spore.

Index

Abies lasiocarpa, 118
Acer, 134
Acetate, 11
Achenes, 146
Achnanthes flexella, 19
Acorn, 151
Acrasiomycota, 25
Adiantum, 96-97
Adventitious roots, 92
Aecia, 64
Aeciospores, 63-64
Aecium, 63-64
Aerial roots, 125
Aerial stem, 82, 86-87
   club moss, 86
   rhizomes, 87
   whisk fern, 82
Agaricus rodmani, 61
Aggregate fruits, 147, 151
Air pore, 71, 72
Air bladder, 38
Akinete, 13
Alfalfa, 8
Alga, 66-67
Algal cells, 67
Algal layer, 67
Algal protists, 18
Algin, 25
Allium, onion, 9, 129
Allomyces, 50
Allophylus, 134
Alternation of generations, 68
Amaranthus, 147
Amaryllis, 142
Amaryllis belladonna, 142
Amoebas, 22, 47
Amoeba proteus, 22
Anabaena, 13
Anaerobes, 11
Anal pore, 24
Anaphase
   lily, 10
   onion, 9
Angiosperms, 119, 121-162
   avocado, 126
   bean, 137, 152, 154, 160
   buttercup, 127
   corn, 126, 130, 152, 159
   dandalion, 150
   geranium, 138-139
   Hibiscus, 142
   lady slipper orchid, 149, 157
   lilac, 148
   lily, 153
   milkweed, 150
   oak, 131, 132
   oleander, 136
   onion, 129
   Panama hat plant, 136
   pea, 145
   peanut plant, 152
   pear, 127, 140, 146
   potato, 129, 160
   rose, 143, 156
   shepherd's purse, 154
   tomato, 153
   tulip, 142
   venus flytrap, 139
   wheat, 126, 131, 154
   yucca, 140
Annular scars, 32
Annulus, 62, 94, 97
   fern, 94, 97
   mushroom, 62
Anobacteria, 11
Anther, 122, 141-146, 148
   angiosperm, 122, 141
   candy tufts, 145
   Family Asteraceae, 144
   gladilus, 143
   grassess, 144
   pear, 146
   pollination, 148
   rose, 143
   tomato, 146
   tulip, 142
Antheridia
   ascomycete, 54
   club moss, 85
   brown alga, 41-42
   fern, 94
   green alga, 31-33, 37
   liverwort, 70, 74
   moss, 77, 79
   water felt, 21
   whisk fern, 82
Antheridial head
   liverwort, 74
   moss, 79
Antheridial receptacle, 71
Antheridium, 20, 21, 32, 37, 42, 54, 70, 74, 77, 79, 82, 94
   water felt, 20
Anthoceros, 75-76
Anthocerotophyta, 68
Anthophyta, 121
Antipodal cells, 153
Antipodals, 148
Apex, 154

Apical meristem
   bean, 137
   pear, 127
   shepherd's purse, 154
Apothecium, 55, 66
Apple, 151, 153
Araucaria excelsa, 111
Araucaria auracana, 111
Araucaria, 111, 113
Archaebacteria, 11
Archegonia
   club moss, 85
   cycad, 102, 105
   fern, 94, 98
   liverwort, 70, 73-74
   moss, 77, 79
   Selaginella, 88
   whisk fern, 82
Archegonial receptacle, 70-73
Archegonial heads, 72,
Archegonium, 70, 73-74, 77, 82, 85, 88, 94, 102, 105
Arcyria, 48
Arrowroot, 148
Asci, 55-56
Asclepias, 150
Ascocarp, 54-56
Ascogenous hyphae, 54
Ascogonium, 54
Ascomycetes, 54-56, 59
Ascospores, 54-57
Ascus, 54, 57
Asexual cycle
   bread mold, 53
   green alga, 29, 31
   Volvox, 27
   water felt, 20
Asexual reproduction, 1
   fission, 9
   vegetative propagation, 9
   water felt, 9
Aspergillus, 58-59
Asplenium, 95-96
Asters, 144
Astragalus oophorus, 155
Astrosclereid, 7
Atriplex hymenelytra, 157
Axillary bud
   bean, 137
   Coleus, 130
   dicot, 121
Azolla, 95

Bacillus, 12
Bacillus megaterium, 17
Bacteria, 11, 12, 14, 16

Baker's yeast, 55
Barberry leaf, 64
Bark, 112, 128
Barley smut, 5
Basal cell, 154
Basal, 32
Basal body, 23
Basidia, 60, 62
Basidiocarp, 60-61
Basidioma, 60
Basidiomycete, 62
Basidiospores
   mushroom, 60, 62
   wheat rust, 63
Basswood, 132
Bean, 137, 152, 154
Begonia, 6
Berry, 151
Blackberry, 151
Blade
   brown alga, 42
   kelp, 41
   Sarassum, 40
   seaweed, 38
Blepharostoma, 69
Blue-green algae, 11
Blueberries, 151
Boletus, 61
Borrelia recurrentis, 17
Botrychium multifidum, 95
Branches, 92
   angiosperm, 128
   conifer, 114
   Equisetum, 93
Bread mold, 52-53
Breadroot, 157
Brevifolia, 6
Brome grass, 64
Brown algae, 25, 38-42
Bryophyta, 68
Bryophytes, 68-80
Bud, 79
Budding, 2
Bulb, 128
Bundle sheath, 137
Burdock, 151
Buttercup, 127

Cabbage, 49
Calyptra
   liverwort, 72-73
   whisk fern, 82
Calyx, 141
Camas, 124
Cambium
   angiosperm, 112

dicot, 127
linden, 132
Canal cells, 73
Candy tuft, 145
Capsella bursa-pastoris, 153, 154
Capsule
   hornwort, 76
   Klebsiella pneumoniae, 17
   liverwort, 72-74
   moss, 77, 80
Carex scirpoidea, 156
Carludovica palmata, 136
Carnegiea gigantea, 156
Carpels, 141
Carpospores, 44-45
Carposporophyte, 45
Carrot, 125
Cell
   cycle, 1
   division, 10
   lumen, 6, 8
   membrane, 1-2, 5, 8, 22-23
   nucleus, 32
   wall, 1-10, 30, 35
Cellular slime molds, 25
Cellulose, 1, 18, 22, 25
Centromere, 9, 10
Ceramium, 43
Ceratium hirundinella, 21
Chalaza, 153
Chara, 37
Chicory, 142
Chlamydamonas, 26
Chlorophyll, 11
Chlorophyta, 25
Chloroplasts, 1-2, 18,
   angiosperm, 134
   Elodea, 4
   Euglena, 23
   green alga, 30
   lycododium, 87
   protista, 25
   Spirogyra, 35
   water felt, 21
   sugar cane, 5
Chondrus crispus, 46
Chromatids, 2, 9-10
Chromosomes, 1, 9-10
Chrysophyta, 18
Cichorium intybus
Cilia, 24
Ciliates, 24
Class Anthocerotae, 75-76
Class Hepaticae, 69-75
Class Musci, 68, 77-80

167

Class Ulvophyceae, 29, 37
Cladophora glomerata, 37
Clavariaceae, 61
Claviceps purpurea, 56
Cleistothecium, 55
Cleistothecium, 36, 59
Closterium, 36
Club mosses, 85-90
Cocci, 12
Cocconeis, 18
Coccus, 11-12
Coconut, 151
Codium, 36
Coenocytic filament, 21
Coleoorhiza, 154
Coleoptile, 147, 154
Coleus, 130
Collenchyma, 3, 6
Columbine, 124
Columella
  bread mold, 53
  hornwort, 76
  moss, 79
  Stemonitis, 49
Comatricha typhoides, 49
Composite flower, 150
Conceptacle, 41-42
Cone axis, 114
Cones, 119-120
Conidia
  ascomycete, 54
  Aspergillus, 59
  Penicillium, 58
  powdery mildew, 57
Conidiophores, 58, 59
Conifers, 109-115, 116-118
Conjugation, 34-36
Contractile vacuole, 22-24
Coprinus, 62
Coral fungus, 61
Coralina, 43
Corm, 90, 128
Corn, 124-126, 130, 151-152
Corn stalk, 64
Corolla, 141, 144
Cortinarius, 61
Cortex
  buttercup, 127
  clover, 131
  club moss, 86-87
  conifer, 112
  cycad, 103
  dicot, 121, 123, 127
  ginkgo tree, 107
  greenbriar, 126
  linden, 132
  maidenhair fern, 96
Cosmarium, 36
Cotyledon, 124, 147, 152, 154
Crustose lichen, 66
Cucurbita maxima, 6
Cuticle, 3, 68, 137, 140
Cyanobacteria, 11, 13
Cyanobacterium, 14
Cyanophyta, 13
Cycads media, 104
Cycas revoluta, 103, 104
Cycas, 105
Cyclotella, 19
Cymbella, 18
Cymopterus purpurascens, 157
Cyperus papyrus, 156
Cypripedium calceolus, 157
Cyrtomium falcatum, 96, 98

Cystocarps, 45
Cytokinesis, 1
Cytoplasm, 1, 4, 8, 14-15, 18

Dandelions, 144, 150-151
Daughter colonies, 27-28
Dead man's fingers, 36
Deciduous leaf, 134
Dermal tissue system, 1, 3
Desmids, 36
Diatoma, 18
Diatoms, 18-21
Dibotryon morbosum, 59
Dicots
  angiosperms, 121-124
  arrowroot, 148
  leaf, 137
  root tips, 127
  sherpherd's purse, 153-154
  woody branchess, 129
Dicotyledonae, 121
Dictyosome, 2
Dikaryotic cells, 60
Dionaea muscipula, 7, 139
Diospyros virginiana,
  persimmon, 8
Diplobacillus, 11
Diplococcus, 11
Diploid cells, 60
Diploid stage, 60
Division Anthophyta, 121, 122, 124
Division Apicomplexa, 23
Division Ascomycota, 54-59
Division Basidiomycota, 60-65
Division Bryophyta, 68-80
Division Chlorophyta, 26, 28-37
Division Chrysophyta, 18-21
Division Ciliophora, 24
Division Coniferophyta, 109-115
Division Cycadophyta, 102-105
Division Ginkgophyta, 101, 106-108
Division Gnetophyta, 118-120
Division Lycophyta, 85-90
Division Myxomycota, 47-49
Division Oomycota, 50
Division Phaeophyta, 37, 41, 42
Division Psilotophyta (*also* Psilophyta), 81-82
Division Pterophyta, 94
Division Pyrrhophyta, 21-22
Division Rhizopoda, 22
Division Rhodophyta, 43-46
Division Sphenophyta, 91-93
DNA, 2
Douglas fir, 117
Downy mildews, 25, 50
Drupe, 151
Duckweed, 155
Dyad, 9
Dyads, 10

Earthstars, 61
Ectoplasm, 22
Ectoplasm, 22
Egg
  brown alga, 41-42
  Bryophyta, 68

club moss, 85
green alga, 31-32, 37
lily, 153
liverwort, 70
moss, 77, 79
pollination, 148
Selaginella, 88
Volvox, 27-28
water felt, 20
water mold, 50
whisk fern, 82, 94
Eichhornia, 137
Elaters, 70, 74-76, 93
Elodea, 4
Elongation region, 125, 127
Embryo, 70, 74, 77, 82, 85, 102, 105, 108-109, 122, 147, 152-154
Endodermis, 86, 113, 126-127
Endoplasm, 22
Endoplasmic reticulae, 2, 8, 18
Endosperm, 8, 122, 147, 153-154
Endospores, 11, 17
Endothelium, 153
Envelope, 2
Enzymes, 11
Ephedra, 101, 118-120
Ephedra fassciculata, 118
Epicotyl, 147, 154
Epidermal hairs, 139
Epidermis,
  angiosperms, 137-140
  avocado, 126
  buttercup, 127
  club moss, 86-87
  conifers, 113
  corn, 152
  cycad, 104
  liverwort, 71, 76
  maidenhair fern, 96
  Mormon tea, 119
  venus flytrap, 7
  wheat rust, 63
Epigynous, 141
Epiphytes, 33
Equisetum, 93
Equisetum arvense, 81, 92
Erysiphe graminis, 57-58
Escherichia coli, 15
Eubacteria, 11
Euglena, 23
Eukaryotic, 18
Eunotic, 19
Euphorbia canariensis, 155
Euphorbia, 155

Family Asteraceae, 144
Family Cactaceae, 155
Family Cupressaceae, 116
Famly Ephedraceae, 119
Family Leguminosae, 155
Family Orchidiaceae, 157
Family Pinaceae, 117-118
Family Poaceae, 152
Family Scrophylariaceae, 157
Fascicular cambium, 131
Ferns, 94-98
Fertilization
  angiosperm, 122
  green alga, 29, 31
  liverwort, 70
  moss, 77

pore, 21
red alga, 44
Volvox, 27
water felt, 20
Fiber, 4, 8
Fibrous root system, 125
Fiddleheads, 81, 94-95
Fig, 151
Filaments
  cabbage, 49
  candy tuft, 145
  cellular, 49
  flowers, 146
  Gladiolus, 143
  green alga, 29-31, 33, 35
  grassess, 144
Filbert, 151
Fir tree, 118
Fission, 2, 9
Flagella, 16, 21
Flagellum, 23
Flax, 8
Floats, 39, 40-41
Floral tube, 146
Floret, 144
Flowers, 121-122, 128, 141-146
Foliose lichen, 66
Food vacuole, 22, 24
Foot, 73-74, 76
Fragaria, 146
Fragmentation, 2
Fronds, 81
Fucus, 38, 41-42
Fruit wall, 152, 153
Fruit, 147
Fruits, 101, 144-146, 154
Fruticose lichen, 66
Fuligo, 48
Fungal hyphae, 67
Fungal layer, 67
Fungi, 25, 51, 64, 66-67
Funiculus, 153

Galls, 132
Gametangia
  club moss, 85
  green alga, 37
  water felt, 21
  water mold, 50
Gametangium, 29
Gamete, donor, 35
Gametes, 1, 29, 36, 47, 50, 68
Gametophyte
  cycad, 102
  club moss, 85
  ginkgo, 105
  hornwort, 75-76
  liverwort, 74
  moss, 77-79
  Selaginella, 88
  red alga, 44-46
  water felt, 20
  whisk fern, 82
Ganoderma applanatum, 61
Gastrointestinal (GI) tract, 12
Geastrum saccutum, 61
Gelidium robustum, 43
Gemmae cup, 71, 75
Geranium, 138, 139
Germ sporangium, 52

Germination, 34, 147-154
Giant kelp, 38, 42
Gills, 60-62
Ginkgo, 106-108
Ginkgo biloba, 101, 107-108
Girdles, 21
Gladiolus grandiflorus, 142
Gladiolus, 128, 142-143
Gland, 7
Glenodinium monensis, 21
Glume, 144
Gnetophytes, 119-120
Gnetum, 101, 120
Golden algae, 18-19, 20-21
Golden-brown algae, 18
Golgi apparati, 2, 18
Gonium, 26
Grana, 5
Grape, 7, 128, 151
Grass, 125, 151
Green alga, 25, 27-37
Ground tissue system, 1, 3, 130-131
Growth, 1
Guard cell, 87, 134, 137, 138
Gullet, 24
Gymnodinium, 21
Gymnosperms, 101-121

Haploid cells, 10
Hedge privet, 137
Helvella, 56
Hemitrichia, 48
Hepaticophyta, 68
Herbaceous stems, 128
Heterocyst, 13
Heterotrophic saprophytes, 12
Heterotrophs, 25
Hibiscus, 142
Hilum, 152
Histones, 1
Holdfast
  brown alga, 38
  green alga, 32-33
  kelp, 39
Holly, 157
Homeostasis, 1
Hook, 54
Hornworts, 75-76
Horsetails, 81, 91-93
Hoya carnosa, 8
Hydration, 147
Hydrodictyon, 37
Hymenium, 57
Hyphae
  bread mold, 52-53
  mold, 59
  morel, 51, 54
  mushroom, 60, 62
Hyphal cell, 63
Hypocotyl, 147, 152, 154
Hypogynous, 141
Hypoxylon, 55

Imbibition, 147
Immature plastid, 8
Indian rice grass, 156
Indusium, 97, 98
Inflorescence, 144
Integument
  cycad, 102
  ginkgo, 108
  lima bean, 152
  lily, 153

Index

Interfascicular region, 131
Internode
    angiosperms, 128-129
    dicot, 121
    horsetail, 91-92
Interphase, onion, 9
Iarvensis, 142
Iris, 124, 128
Irish moss, 46
Isoetes melanopoda, 90
Isogametes, 29

Juniper, 116
Juniperus, 116

Karyogamy, 47, 52, 54, 60
Keel, 145
Kelp, 39
Klebsiella pneumoniae, 17
Kingdom Fungi, 51
Kingdom Monera, 11
Kingdom Plantae, 68-98, 101-162
Kingdom Protista, 18, 25

Lady slipper orchid, 149
Lady Slipper, 157
Lamina
    brown alga, 39
    dicot, 121, 134
    leaves, 135, 136
Laminara, 38-39
Larch, 118
Larix, 113
Larix dahurica, 118
Lateral roots, 154
Lateral bud, 128-129
Leaves
    club moss, 85-86
    dicot, 121
    Ephedra, 119
    gap, 96, 130
    ginkgo tree, 107
    peat moss, 78
    pitcher plant, 139
    plant organization, 3
    sear, 128
    sheaths, 92-93
    trace, 86
    veins, 124
Legumes, 12, 151
Lemma, 144
Lemna, 155
Lenticels, 128-129
Lepidozia, 69
Lichens, 66-67
Ligule, 88-89
Ligustrum, 137
Lilac, 148
Lilium, 146, 153
Lilium columbianum, 142
Lily, 10, 142, 146, 149, 153
Lima bean, 151
Linden, 132
Linum, 8
Liverworts, 69-75
Lobe, 97
Lobularia, 145
Loco weed, 155
Locule, 153
Lodicule, 144
Lycogala, 48
Lycoperdon ericetorum, 61
Lycoperscion esculentum, 146

Lycophyte, 81
Lycopod, 86, 90
Lycopodium clavatum, 81, 85-87

Macronucleus, 24
Macrosclerids, 152
Maidenhair tree, 106
Maidenhair fern, 96-97
Maple, 151
Marchantia, 70-75
Margin, 121, 135
Marigolds, 1 88, 94, 102, 122, 44
Maturation region, 125, 127
Megagametophyte
    angiosperm, 122
    ginkgo tree, 108
    pine, 109
    Selaginella, 88-89
Megasporangium, 88-89, 105
Megaspore, 88-90, 122
Megasporophyll, 88-89, 102, 105
Meiosis, 10, 20, 25, 27, 29, 31, 41, 47, 52, 54, 60, 70, 77, 85, 88, 94, 102, 122
Meiospores
    liverwort, 70, 74
    moss, 77, 79
    pine, 109
    slime mold, 47
Merismopedia, 13
Meristem, 130
Meristematic region, 125
Merozites, 23
Mesophyll, 104, 134
Metabolism, 1, 11
Metaphase, 10
Methanobacteria, 11
Methanogens, 11
Micrococcus luteus, 16
Microgametophyte, 88
Micronuclei, 24
Micronucleus, 24
Microphylls, 85-86, 102
Micropyle
    cycad, 102
    ginkgo, 108
    lily, 153
Microsporangia, 88, 90, 102, 104-105, 108-109, 114, 122
Microsporangiate cones, 109, 114, 119-120
Microspore, 10, 88-90, 109
Microsporophyll
    cycad, 104-105
    ginkgo, 108
    Selaginella, 88
Microtubles, 1
Midrib, 121, 134, 136
Milkweed, 150
Mimosa pudica, 138
Mitochondria, 18
Mitochondrion
    alfalfa, 8
    barley smut, 5
    sugar cane, 2, 5, 8
Mitosis, 1, 9, 54
Mnium, 79
Modified taproot, 125
Molds, 54-59
Monocots, 122-126, 130-131
Monocotyledonae, 121

Monolina fructicola, 56
Morchella, 56, 57
Morels, 54-59
Mormon tea, 118-119
Morning Glory, 142
Mosquito, 23
Mosses, 68, 77-80
Multiple fruits, 147, 151
Mushrooms, 60-65
Mussel shell, 39
Mycelium, 57
Myxomycota, 25

Navicula, 18, 19
Neck, 79
Nemalion, 43
Nereocystis, 39
Nerium oleander, 136
Nitrobacter, 12
Nitrogen, 12
Nitrosomonas, 12
Node
    angiosperm, 128, 134
    dicot, 121, 129
    Equisetum, 92
    fern, 93
Norfolk Island pine, 111
Nostoc, 13
Nucellar tissue, 153
Nucellus, 108-109
Nuclei, 37
Nucleic acid, 1
Nucleolus, 1-2, 8, 35
    alfalfa, 8
Nucleus, 1, 4, 8, 18, 22-23
    alfalfa, 8
    barley smut, 5
    Elodea, 4
    envelope, 2
    membrane, 1
    pores, 2, 5
    potato, 4
    sugar cane, 5
Nuphar, Pondlily, 7
Nuts, 151

Oak, 132
Oedogonium, 31-33
Oleander, 136
Onion, *Allium*, 9
Oogonia, 32, 42
Oogonium
    brown alga, 41-42
    green alga, 31-32, 37
    Volvox, 27
    water felt, 20
    water mold, 50
Oomycota, 25
Operculum, 79-80
Opuntia, 140
Oral cavity, 24
Orchid, 125
Organ, 3
Organelles, 1
Organism, 1, 3
Organs, 1
Oscillatoria, 13, 14
Ostiole, 42
Ovary, 122, 141-146, 148, 150, 153
Ovulate cones, 109, 114
Ovule, 102, 105, 107, 114, 122, 141, 143-144, 146, 148, 153
Ovulferous scale, 114

Palea, 144
Palisade mesophyll, 137
Panama hat plant, 136
Paphiopedilum venustii, 149
Pappus, 150
Pappus scale (plume), 144
Papyrus, 156
Paramecium, 24
Paramecium caudatum, 24
Paramylon granule, 23
Paraphyses, 42, 79
Parenchyma cell
    buttercup, 127
    corn, 130
    Pinus, 112
    plant tissue, 3-4
    flax, 8
    pondlily, 7
    squash, 6
    yucca, 7
Pea, 145
Peach, 151
Peanut plant, 152
Pear, 127, 146
Peat moss, 78
Pedicel
    flowers, 145-146
    fruit, 153
    pollination, 148
Peduncle, 144
Pelia, 75
Pellicle, 23, 24
Penicillium, 58
Pennate diatoms, 18
Peptidoglycan, 11
Pericarp, 45, 154
Pericycle
    buttercup, 127
    club moss, 86
    greenbriar, 126
Periderm, 3, 112, 128, 132
Peridinium wisconsiense, 21
Perigynous, 141
Peristoma, 80
Perithecia, 54-56
Persea, 126
Persimmon, *Diospyros virginiana*, 8
Petal, 122, 141-142, 146
Petiole, 121, 123, 134, 135
Peudomonas, 12
Peziza repanda, 56-57
Phaeophyta, 25
Phagocytosis, 18, 25
Phaseolus, 137, 152, 154
Phloem, 3, 7, 32, 137, 86-87, 89, 96, 112-113, 121, 123, 126-128, 130-131,
Photoautotrophs, 25
Photoreceptor, 23
Photosynthesis, 11, 134
Photosynthetic, 18
    cells, 78
    bacteria, 11
    mesophyll, 113
    tissue, 71, 76, 79
Phycodrys rubens, 46
Phylloceladus alpinus, 115
Phylloceladus, 115
Physarum, 49
Picea pungens, 110-111
Pigweed, 147
Pileus, 60, 62
Pine needles, 114

Pineapple, 151
Pinna, 96, 98
Pinnae, 96
Pinnularia, 19
Pinnule, 94, 97-98
Pinocytosis, 18
Pinus, 5, 109-110, 112-115, 117
Pinus edulis, 113
Pinus flexilis, 117
Pinus lambertiana, 113
Pistil, 141, 145
Pisum, 145
Pitcher plant, 139
Pith, 103, 107, 112, 119, 121, 126, 128, 131-132
Plankton, 18-19, 25
Plant cells, 4
Plant organization, 3
Plasma, 1-2
Plasma membrane, 11
Plasmodesmata, persimmon, 8
Plasmodial slime molds, 25, 47-49
Plasmodiophora brassicae, 49
Plasmodium, 25, 47
Plasmodium vivax, 23
Plasmogamy, 47, 52, 54, 63
Platanus, 134
Platycerium alcicorne, 96
Plectocolea, 69
Plume, 144
Plumule, 152
Podocarpus, 111
Polar nuclei, 148, 153
Pollen
    cycad, 102
    grains, 124, 141, 143
    pigweed, 147
    pine, 109
    pollination, 148
    tube, 122, 149
    wings, 115
Pollen tube, 153
Pollen tube, 122, 145, 149
Pollination, 141, 148-149
Polymerases, 11
Polypodium, 97
Polysiphonia, 43-46
Polystichum, 98
Pome, 151
Pond cypress, 116
Pondlily, *Nuphar*, 7
Poppy, 151
Porella, 74
Pores, 78
Porphyra, 43
Postelsia palmaeformis, 38
Potato, *Solanum tubersum*, 4, 6, 25, 128-129
Powdery mildew, 57, 58
Prickly pear, 140
Primordium, 137
Procambium, 130
Prokaryotic cells, 1, 11, 12
Prop roots, 125-126
Prophase, 9-10
Prothallus, 81
Protists, 18, 25
Protonema, 79
Protozoa, 18, 22
Pseudomonas, 15
Pseudopodia, 22, 78

Pseudotsuga menziesii, 117
Psilotum, 81-82
Puccinia graminis, 63
Puffball, 61-62
Pycnium, 63-64
Pyrenoid, 35
Pyrrhophyta, 18
Pyrus, 127

Quercus, 131-132
Quillworts, 85-90

Rachilla, 144
Radicle, 147, 152, 154
Ranunculus, 127
Raphe, 19
Ray parenchyma cells, 4
Ray flower, 144
Receptacle, 38, 41-42, 143-144, 146, 148, 153
Red blood cells, 17, 23
Red algae, 25, 43-46
Reservoir, 23
Resin duct, 112-113
Reticulum, 8
Rheumatic fever, 15-16
Rhizobium, 12
Rhizoids
 bread molds, 52-53
 fern, 98
 liverwort, 70-71
 mosses, 68
Rhizomes, 81-82, 91-92, 94, 128
Rhodophyta, 25
Ribosomes, 2, 8
 alfalfa, 8
Rhizopus, 52-53
Riccia fluitans, 69
Rockweed, 38, 42
Root cap, 121, 125, 127
Root tip, 154
Root hair, 121, 125-127
Root systems, 125
Root tip, 127
Roots, 3, 85, 88-89, 121, 123-124
Rosa woodsii, 142, 156
Rose, 124, 142-143, 156
Ruscus aculeatus, 133
Rusts, 60-65

Saccharomyces cerevisiae, 55
Sage, 155
Saguaro cactus, 156
Salvia dorrii, 155
Saprolegnia, 50
Sargassum, 38, 40
Sclereids, (stone cells), 8
Sclerenchyma, 3, 6-7, 130
Scutellum, 154
Scytonema, 14
Sea lettuce, 36
Sea palm, 38
Seaweeds, 25, 38, 39
Sedge, 156
Seed,
 coat, 102, 105, 108-109, 114, 122, 152, 154
 dispersal, 151
Seedless vascular plants, 81
Selaginella, 81, 88-90
Sepal, 122, 141, 145-146
Sequoia, 110, 117
Sequoiadendron giganteum, 110, 117
Seta, 73-74

Sexual cycle
 bread mold, 52
 green alga, 29, 31
 Volvox, 27
 water felt, 20
Sexual reproduction, 2
Shelf fungus, 61
Shepherd's purse, 153-154
Shrub, 125
Sieve cell, 112
Sieve plates, grape, 7
Sieve tube, 130
Simple fruits, 147, 151
Smilax, 126
Solanum tuberosum, potato, 6
Smuts, 60-65
Soredia, 67
Sori, 94, 96-98
Sperm, 20, 27, 31-33, 37, 41, 68, 70, 77, 82, 85, 88, 94, 102, 148
 nuclei, 148
 packet, 27-28
Spermatangia, 44, 46
Spermatia, 44, 63
Spermatogenous tissue, 74, 79
Speroidal, 55
Sphagnum, 78
Spike mosses, 85-90
Spikelet, 144
Spindle, 1
Spindle fibers, 10
Spirillum, 11-12
Spirillum volutans, 16
Spirochaete, 11
Spirochete, 17
Spirogyra, 34, 35
Spongy mesophyll, 137
Sporangia, 47-48, 52-53, 68, 86-87, 89, 93-94, 97-98
Sporangiophore, 52-53, 93
Sporangium, 47-48, 52-53, 68, 70, 74, 76, 85, 87, 89, 93-94, 97
Spores, 48-49, 53, 72, 74-76, 81-82, 85, 93-94, 97-98
Sporogenous tissue, 73, 82
Sporophyll, 85-87, 114
Sporophyte, 68-70, 72-78, 80, 82, 85, 88, 94, 98, 102, 109, 115, 122
Sporozoans, 23
Sporulation, 2
Squash, *Cucurbita maxima*, 6
Staghorn fern, 96
Stalk, 68, 79
Stamen, 122, 141
Standard, 145
Staphylococcus, 11, 16
Starch grains
 potato, 6
Stele, 86, 126, 127
Stemonitis, 49
Stem, 3, 88-89, 92-93, 119-121, 123, 128-129, 132-133, 134
Stephanodiscus, 19
Sterigma, 62
Stigma, 122, 141-146, 148, 153
Stigonema, 14
Stipa hymenoides, 156
Stipe, 38-40, 42, 60, 62
Stipules, 145
Stolon, 9, 52
Stoma, 76, 87, 113, 137, 138
Stomata, 68, 134, 138
Stomatal chamber, 87
Stone cells, wax plant, 8

Storage tissue, 71
Strawberry, 146, 151
Strep throat, 15
Streptobacillus, 11
Streptococcus, 11
Streptococcus pyogenes, 15-16
Striae, 19
Striated pellicle, 23
Strobili, 85-89, 91-93, 107
Stroma, sugar cane, 5, 56
Stropharia semiglobata, 62
Style, 122, 141, 143, 145-146, 148, 153
Subepidermal sclerids, 152
Subsidiary cells, 138
Sugar cane, 5
Sulfur, 14-15
Sunflower, 124, 144
Surirella, 18
Symbiont protists, 25
Synapsis, 10
Synedra, 18
Synergid cells, 148, 153
Syngamy, 20, 25, 27, 29, 31, 34, 41, 44, 70, 77, 82, 85, 88, 94, 102, 109
Synthesis, 1
Syringa, 148

Taproot system, 125
Taraxacum, 150
Taxodium, 110
Taxodium ascendens, 116
Taxodium distichum, 115
Telia, 64
Teliospore, 63-65
Telium, 63
Telophase
 lily, 10
 onion, 9
Tendril, 128, 145
Terminal bud, 121, 128-129
Terpenes, 155
Tetraspores, 44-45
Tetrasporophyte, 44-45
Thallus, 66-67, 71-72
Thermoacidophiles, 11
Thiothrix, 14, 15
Thuja orientalis, 116
Thylakoid membrane, 5
Tissues, 1, 3
Tmesipteris, 81
Toadstools, 60-65
Tomato, 146, 153
Touch-me-not, 151
Tracheid, 4, 5, 112
Tradescantia, 138
Transfusion tissue, 113
Transverse groove, 22
Trichocyst, 24
Trillium, 124
Triticum aestivum, 154
Trophozoite, 23
True bacteria, 11
Truffles, 54-59
Trychome, 130, 137
Tube nucleus, 148
Tuber, 128
Tulip, 142
Twigs, 128

Ulothrix, 29-30
Ulva, 36
Unarmored dinoflagellates, 21

Uredium, 63
Uredospore, 63
Ustilago maydis, 64

Vacuole
 barley smut, 5
 Elodea, 4
Valve, 19
Vascular cambium cells, 3, 123
Vascular tissue, 1, 3, 103, 107, 130, 136, 154
Vascular bundle, 104, 124, 130, 131
 yucca, 7
Vaucheria, water felt, 20-21
Vegetative cell, 13, 33
Vegetative propagation, 9
Vegetative filament, 30
Vegetative cell, 28, 32, 33
Veins, 121, 134, 135, 136
Venation, 106, 135
Venter, 79
Venus flytrap, *Dionaea musciipula*, 7, 139
Verbascum thapsus, 157
Vesicle, 2
Vessel, squash, 6
Vessel, 4
Vibrio, 11
Violet, 124
Virginiana, 8
Vitis vinifera, grape, 7
Volvox, 27, 28

Water felt, 20-21
Water hyacinths, 137
Water lily, 124
Water molds, 50
Wax plant, 8
Welwitschia, 101, 120
Wheat, 124, 154
Whisk ferns, 82
White rusts, 25, 50
White blood cells, 17
Wing, 145
Wood fungus, 62
Wood, 128
Woody dicot, 132
"Woody" monocot, 132
Woody stems, 128

Xylem, 3, 5, 7, 86-87, 89, 96, 112-113, 121, 123, 126-128, 130-132, 137

Yucca, 140
Yucca brevifolia, yucca, 6, 7
Yucca epidermis, venus flytrap, 7
Yeasts, 54-59
Yellow-green algae, 18
Yucca, *Yucca brevifolia*, 6-7
 epidermis, 6
 phloem, 6-7
 sclerenchyma, 6

Zamia, 103-105
Zamia pumila, 103
Zea mays, 126, 130, 152
Zoosporangium, 33, 52
Zygospores, 20, 27-31, 34, 44
 water felt, 20
Zygotes, 28-29, 31, 34-36, 41, 47, 54, 63, 68, 70, 74, 81-82, 85, 88, 109, 122
Zygnema, 35